TREATISE ON BASIC PHILOSOPHY

Volume 3

ONTOLOGY I: THE FURNITURE OF THE WORLD

TREATISE ON BASIC PHILOSOPHY

1

SEMANTICS I *Sense and Reference*

2

SEMANTICS II *Interpretation and Truth*

3

ONTOLOGY I *The Furniture of the World*

4

ONTOLOGY II *A World of Systems*

5

EPISTEMOLOGY I *The Strategy of Knowing*

6

EPISTEMOLOGY II *Philosophy of Science*

7

ETHICS *The Good and the Right*

MARIO BUNGE

Treatise on Basic Philosophy

VOLUME 3

Ontology I:

THE FURNITURE OF THE WORLD

D. REIDEL PUBLISHING COMPANY
DORDRECHT-HOLLAND/BOSTON-U.S.A.

Library of Congress Cataloging in Publication Data

Bunge, Mario Augusto.
The furniture of the world.

(His Ontology; 1) (His Treatise on basic philosophy; v. 3)
Bibliography: p.
Includes indexes.
1. Substance (Philosophy). 2. Form (Philosophy). 3. Change. 4. Space and time. I. Title.
BD311.B84 vol. 1 [BD331] 111s [110] 77-1756
ISBN 90-277-0780-4
ISBN 90-277-0785-5 pbk.

Published by D. Reidel Publishing Company,
P.O. Box 17, Dordrecht, Holland

Sold and distributed in the U.S.A., Canada, and Mexico
by D. Reidel Publishing Company, Inc.
Lincoln Building, 160 Old Derby Street, Hingham,
Mass. 02043, U.S.A.

Printed in The Netherlands

GENERAL PREFACE TO THE *TREATISE*

This volume is part of a comprehensive *Treatise on Basic Philosophy*. The treatise encompasses what the author takes to be the nucleus of contemporary philosophy, namely semantics (theories of meaning and truth), epistemology (theories of knowledge), metaphysics (general theories of the world), and ethics (theories of value and of right action).

Social philosophy, political philosophy, legal philosophy, the philosophy of education, aesthetics, the philosophy of religion and other branches of philosophy have been excluded from the above *quadrivium* either because they have been absorbed by the sciences of man or because they may be regarded as applications of both fundamental philosophy and logic. Nor has logic been included in the *Treatise* although it is as much a part of philosophy as it is of mathematics. The reason for this exclusion is that logic has become a subject so technical that only mathematicians can hope to make original contributions to it. We have just borrowed whatever logic we use.

The philosophy expounded in the *Treatise* is systematic and, to some extent, also exact and scientific. That is, the philosophical theories formulated in these volumes are (*a*) formulated in certain exact (mathematical) languages and (*b*) hoped to be consistent with contemporary science.

Now a word of apology for attempting to build a system of basic philosophy. As we are supposed to live in the age of analysis, it may well be wondered whether there is any room left, except in the cemeteries of ideas, for philosophical syntheses. The author's opinion is that analysis, though necessary, is insufficient – except of course for destruction. The ultimate goal of theoretical research, be it in philosophy, science, or mathematics, is the construction of systems, i.e. theories. Moreover these theories should be articulated into systems rather than being disjoint, let alone mutually at odds.

Once we have got a system we may proceed to taking it apart. First the tree, then the sawdust. And having attained the sawdust stage we should move on to the next, namely the building of further systems. And this for three reasons: because the world itself is systemic, because no idea can become fully clear unless it is embedded in some system or other, and because sawdust philosophy is rather boring.

The author dedicates this work to his philosophy teacher

Kanenas T. Pota

in gratitude for his advice: "Do your own thing. Your reward will be doing it, your punishment having done it".

CONTENTS OF *ONTOLOGY I*

PREFACE TO *ONTOLOGY I*

This book and its companion, namely Volume 4 of our *Treatise*, concern the basic traits and patterns of the real world. Their joint title could well be *The Structure of Reality*. They constitute then a work in ontology, metaphysics, philosophical cosmology, or general theory of systems. Our work is in line with an old and noble if maligned tradition: that of the pre-Socratic philosophers, Aristotle, Thomas Aquinas, Descartes, Spinoza, Leibniz, Hobbes, Helvetius, d'Holbach, Lotze, Engels, Peirce, Russell, and Whitehead. But at the same time it departs from tradition in the matter of method. In fact our aim is to take the rich legacy of ontological problems and hints bequeathed us by traditional metaphysics, add to it the ontological presuppositions of contemporary scientific research, top it with new hypotheses compatible with the science of the day, and elaborate the whole with the help of some mathematical tools.

The end result of our research is, like that of many a metaphysical venture in the past, a conceptual system. It is hoped that this system will not be ridiculously at variance with reason and experience. It is intended moreover to be both exact and scientific: exact in the sense that the theories composing it have a definite mathematical structure, and scientific in that these theories be consistent with and moreover rather close to science – or rather the bulk of science. Furthermore, to the extent that we succeed in our attempt, science and ontology will emerge not as disjoint but as overlapping. The sciences are regional ontologies and ontology is general science. After all, every substantive scientific problem is a subproblem of the problem of ontology, to wit, *What is the world like?*

After a long period underground, talk about metaphysics has again become respectable. However, we shall not be talking at length about ontology except in the Introduction. We shall instead do ontology. In the process we shall attempt to exhibit the mathematical structure of our concepts and we shall make the most of science. Being systematic our ontology may disappoint the historian. Being largely mathematical in form it will be pushed aside by the lover of grand verbal (but sometimes deep and fascinating) systems – not to speak of the lover of petty verbal matters. And being science-oriented it will fail to appeal to the friend of the esoteric. Indeed we shall be concerned with concrete objects such as atoms, fields, organisms, and societies. We shall abstain from talking about items that are neither concrete things nor properties, states or changes thereof. Any fictions entering our system will be devices useful in accounting for the structure of reality. (Constructs were dealt with in Volumes 1 and 2 of this work.)

The first ideas for this work dawned upon me when I was engaged in axiomatizing some basic physical theories involving ontological concepts such as those of thing, property, possibility, change, space, and time, none of which are the exclusive property of physics but all of which belong to the metaphysical background of this science, or protophysics (Bunge 1967b). And the earliest plan for this work occurred to me a bright day of June 1966 when travelling from Freiburg im Breisgau to Geneva at the invitation of Jean Piaget. I have been working on this project ever since, on and off, stimulated by what seemed a grand design and occasionally inhibited by the difficulties met with in carrying it out. The result is a system but not a closed and final one: there is much room for improvement and of course also for divergent developments.

This volume deals with the concepts of substance, form (or property), thing (or concrete object), possibility, change, space, and time. The companion volume, *A World of Systems*, will tackle the concepts of system, novelty, biosystem, psychosystem, and sociosystem.

ACKNOWLEDGEMENTS

It is a pleasure to thank the various generations of curious, bewildered and merciless students who survived my ontology courses at McGill University and the Universidad Nacional Autónoma de México from 1969 to 1976. I am also indebted to Professors Rutherford Aris (Chemical Engineering, Minnesota), Thomas A. Brody (Physics, UNAM, México), Máximo García Sucre (Physics, IVIC, Caracas), Tomás Garza (IIMAS, UNAM, México), Andrés J. Kálnay (Physics, IVIC, Caracas), and Roberto Torretti (Philosophy, Puerto Rico) for critical remarks and suggestions. My former research assistants Dr David Salt and Dr Gerhard Vollmer made a number of useful remarks. My research assistant Mr Robert Blohm enhanced the clarity, improved the grammar, and asked questions still to be answered. But my greatest debt is to my former research associates Professors Adalberto García Máynez (Mathematics, IPN, México), William E. Hartnett (Mathematics, SUNY, Plattsburgh), and Arturo A. L. Sangalli (Mathematics, Ottawa), none of whom understood friendship to be in the way of rigor.

The Alexander von Humboldt-Stiftung covered my first incursions into the intersection of physics and metaphysics (1965-66). The Canada Council contributed to this project by awarding me research grants (1969–72, 1974–76), one of them on behalf of the Killam Foundation. And the John Simon Guggenheim Memorial Foundation awarded me a fellowship that gave me a happy and fruitful year (1972–73). It was during the tenure of that fellowship, at the ETH Zürich, that I started to write this book. I am grateful to all those organizations for their support, and to Professors Gerhard Huber and Peter Huber for their hospitality at the ETH. Last, but not least, I thank my guide in the fascinating and puzzling Mexican labyrinth, Professor Fernando Salmerón, director of the Instituto de Investigaciones Filosóficas of the UNAM, where this volume acquired its final shape during the 1975–76 academic year.

MARIO BUNGE

SPECIAL SYMBOLS

$x \rhd y$	*x acts* on *y*
$c = a \circ b$	*c* is the *association* of individuals *a* and *b*
$A \times B$	the *cartesian product* of sets *A* and *B*
$\langle s, s', g \rangle$	the *change* from state *s* to state *s'* along curve *g*
C	set of constructs
$\mathscr{C}(x)$	the *composition* of thing *x*
$x \rceil y$	*x* and *y* are *detached*
$E_{\mathbb{L}}(x)$	the *event space* of thing *x*
$E_A x$	*x exists* in *A*
$\mathscr{E}(A)$	the *extension* of attribute (predicate) *A*
$f: A \to B$	*function f* maps set *A* into set *B*
F	set of *facts*
$\mathbb{F} = \langle F_1, F_2, \ldots, F_n \rangle$	*state function*
$h(x)$	the *history* of thing *x*
$a \vert c \vert b$	*c interposes* between *a* and *b*
$c = a \dotplus b$	*c* is the *juxtaposition* of *a* and *b*
$k(\mathbb{R})$	the *kind* of things sharing all properties in $\mathbb{R}$
$G_{\mathbb{L}}(x)$	the set of *lawful transformations* of the states of *x*
$\mathbb{L}(x)$	the *laws* of thing *x*
$\langle x, y \rangle$	the *ordered pair* of *x* and *y*
$\Box$	the *null thing*
$x \sqsubset y$	*x* is a *part* of *y*
$\mathscr{P}(S) = 2^S$	the *power set* of *S*
$P \leqslant Q$	property *P precedes* property *Q*
$\mathbb{P}$	the set of *all properties*
$p(x)$	the collection of *properties of thing x*
Pr	*probability* function
$\mathbb{R}$	the *real line*
$S_{\mathbb{L}}(x)$	*lawful state space* of thing *x*
$c = a \dot{\times} b$	*c* is the *superposition* of things *a* and *b*
S	the set of *all substantial* (*concrete*) *individuals*
$\mathscr{S}(P)$	the *scope of property P*
$[T] = \inf T$	the *additive aggregation* of all things in *T*
$(T) = \sup T$	the *multiplicative aggregation* of all things in *T*
Θ	the *set of all things*
▯	the *world* or *universe*

INTRODUCTION

In this Introduction we shall sketch the business of ontology, or metaphysics, and shall locate it on the map of learning. This has to be done because there are many ways of construing the word 'ontology' and because of the bad reputation metaphysics has suffered until recently – a well deserved one in most cases.

1. Ontological problems

Ontological (or metaphysical) views are answers to ontological questions. And ontological (or metaphysical) questions are questions with an extremely wide scope, such as 'Is the world material or ideal – or perhaps neutral?', 'Is there radical novelty, and if so how does it come about?', 'Is there objective chance or just an appearance of such due to human ignorance?', 'How is the mental related to the physical?', 'Is a community anything but the set of its members?', and 'Are there laws of history?'.

Just as religion was born from helplessness, ideology from conflict, and technology from the need to master the environment, so metaphysics – just like theoretical science – was probably begotten by the awe and bewilderment at the boundless variety and apparent chaos of the phenomenal world, i.e. the sum total of human experience. Like the scientist, the metaphysician looked and looks for unity in diversity, for pattern in disorder, for structure in the amorphous heap of phenomena – and in some cases even for some sense, direction or finality in reality as a whole. Metaphysics and science have then the same origin. However, they can be distinguished up to a certain point, namely by the scope of their problems. Whereas the scientific specialist deals with rather specific questions of fact, the ontologist is concerned with all of the factual domains: he is a generalist not a fragmentarian. His enterprise is more ambitious, hence also riskier, than any one scientific project. But the two concerns are not mutually exclusive and, in fact, sometimes they are indistinguishable: an extremely general scientific question may be a special ontological one.

Ontological questions are not easy to characterize or even to recognize as meaningful in isolation from ontological frameworks or ontological theories. Consider for example the following questions:

(i) Why is there something rather than nothing?
(ii) Does essence precede existence?
(iii) What is being?
(iv) Where is one?
(v) What is there?

The first question makes sense in any creationist system of theodicy, such as Leibniz', but it makes no sense elsewhere. The second, at first blush unintelligible, makes good sense in a Platonic metaphysics, where essences are ideal and prior to physical existents. The third question becomes meaningful if reformulated as 'What are the features common to all existents?'. The fourth question was asked by my son Eric when he was 18 months old. It must have made some sense in his own *Weltanschauung* and, rewriting 'one' with a capital O, it might be safely attributed to Parmenides. The fifth – which is how Quine seems to understand the task of ontology – calls for either an exhaustive inventory of existents – a job for scientists of the Baconian persuasion – or a simplistic answer such as 'There are bodies and persons' (Strawson, 1959).

The preceding questions make hardly any sense in the system to be developed in this book. On the other hand the following problems do make sense in it and moreover can be given definite answers:

(vi) Are things bundles of properties? (No.)
(vii) Are there natural kinds? (Yes.)
(viii) Is change possible without an unchanging substrate? (Yes.)
(ix) How do emergent properties come about? (Wait for Volume 4.)
(x) What is the mind? (Ditto.)

These questions are both fundamental and extremely general. Moreover they are factual questions – only, comprehensive or cross-disciplinary rather than special. There are as many such questions as we care to ask. The more we come to know, the more problems we can pose and the less final our solutions prove to be. In both regards then – factual content and open-endedness – ontological questions are no different from scientific ones. (For the latter see Bunge, 1967a, Vol. I, Ch. 4.) They differ only in scope. And even this difference is often non-existent, as will be seen in Sec. 7.

2. The business of ontology

At least the following ten conceptions of the concern of ontology (or metaphysics) have followers nowadays:

(i) Metaphysics is a discourse (in either ancient Greek or modern German) *on Being, Nothingness, and Dasein* [human existence] (Heidegger, 1953). Objection: impossible, because the said discourse is unintelligible and moreover avowedly irrational. If in doubt try to read Heidegger or Sartre.

(ii) Metaphysics is a collection of *instinctive* (as opposed to intellectual) *beliefs* (Bergson, 1903). Objection: if metaphysics is to be a discipline then it cannot collect blindly any received ideas, be they "instinctive" or gotten from tradition. The fact that cave men held "instinctive" beliefs without subjecting them to methodic criticism does not justify us in upholding an attitude that would never have taken us out of the cave.

(iii) Metaphysics is the *justification of instinctive beliefs*: "the finding of reasons, good, bad or indifferent, for what we believe on instinct" Strawson, 1959, p. 247). Objection: our least educated tenets must surely be studied – by the cultural anthropologist. Scholars, whether scientists or metaphysicians, are supposed to examine, refine or reject that starting point – and above all to propose new ideas. Reassuring the cave man in his primitive metaphysics is worse than sharing it.

(iv) Metaphysics is "*the science of absolute presuppositions*" (Collingwood, 1940). That is, metaphysics is the study of all presuppositions, of any discipline, in so far as they are absolute, i.e. lurk behind every question and every answer, and are moreover beyond question. This is a respectable view. However, it is open to the following objections: (*a*) most presuppositions are not absolute but are bound to rise and fall with the special theory concerned: look at the history of ideas; (*b*) even though metaphysics does study some of the presuppositions of science, it does not handle them all, as some of them are purely formal (logical or mathematical) and others are methodological.

(v) Metaphysics deals with *everything thinkable*, whether or not it actually exists, whether reasonable or absurd: it is concerned with "the totality of the objects of knowledge" (Meinong, 1904, in Chisholm, 1960, pp. 78–79). Objections: (*a*) no theory is possible that will encompass both concrete objects and conceptual ones; in particular, logical truths may refer to anything but they do not describe or represent any objects except the logical concepts ("or", "all", etc.); (*b*) objects known to be fantastic, such as Pegasus, may be imagined but cannot constitute the subject of any discipline: only our beliefs about such mythological objects can be studied scientifically.

(vi) Metaphysics is *the study of objects neither physical nor conceptual* – i.e. of spiritual beings, and of God and his celestial court in the first place. This opinion is quite popular and was occasionally voiced by Thomas himself (1259 Bk. I, Ch. IV). Objection: that is the proper subject of theology, which is no longer recognized as a part of philosophy.

(vii) Metaphysics is *the science of being as such*: unlike the special sciences, each of which investigates one class of being, metaphysics investigates "all the species of being *qua* being" and "the attributes which belong to it *qua* being" (Aristotle, *Metaphysics* Bk. IV, Chs. 1 and 2). This is what nowadays one would call *general* ontology by contrast to the various *special* or *regional* ontologies (of the biological, the social, etc.). Certainly the Philosopher had a correct grasp of the relation between metaphysics (general) and the sciences (special). Still, the following objections must be raised: (*a*) the formulation is too imprecise, so much so that it has suggested to some that becoming is not within the purview of metaphysics – an opinion certainly not shared by the Stagirite, who was centrally concerned with change; (*b*) a science of pure being is a contradiction in terms because it has no definite subject matter (Collingwood, 1940, pp. 10–11).

(viii) Metaphysics is *the study of change*: of events and processes – because this is what things are (Whitehead, 1929). Objection: an event is a change in the condition (state) of some thing and therefore cannot be studied apart from it any more than things can be studied apart from their changes.

(ix) Metaphysics concerns *all possible worlds*: it is an ontological interpretation of logic. A system of metaphysics is a set of statements satisfying two conditions: (*a*) "The horizon [set of referents] of a significant metaphysical statement must surpass in an unambiguous way the horizon of a physical statement", and (*b*) "A metaphysical statement must not lag behind a physical statement as far as exactness and stability [*Standfestigkeit*] are concerned" (Scholz, 1941, pp. 138–139). While I have no quarrel with the exactness condition, I dispute the others. My objections are: (*a*) the fact that logic may refer (apply) to anything does not render it a theory of all possible worlds; (*b*) while some metaphysical statements do concern all concrete things, others refer to things belonging to certain genera such as physical objects or organisms or societies; (*c*) metaphysical statements cannot be any less fallible than scientific (e.g. physical) statements.

(x) Metaphysics is *general cosmology or general science*: it is the science concerned with the whole of reality – which is not the same as reality as a whole. "Its business is to study the most general features of reality and real objects" (Peirce, 1892–93, p. 5). It "is concerned with all questions of a general and fundamental character as to the nature of the real" (Montagu, 1925, p. 31; see also Woodger, 1929; Williams, 1937; and Quinton, 1973). In other words, metaphysics studies the generic (nonspecific) traits of every mode of being and becoming, as well as the peculiar features of the major genera of existents. This is the task Hegel (1812–16) assigned to "objective logic" and Engels (1878) to what came to be known as dialectical materialism.

We adopt the latter position: we maintain that the ontologist should stake out the main traits of the real world as known through science, and that he should proceed in a clear and systematic way. He should recognize, analyze and interrelate those concepts enabling him to produce a unified picture of reality. (The word 'reality' is here understood in a strict and non-Platonic sense, namely as the concrete world.) In this sense the reader is real and so is any utterance of the word 'reader'; but the concept designated by this word is unreal.

Because unreal objects have nonphysical properties, they satisfy nonphysical laws if any. For this reason it is impossible to make any nontautological statements applying to all objects: ontology, as conceived by Meinong and Leśniewski, i.e. as a general theory of objects of any kind, and yet different from logic, is impossible. So is the modern version of this doctrine, namely general systems theory construed as a mathematical theory "dealing with the explanations of observed phenomena or conceptual constructs in terms of information-processing and decision-making concepts" (Mesarović in Klir, 1972, p. 253). If the "system" is purely conceptual, as is the case with a number system, then it cannot combine with material systems to form supersystems, it cannot interact with them, it does not obey laws of the same kind and therefore it cannot be studied with the special methods of factual science. Whatever we may think of the construct-thing duality, whether we take a Platonist or a materialist stand, or whether or not we wish to reduce objects of one kind to objects of another, we should keep the duality at the methodological level. (Recall Vol. 1, Ch. 1, Sec. 3.)

We leave formal science, i.e. logic, mathematics, and semantics, the task of studying (and creating) formal or ideal objects of the law-abiding kind, such as sets and categories. (More in Sec. 6.) We take factual (natural or social) science and ontology to be the only disciplines concerned with concrete objects. And we assign ontology the task of constructing the most general theories concerning such and only such objects. But is this task possible?

3. Is ontology possible?

Up until the first scientific revolution, in the early 17th century, metaphysics was taken for granted and was usually regarded as an adjunct to theology. Thereafter the formidable successes of science, jointly with the failure of the grand metaphysical systems – chiefly those of Aristotle, Aquinas, Descartes and Leibniz, and later on Hegel's and Heidegger's – suggested, nay "proved", that metaphysics had become impossible and had been replaced with science.

Those who deny the possibility of metaphysics as an authentic and creative intellectual endeavor hold that metaphysical sentences are either meaningless (hence neither true nor false) or meaningful but untestable; or, if meaningful and testable, then false. Or they hold that, aside from questions of meaning and truth, such statements have no value either practical or heuristic. That many metaphysical statements – e.g. most of Hegel's and all of Heidegger's – are hardly meaningful, is true and sad enough but speaks only against bad metaphysics. (Likewise the fact that most statements about the mental are nonscientific does not disqualify psychology as a science.) That metaphysical propositions cannot be checked for truth the same way scientific statements are tested, is correct but in no way proves that they are beyond criticism and confirmation. That many, perhaps most metaphysical propositions have turned out to be false, and many more will be refuted, is again true but does not render metaphysics any less possible than science, which is in the same predicament. Finally, it is not quite correct that metaphysics lacks in heuristic or even practical import. For one thing scientific research makes use of a number of ontological hypotheses, as will be seen in Sec. 7. For another the philosophic core of every world view and every ideology is, for better or (usually) worse, some ontological system or other together with some value system. In short, it is not true that ontology has become impossible after the birth of modern science. What has become impossible *de jure* – though unfortunately not *de facto* – is nonscientific ontology.

The sceptic may not deny the possibility of ontology but will regard it as inconclusive. With this we concur: metaphysics is indeed inconclusive – but so is factual science. Genuine knowledge is characterized not by certainty but by the possibility of correction in an endless effort at attaining truth and depth. This holds for ontological knowledge no less than for scientific knowledge. Let the sceptic criticize every single metaphysical system: let us keep him busy by offering him new metaphysical theories.

Ontology is not just possible but very much alive these days. (As a matter of fact it has never ceased to exist: it just went underground for a while.) A casual look at the literature will bear out this contention. Moreover, metaphysics only recently has undergone a revolution so deep that nobody has noticed it: indeed ontology has gone mathematical and is being cultivated by engineers and computer scientists. As a matter of fact a number of technologists have developed, over the past three decades, certain exact theories concerning the most basic traits of entities or systems of various genera. Switching theory, network theory, automata theory, linear systems theory, control theory, mathematical machine theory, and information theory are among the youngest metaphysical offspring of contemporary technology (Bunge, 1971, 1974d, 1977a). This kind of ontology, both exact and scientific, is the one we wish to develop and systematize. Its place in the map of metaphysics is shown in the following table. (Nonacademic varieties of metaphysics, mostly esoteric, have been left aside.)

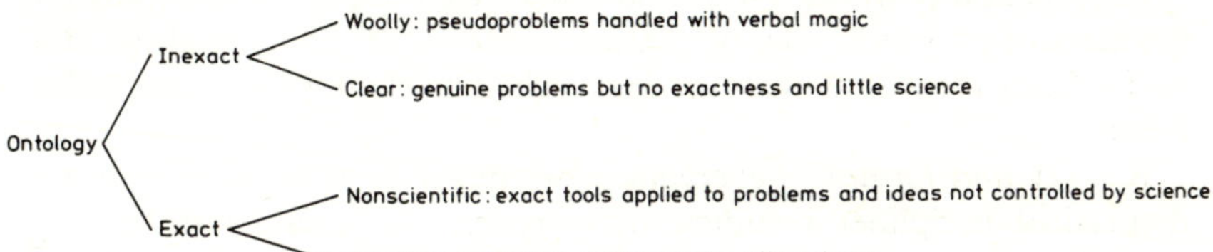

4. The Method of Scientific Ontology

The ontology we wish to build is both exact and contiguous with science. 'Exact' means of course logical or mathematical in form. Exact metaphysics is then the set of metaphysical theories built with the explicit help of logic or mathematics. For example, a mathematical theory of the synthesis of wholes out of lower level units would qualify as a piece of exact metaphysics. Not so a theory, however interesting, tackling the same problem in ordinary knowledge terms: it would remain at the level of inexact metaphysics because of the ambiguity and imprecision of ordinary language.

We are interested in the scientific variety of exact metaphysics, that is, in the ontological theories that, in addition to being exact, are scientific. The difference between 'exact' and 'scientific' is this: the latter implies the former but the converse is false. That is, there are systems of exact metaphysics that are out of touch with factual (natural and social) science. For example, Leibniz, Bolzano, Scholz and Montague were exact metaphysicians but they conceived of metaphysics as an a priori science (see e.g. Bolzano, 1849, p. 29); hence their work is not in tune with the science of their day. Likewise most of the contemporary essays on possible worlds, temporal logic, and causality, though often exact, are far removed from science and sometimes even incompatible with it. There is little if any philosophical meat in these theories; and when there is some, it turns out to be stale.

As for scientific metaphysics, it is still largely a program. Even Peirce (1892–93), perhaps the first to employ the expression, did not advance beyond some programmatic remarks. To implement that program we need some guidelines. Here are some of the *regulae philosophandi more geometrico et scientifico* we shall try to abide by:

R1 *Take the metaphysical tradition into account but do not stick to it*: go ahead revising the traditional stock of problems and solutions, posing new problems, and trying new solutions to both old and new questions.

R2 *Avoid any words that fail to convey clear ideas*: obscurity is not the mark of profundity but of confusion or even of intellectual swindle. As for fuzzy ideas – all ideas are fuzzy when newly born – try and refine them.

R3 *Try and formalize everything*: whatever is worth saying in any theoretical discipline, including metaphysics, can be said with the help of mathematics and ought so to be said for the sake of clarity and systematizability.

R4 *Do not mistake symbolization for mathematization*: stenography does not elucidate and does not systematize. Unless a construct is assigned a definite mathematical status (as set, relation, function, group, topological space, or what have you) it is not exact and may be a fake, i.e. a *flatus vocis* rather than a genuine concept.

R5 *Strive for rigor but do not allow it to curtail vigor*: exactness is a means not an end – a means to attain clarity, systemicity, cogency and testability. Insisting on rigor for its own sake and at the price of giving up deep intuitions is a mark of sterility.

R6 *Explain the concrete by the abstract* rather than the other way around. Invert Russell's advice to replace "inferred" (unobserved, hypothetical) entities with logical constructions out of sense impressions. Imitate the atomists and the field theorists. Admit observable, ostensive and secondary properties only if they can be analyzed in terms of primary properties.

R7 *Keep clear of subjectivism* – e.g. avoid definitions in terms of personal experience. Example of the strategy to be avoided: Whitehead (1920, p. 13) defined the concept of a fact as "the undifferentiated terminus of sense-awareness". Every ontological elucidation, even of concepts concerning subjective phenomena, should be couched in purely objective (subject-free) terms.

R8 *Do not reify whatever is not a thing*, and do not treat as an autonomous entity what is but the result of abstraction. E.g. do not talk about events apart from or even as constituting changing things.

R9 *Strive for systemicity*: try to build theories and establish links among them. Do not introduce – except by way of extrasystematic remarks – concepts that have not yet been elucidated: proceed in orderly fashion, and if necessary axiomatically. Philosophical analysis is indispensable but not enough, and is best done in the context of a system anyway.

R10 *Check* your metaphysical hypotheses and theories not only for internal consistency but also for compatibility and even contiguity with contemporary science.

Abiding by these rules should help us in building science-oriented ontological frameworks and even theories, though it won't secure them.

5. The Goals of Scientific Ontology

Every philosophical activity has two possible intellectual goals: analysis or synthesis – i.e. taking apart or building. Philosophical analysis examines certain concepts and propositions in order to clarify them. Philosophical synthesis creates frameworks and theories in order not only to elucidate notions and statements but also to understand what goes on in the world. The two operations are complementary rather than mutually exclusive. In fact the best analysis is that performed in a definite theoretical context, just as the best synthesis is that which articulates analyzed notions. In any case it is synthesis, not analysis, that provides some understanding of a field of reality or of human experience. Thus only a theory of ideation, rather than an analysis of the notion of ideation – let alone an analysis of the word 'ideation' as used in a given community – can account for ideation.

In the case of ontology, analysis bears on any metaphysical concepts or propositions – or candidates for either role. Ontological analysis bears, in particular, on ontological categories – such as those of quality and society – and ontological principles – such as the hypothesis that every concrete thing is in flux. The analysis we expect from scientific ontology concerns, in particular but not exclusively, the ontological categories and hypotheses that occur, either in a heuristic or in a constitutive capacity, in scientific research. Some such categories are those of thing, property, fact, and value. As for the ontological principles inherent in science, suffice it to mention the assumption that a society, far from being either an amorphous set of individuals, or a totality transcending individuals, is a system of interacting persons. More in Secs. 7 and 8.

However, the ontological analyst need not confine himself to listing and classing the ontological categories and principles actually employed – usually with neither acknowledgement nor apology – in scientific inquiry. He may be more ambitious and criticize certain ontological theses, be they proposed by philosophers or by scientists, and he may even propose his own theses on the matter. Take for example the generic or ontological notion of a mental event. The ontological analyst may not only examine the various construals of the expression 'mental event' but also have an axe to grind – or ought to if he is a philosopher at all. But he may not be able to accomplish much without the help of some general (ontological) theory of events (of all kinds) and some general (ontological) theory, or at least framework, of the mental. That is, ontological analysis, if profound, calls for ontological synthesis or theory, or at least a framework for such.

Ontological theories are of course those involving ontological categories. A general theory of changes of any kind, physical or mental, quantitative or qualitative, is an ontological theory. And a theory of mental events of any kind – feeling visceral processes, perceiving physical stimuli, forming images or concepts, etc. – is an ontological theory too, though a regional one instead of a universal theory. A complete ontology should include both *universal* and *regional* ontological theories. The former serve as frameworks for the latter, which will in turn illustrate and in a way test the former. Take again the case of mental events. In our own ontology the dualistic tenet concerning the mind-body problem makes hardly any sense because it presupposes that mental events are exceptional in that, unlike all others, they fail to be changes in some concrete thing. In fact in our ontology all and only things change, and every change (event of process) is a modification in the state of some concrete thing or other. Once this general ontological view has been adopted and systematized (made into a theory or at least a framework) there is no room left for changes that fail to be changes in or of some concrete thing, be it an electron or a neuron, a brain or a society. Only one view of mental events is consistent with that general ontological theory of change, namely that according to which the mind is a certain activity of the brain (Hebb, 1949). Moreover it is clear that any alternative view should be shown to be consistent with some general theory of change. This is certainly possible: a mentalistic (e.g. idealistic) ontology will include a mentalistic view of the physical. But of course such an ontology will be at variance with physics, chemistry and biology, none of which is animistic. We shall discuss these matters in detail in Vol. 4, Ch. 10. The point of bringing them up here was to defend the view that a full fledged ontology must consist of theories – or at least theory embryos – some universal and some regional, and that these theories should be mutually consistent. There is no salvation outside a system.

Needless to say, when using the word 'theory' we mean a hypothetical-deductive system rather than a stray opinion or an unsystematic set of opinions. In particular, an *ontological theory* is a theory that contains and interrelates ontological categories, or generic concepts representing components or features of the world. (Our own set of categories includes those of thing, property, law, possibility, change, spacetime, life, mind, value, and society.) Ideally, an ontological system or theory is a system, not just a set, of interrelated ontological categories. But we shall not always succeed in building theories proper. Sometimes we shall offer just ontological frameworks, which are constructs intermediate in structure between shapeless views and closed hypothetical-deductive systems. (For the general concept of a framework see Vol. 1, Ch. 2, Sec. 3.4, Def. 2.10.)

We shall agree to call an ordered triple $\mathbb{C} = \langle S, \mathbb{P}, D \rangle$ an *ontological framework* iff S is a set of statements in which occur only the predicate constants in the predicate family $\mathbb{P}$, which includes a nonempty set $\mathbb{O}$ of basic ontological concepts (i.e. categories), and the reference class of every P in $\mathbb{P}$ is included in the universe or domain D of hypothesized entities, or objects assumed to exist (in our case concrete objects or things). Though no substitute for an ontological theory, an ontological framework serves as a matrix for any number of ontological systems or theories, and has thus a guiding or heuristic power. (For the difference between theory and framework, and the heuristic role of the latter, see Bunge (1974e).) Indeed the mere display of D's and their properties (i.e. the members of $\mathbb{P}$) invites the formulation of hypotheses (statements in S) concerning the mode of being and becoming of those alleged entities.

An ontological framework will be *monistic* or *pluralistic* according as both the domain D of individuals and the collection $\mathbb{P}$ of their properties (and relations) are taken to be a single genus each or the union of two or more genera. (In short, monism is reductionistic in some sense or other, pluralism is not.) And, whether monistic or pluralistic, an ontological framework can be either *spiritualistic* or *naturalistic*, according as it postulates that D includes, or fails to include, a set of spiritual (disembodied) entities such as immaterial souls. But whereas according to spiritualistic monism all entities are spiritual and of the same kind (e.g. sensations, or ideas), according to pluralistic spiritualism all entities are spiritual but there are several types of them and these kinds are not reducible to one basic kind. And whereas according to monistic naturalism all entities are concrete and of the same kind (e.g. physical), pluralistic monism asserts that, although all entities are concrete, there are several irreducible kinds of them. The ontological frameworks and theories to be developed in the present volume and its sequel, Vol. 4, will turn out to be both *naturalistic and pluralistic*: we will assume only concrete existents but will assert their qualitative variety.

In summary, the goals of scientific ontology are *to analyze and to systematize the ontological categories and hypotheses germane to science.* These are the aims served by the construction of frameworks and theories close to science rather than closed to science.

6. Ontology and Formal Science

Formal science, or at least some of it, constitutes both the language and the formal skeleton of scientific ontology. In particular, scientific ontology presupposes *abstract mathematics*, including deductive logic. The choice of abstract mathematics as the formal backbone of metaphysics is only natural, for only abstract systems can be assigned alternative interpretations, in particular ontological interpretations. The theories of semigroups and of lattices, which are abstract, have far more potential use for metaphysics, and for philosophy in general, than number theory and the infinitesimal calculus, which are fully interpreted (in mathematical terms), hence nonabstract. In any case, metaphysics can make use of mathematics and must employ it if it is to become exact and contiguous with science. Is there more to it? Let us see.

Many philosophers, from Parmenides through Leibniz and Hegel to Gonseth and Scholz, have believed that logic is a sort of universal physics, i.e. the most general theory of being or becoming. The Leibnizians put it this way: (*a*) the real world is merely one among many equally possible worlds; (*b*) logic is true of all possible worlds, hence also of ours. (See e.g. Lewis, 1923; Scholz, 1941; Hasenjäger, 1966.) The rationale for this view is that a tautology of ordinary logic, such as $\ulcorner(x)(Px \vee \neg Px)\urcorner$, is true of anything, i.e. no restrictions are placed on the individual variable or the predicate variable. This is certainly true: logic refers to everything and holds for anything in any respect. It would seem therefore that logic is a sort of minimal ontology, a *physique de l'objet quelconque* (Gonseth, 1938).

However, although logic is indeed indifferent to the kind of referent, it does not describe or represent, let alone explain or predict, any factual items. The statement "Every organism is either alive or dead" concerns organisms but does not say anything definite about them; hence it is not a biological statement or even an ontological one. Rather, the proposition says something about the particular combination of "or" and "not" that occurs in it. Logic is the set of theories describing the properties of logical concepts – the connectives, the quantifiers, and the entailment relation. So much so that none of these concepts has a real counterpart or at least a unique one. Indeed, there are neither negative nor alternative things, properties, states, or events. Generalization, whether existential or universal, is strictly a conceptual operation. And the entailment relation has no ontic correlate either, although it has sometimes been likened to the causal relation. Further, if it be admitted that tautologies are analytic, hence a priori, they cannot be also synthetic a priori truths. In sum logic is not ontological.

The relation between logic and ontology is that of presupposition or logical priority: any cogent metaphysics presupposes logic. (This may be why Heidegger (1953) pushed logic aside with contempt and hatred.) In our work we shall take second order ordinary logic for granted. *Ordinary* logic because it is the one employed in mathematics, hence in theoretical science, neither of which has found use for any deviant logics. And *second order* logic because we want to analyze ontological statements such as "Everything has some property or other", and "Every property of a thing is related to some other properties of it".

The relation between ontology and mathematics is similar: if exact, ontology presupposes mathematics, but the latter is in turn free from any ontological commitments. Surely certain mathematical structures can be assigned ontic interpretations. For example group theory can be construed as describing all possible reversible transformations. But such interpretations, like those inherent in theoretical physics, go beyond pure mathematics. The latter stands on its own feet. If anyone wants to make use of it he is welcome: mathematics (including logic) is a service discipline. If a given field of research finds no use for mathematics, then this suggests only that it is in a backward state – that it has few clear notions and general propositions. The possibility of mathematizing a field of knowledge does not depend upon the subject matter but upon the state of the art, for what get mathematized are not facts but our ideas about them. As long as the latter are confused they are impregnable to mathematization, as was the case with the much overrated *impetus* concept (Koyré, 1957), later on with that of energy, and nowadays with that of the "neural correlate" of a mental event. So much for the relations between ontology and mathematics.

The relation between semantics and metaphysics is another matter. Tarski (1944, p. 363) held that ontology "has hardly any connections with semantics." However, the correspondence theory of truth presupposes that there is a real world – and this is a metaphysical assumption. Moreover the applications of any theory of reference calls for definite assumptions about the furniture of the world. Consider e.g. "The plane is 1 hr late". Does it refer to the plane, to both the plane and the airport, or only to the event of its arrival in the airport, or to both and the duration of the delay? The answer to this question depends on one's cosmology – on the assumed furniture of the world. If things are regarded as the only constituents of reality, the answer is that the statement concerns the plane and the airport; if events are declared the stuff of reality, then the referent will be the plane's arrival; and if time is assumed to have an absolute existence then it, too, will count among the referents. In sum semantics, or at least its application, does have metaphysical presuppositions. (More in Bunge, 1974c, and Vol. 2, Ch. 10, Sec. 4.3.) On the other hand ontological theories seem to have no semantic presuppositions although they do employ semantic concepts such as those of designation, reference, and representation.

To sum up. Deductive logic and pure mathematics, in particular abstract mathematical theories, are ontologically neutral. Precisely for this reason they can be used in building ontological theories. There is no a priori limitation on the variety of mathematical theories that can be employed in metaphysical research. The choice will depend largely upon the metaphysician's background and preferences. Some (e.g. Scholz, 1941) will try to make do with formal logic and model theory only. Others would like to extract an entire cosmology out of modal logic (see e.g. Munitz (ed.), 1971 and 1973). Suppes (1974) seeks instead "to replace the concept of logical empiricism by probabilistic empiricism" in constructing a probabilistic metaphysics. Finally, other philosophers will adopt what may be termed *mathematical opportunism*, or the strategy of using whatever mathematical theories seem promising. The present work exemplifies this strategy.

So much for the relation of metaphysics to formal science. (More in Bunge 1974c.) Let us now look at the other component of science.

7. The Ontology of Science

Metaphysics has often been contrasted with science for allegedly being speculative rather than empirical, hence irrefutable. So it is in many cases. However this is far from necessary: ontology can be consonant with science and just as scientific as physics, even though there will never be any metaphysical laboratories. Moreover we shall argue that (*a*) scientific research is guided or misguided by metaphysical principles – some good, others bad; (*b*) both basic science and technology have produced theories that are scientific as well as metaphysical, and (*c*) it is possible to build systems of scientific ontology. If this is so then there need not be any hostility between science and (scientific) metaphysics. There is not even a gap, let alone an abyss, between them: *ontology is general science* and *the factual sciences are special metaphysics*. In other words, both science and ontology inquire into the nature of things but, whereas science does it in detail and thus produces theories that are open to empirical scrutiny, metaphysics is extremely general and can be checked solely by its coherence with science.

That scientific research proceeds on a number of metaphysical hypotheses has been pointed out from time to time – e.g. by Woodger (1929, p. 228), Collingwood (1940), Russell (1948, pp. 506 ff.), Margenau (1941, 1950), Bunge (1961, 1967a I pp. 291 ff.), Harré (1961, 1972), Agassi (1964, 1975), Lakatos (1969), Harvey (1969), Körner (1970) and Rosenblueth (1970, Chs. 7 and 10). The following list of ontological principles occurring in scientific research (Bunge, 1974d) must suffice here:

M1 There is a world external to the cognitive subject. If there were no such world it would not be subject to scientific inquiry. Rather we would resort to introspection or to pure mathematics instead of attempting to discover the unknown beyond the self.

M2 The world is composed of things. Consequently the sciences of reality (natural or social) study things, their properties and changes. If there were real objects other than things it would be impossible to act upon them with the help of other things.

M3 Forms are properties of things. There are no Platonic Forms in themselves flying above concrete things. This is why (*a*) we study and modify properties by examining things and forcing them to change, and (*b*) properties are represented by predicates (e.g. functions) defined on domains that are, at least in part, sets of concrete objects. (Think of fertility, defined on the set of organisms.)

M4 Things are grouped into systems or aggregates of interacting components. There is no thing that fails to be a part of at least one system. There are no independent things: the borders we trace between entities are often imaginary. What there really is, are systems – physical, chemical, living, or social.

M5 Every system, except the universe, interacts with other systems in certain respects and is isolated from other systems in other respects. Totally isolated things would be unknowable. And if there were no relative isolation we would be forced to know the whole before knowing any of its parts.

M6 Every thing changes. Even the so-called ultimate components of matter end up by changing in the course of their transactions with other things. Even the supposedly stable particles can be absorbed by other systems or merge with their corresponding antiparticles to form photons that may in turn be absorbed.

M7 Nothing comes out of nothing and no thing reduces to nothingness. If this were not so we would make no effort to discover both the origin of new things and the traces left by things that have been destroyed.

M8 Every thing abides by laws. Whether natural or social, laws are invariant relations among properties, and they are just as objective as properties. Moreover a law is a property. If there were no laws we would never discover any and would not utilize them in order to explain, foresee and do. In particular the experimental method would not be feasible, for its essence is the deliberate and controlled wiggling of a component or a variable of some system to find out what effect that change may have on some other feature of the system. Our unrelenting attempt to find out facts of this kind presupposes that there are lawful relations among the items concerned.

M9 There are several kinds of law (nomological pluralism). There are causal (or predominantly causal) laws and stochastic laws, as well as laws exhibiting these two and even further modes of becoming. There are same-level (e.g. biological) laws and cross-level (e.g. psychosocial) laws.

M10 There are several levels of organization: physical, chemical, biological, social, technological, etc. The so-called higher levels emerge from other levels in the course of processes; but, once formed – with laws of their own – they enjoy a certain stability. Otherwise we would know nothing about organisms and societies before having exhausted physics and chemistry – which are inexhaustible anyway.

There are certainly further ontological principles utilized more or less overtly in the selection of problems, in the formation of concepts and hypotheses, in the design of techniques, and in the evaluation of results of scientific research. However, the ones we have listed are enough to establish the thesis that *all science presupposes some metaphysics.* Surely most scientists are not aware of this fact or else they call such principles 'scientific' (e.g. Rosenblueth, 1970). But the fact is there and once in a while we come across some great scientist who does realize that he is upholding certain metaphysical hypotheses, i.e. who has a definite and conscious world view that guides his research. Galilei, Descartes, Leibniz, Newton, Euler, d'Alembert, Priestley, Faraday, Darwin, Maxwell, Einstein, Born, Heisenberg, Schrödinger, J. B. S. Haldane, Bernal, and Dobzhansky – to cite but a few – were among the great scientists aware of some of their own ontological principles. And if scientific research is guided – or misguided – by ontological axioms then it behoves the historian of science to dig them up, and the philosopher of science to formulate them clearly, to justify or criticize them, and eventually to systematize them. This is in fact the task of the *metaphysics of science* (Bunge, 1974d, 1974f). Some of it will motivate, and some will be incorporated into, our system of scientific metaphysics.

To be sure not all of the metaphysical hypotheses proposed or adopted by scientists are reasonable or fertile: some are unreasonable or obstructive. To cite but a few examples: (*a*) the Copenhagen interpretation of quantum mechanics claims that, whereas macrophysical objects are objectively real, their atomic components come into being only as a result of observations; (*b*) the many-worlds interpretation of quantum mechanics holds that at each instant the world we live in branches off into countless parallel universes, mutually disconnected and therefore unknowable; (*c*) a number of scientists deny that chance is an objective mode of being and are therefore committed to the subjectivist or personalistic interpretation of probability: they are die-hard classical determinists; (*d*) according to a widespread interpretation of the theory of evolution, natural selection leads to gradual perfection, which in turn is what lends "meaning" to life; (*e*) behaviorism rests on the Aristotelian principle that nothing can change spontaneously: that the organism is a deterministic automaton every output of which is caused by some environmental input; (*f*) many eminent neurophysiologists and neurosurgeons believe that the mind is something separate from the body that does the minding; (*g*) while some social scientists claim that there are no societal properties and laws above those characterizing the individuals, others refuse to recognize that societal emergents are rooted to the individuals and their interplay. In sum, science is shot through with metaphysics – good and bad, fertile and sterile.

Let us now turn to some areas of scientific research where metaphysics makes itself felt with particular acuteness.

8. Ontological Inputs and Outputs of Science and Technology

We shall presently defend the thesis that ontology can be found not only among the heuristic crutches that guide scientific research, but also (*a*) in some of the very *problems* handled by research, (*b*) in the *axiomatic reconstruction* of scientific theories, and (*c*) among *extremely general theories* in both basic and applied science.

That some of the most interesting scientific problems are at the same time metaphysical, is shown by the following examples:

(i) Is there an ultimate matter (*ápeiron* or *Urmaterie*)? This question triggered Heisenberg's 1956 theory of elementary particles.

(ii) Is life characterized by an immaterial principle (spirit, entelechy, *élan vital*, or any other substitute for ignorance), or has it emerged in a strictly natural evolutionary process? This question has prompted the current experimental studies on the origin of organisms.

(iii) Are biological species embodiments of Platonic archetypes, or just concrete populations, or something else? This question is being asked every day by almost every thoughtful taxonomist.

(iv) Is the mental anything disjoint from and beyond the neural? Are the two correlated, or do they interact, or finally is the mental a function of the nervous system? The latter alternative is no less than the motor of physiological psychology.

(v) Is a society anything beyond and above the individuals that compose it or are there societal laws in addition to the laws concerning individual behavior? This is one of the central disputes in today's methodology and philosophy of the social sciences.

The above questions have proved fruitful: they have triggered entire research lines. Alongside them there are other questions that result from a dubious metaphysics often at work in science, one that is a source of confusion and a sink of efforts. Here is a sample:

(vi) Why are μ-mesons different from electrons? (Why should they not be?)

(vii) Why are there no superluminal particles? (Why should there be?)

(viii) Is it possible to observe time reversal? (Inverting the sign of t in an equation of motion has no "operational" counterpart.)

(ix) Why should the study of behavior be hitched to an investigation of the behaving thing, i.e. the organism? (Because there is no change apart from a changing thing.)

(x) What is the liaison officer between body and mind? (Why presuppose that body and mind are separate entities?)

In sum, ontology inspires or blocks the most interesting research problems and plans. (This being so, the intervention of ontological principles in the design of research policies ought to be studied.)

A second area where ontology overlaps with science is the axiomatic foundation of scientific theories. If a scientific theory is axiomatized, some of the following concepts are likely to occur in it in an explicit fashion: part, juxtaposition, property, possibility, composition, state function, state, event, process, space, time, life, mind, and society. However, the specific axioms of the theory will usually not tell us anything about such fundamental and generic concepts: science just borrows them leaving them in an intuitive or presystematic state. Only ontology is interested in explicating and systematizing concepts which, since they are used by many sciences, are claimed by none. For example, physics asks not 'What is time?', biology 'What is life?', psychology 'What is the mental?', and sociology 'What is sociality?'. It is the task of ontology, jointly with the foundations of science, to try and supply answers to such questions and, in general, to clarify whatever idea science takes for granted or leaves in the twilight. That is, the metaphysician must fill some of the gaps in science.

There are such gaps in scientific knowledge not because certain facts are impregnable to the scientific method and require an esoteric intuition but because there are certain ideas which scientists use freely without bothering to examine, perhaps because they look obvious. The ontological theories that clarify and articulate the general ideas underlying a scientific theory T may be called the *ontological background* of T. This background becomes part and parcel of the axiomatized scientific theory (Bunge, 1967b, 1973b). That is, the metaphysical hypotheses contained in that background become blended with the scientific hypotheses proper. Note that such metaphysical ideas are now *constitutive* not just *regulative* or heuristic like those listed in Sec. 7. Whereas the latter may go down with the scaffolding the former remain in the theory.

In short, the axiomatic reconstruction of any fundamental scientific theory is bound to dig up, exactify and systematize certain ontological concepts. In this way the line between ontology and science is erased.

A third region where science merges with ontology is constituted by certain extremely general theories such as Lagrangian dynamics, the classical theory of fields (of any kind), and the quantum theory of fields (misnamed 'axiomatic field theory'). All these theories are generic in the sense that, far from representing narrow species of things, they describe the basic traits of whole genera of things. For example Lagrangian dynamics, which started out as a reformulation of classical particle mechanics, was eventually generalized to cover mechanical, electrical, and even biological and social systems. As long as the Lagrangian function is not specified, and its independent variables are not interpreted, the theory remains so general as to be indistinguishable from metaphysics. In fact Lagrangian dynamics may have been the earliest member of scientific metaphysics.

In sum, science itself has produced ontological theories by a process of generalization. Here again there is no border to be drawn.

A fourth and last area where ontology mingles with science is contemporary technology. In fact some of the theories included in the so-called information sciences and in systems theory are so general, and at the same time so precise, that they qualify as theories in scientific metaphysics. For example a general control (or cybernetic) theory will apply not only to machines with feedback loops but also to goal-seeking systems (members of higher animal species). Being so general, a theory of this kind will not represent any details of the system of interest: for example it will be insensitive to the nature of the materials and mechanisms involved. Hence it will not replace any of the special theories built by scientists and technologists. But such general theories provide insights and they link previously isolated fields of research. (Cf. Bunge, 1977a.)

In short, post-war technology has produced a number of ontological theories. In fact it has been more fertile than academic metaphysics.

The upshot of our discussion so far is this: science, whether basic or applied, is permeated with ontological ideas, now heuristic, now constitutive – so much so that some scientific theories are at the same time metaphysical. More precisely: every scientific theory, if extremely general, *is* ontological; and every ontological theory, if exact and in tune with science, *is* scientific. Therefore the expression *scientific metaphysics*, though perhaps shocking at first, designates an existing field and one that is hardly distinguishable from science.

True, ontological theories, whatever their degree of scientificity, cannot be tested empirically. Nor, for that matter, can extremely general scientific theories. (For example the basic equations of continuum mechanics cannot even be solved, let alone put to the test, unless they are adjoined special assumptions concerning the forces, the mass distribution, and the internal stresses.) Like the generic theories of science, those of ontology are *vicariously testable*, i.e. through the checking of more special theories gotten from the general ones by conjoining them with subsidiary assumptions (Bunge, 1973a, Ch. 2). In sum, concerning testability as much as concerning reference, there is no distinction between an ontological theory and an extremely general scientific one.

Because there is no clear-cut border between science and ontology, it is pointless to try to find the correct criterion for the strict demarcation between these two fields. The demarcation problem, which has occupied so many eminent brains from Kant on (see e.g. Popper, 1963), has disappeared. Another, far more interesting and rewarding task, has replaced it: namely that of digging up the metaphysics of science and technology, and building a scientific ontology.

9. USES OF ONTOLOGY

Metaphysics is a traditional branch of philosophy and as such it need not be justified in the eyes of the philosopher provided he is either pre-Kantian or post-positivist. (It has been remarked many times, and rightly so, that an antimetaphysician is one who holds primitive and unexamined metaphysical beliefs.) A philosopher may dislike the word and prefer some substitute for it, such as 'ontology', 'philosophical cosmology', 'general science', or 'scientific outlook'. But, since he acknowledges the existence of certain problems that have traditionally been called 'metaphysical', unless he is an irrationalist he won't deny that those problems deserve an answer and, moreover, that the best answer to any theoretical problem is supplied by a theory.

If metaphysics is a part of philosophy then it must be related to the other branches of this field: otherwise it would not constitute such a part. It can certainly be distinguished from the other branches of philosophy – mainly logic, semantics, epistemology, value theory, and ethics – but not separated from them. Thus it makes a lot of difference for ontology if one adopts a realist epistemology or a phenomenalist one: with the former one will attempt to disclose the structure of reality, with the latter only the structure of appearance. In turn, one's epistemology depends to a large extent upon one's ontology: if the latter is naturalist one will conceive of epistemology as the ontology of knowing; but if one adopts a spiritualist ontology then the way to a subjectivist epistemology is open. To generalize: ontology is closely linked to the rest of philosophy.

However, the close interactions between ontology and its sister disciplines should not blind us to their differences. In particular we must not mistake the categories of knowing (e.g. truth) with those of being (e.g. property). And yet this distinction is sometimes overlooked. For example Strawson (1959) has claimed that material bodies are basic among particulars, not because they are the ultimate constituents of everything, but because every particular is allegedly identified or reidentified in terms of material bodies. Surely material bodies are a condition for our knowledge of the external world. Yet so are our senses and our intellect and yet these are not basic particulars. In sum, while defending the unity of philosophy, i.e. the interrelations among its branches, we must not forget their differences.

In any case ontology hardly needs a defense among contemporary philosophers. It has even become fashionable among them to talk about ontology. On the other hand scientists will not be impressed by the statement that metaphysics is worthy because it happens to be part and parcel of philosophy. (After all, they were told by philosophers, over a century ago, that metaphysics is balderdash.) But scientists might learn to tolerate ontology if shown that, though at first blush useless (or even nonsense), it can be of some use, namely in the following ways:

(i) *Metaphysics can help dissolve pseudoquestions* that arise in science and originate in misconceptions. Example: "What is the root of the arrow of time: irreversibility, ignorance of microscopic details, experiment, the expansion of the universe, or what?" This question dissolves upon realizing that time has no objective arrow: that only processes can have an "arrow", i.e. be irreversible. But this realization requires a full-fledged theory of time, and this is a typical task for ontology.

(ii) *Metaphysics can dig up, clarify, and systematize* some basic concepts and principles occurring in the course of scientific research and even in scientific theories – constructs that are common to a number of sciences, so that no single science takes the trouble to regiment them, as is the case with "property" and "space". Some of these constructs are included in the ontological background of scientific theories: recall Sec. 8.

(iii) *Metaphysics can suggest new scientific problems, research plans, and even theories.* Example: the mechanistic metaphysics born in the early 17th century inspired scientific research for two centuries. Another: Nobel laureate Yukawa (1973) blames the current cult of blind computation and measurement, and the concurrent lack of a deep understanding of high energy physics, on the lack of an inspiring metaphysics.

(iv) *Metaphysics can render service by analyzing fashionable but obscure notions* such as those of system, hierarchy, structure, event, information, and possible world, as well as by criticizing popular misconceptions such as that the world is a collection of facts and the mind the control of the body.

(v) *Metaphysics can perform another public service by examining the ontological kernel of the current ideologies* and finding out whether they live up to the standards of contemporary intellectual work.

Briefly, ontology is far from useless: sometimes it fuels good work, at other times it is an obstacle to research, and at all times it is there, in the very midst of our speculations about the world.

10. Concluding remarks

By way of summary:

(i) Ontology can be exact or inexact, according as it uses any formal tools or fails to employ them.

(ii) Exact (or mathematical) ontology may be meaty or hollow, according as it tackles or evades the great problems concerning the nature of reality.

(iii) Exact and meaty metaphysics may or may not be scientific, according as it is contiguous with science or alien to it.

(iv) Scientific ontology is a collection of general or cross-disciplinary frameworks and theories, with a factual reference, mathematical in form and compatible with – as well as relevant to – the science of the day.

(v) There is no gap between good metaphysics and deep science: every scientific theory with a wide scope may be regarded as metaphysical, and every ontological theory that brings together and generalizes scientific results, or else occurs in the background of an axiomatized scientific theory, qualifies as scientific.

(vi) Science, both basic and applied, teems with metaphysical concepts and hypotheses: it presupposes certain ontological principles of both the heuristic and the constitutive kind, and it is a powerful source of metaphysical conjectures. In fact some theories are both metaphysical and scientific. Since there is no definite border line between the two fields, there is no demarcation problem left.

Enough of meta-metaphysical discourse. Let us build a new ontology.

CHAPTER 1

SUBSTANCE

One of Descartes' *regulae philosophandi* counsels us to start by simplifying, and one of our own maxims (Introduction, Sec. 4) enjoins us to hypothesize unobservables in order to account for appearances. A radical joint implementation of these two rules consists in stripping real things of all their properties – only for a start of course. What remains is the qualitatively indeterminate particular, the *bare individual.* Our bare individual will turn out to be similar to, though not identical with, Plato's formless matter and Aristotle's primary substance. Similar because neither has inherent properties (except possibly composition in our case), different because our bare individuals are endowed with the capacity of associating, i.e. of forming composite entities. Whereas each bare individual is nondescript, an aggregate of bare individuals has the definite property of being composed, i.e. of having parts. The association of bare individuals is then a beginning of complexity and thus a step towards realism.

Surely every real thing consists of some definite stuff endowed with definite properties: there are no bare individuals except in our imagination. This is a tacit assumption of factual science. In other words the Platonic concept of formless matter and the Aristotelian notion of unchanging substratum have been shelved by science. Still, although we no longer believe in the actual existence of nondescript things, we find it methodologically convenient to feign them – only to endow them with further properties later on, namely in Ch. 2. A fully qualitied individual, if substantial or concrete, will be called a *thing* in Ch. 3. And a complex thing with coupled components will be termed a *system* in Ch. 5. This is how we shall tackle the basic problem of traditional metaphysics, i.e. that of substance and attribute. First, then, we address ourselves to the concept of substance.

1. Association

1.1. *Concatenation and Its Ontological Interpretation*

Individuals can associate to form further individuals. When neither the kind of individual nor the manner of association are specified, we have the (bare) association of (bare) individuals. This notion of association can be formalized by the algebraic concept of concatenation, which is elucidated in the theory of semigroups. (Cf. Ljapin, 1963.)

A semigroup is a set S together with a binary, internal and associative operation $\circ$ of concatenation. The operation is said to be internal because it is defined in S, i.e. because S is closed under $\circ$. (I.e., concatenation does not give rise to individuals alien to S.) And concatenation is associative because it satisfies the law of associativity: if $x, y, z \in S$, then $x \circ (y \circ z) = (x \circ y) \circ z$. More succinctly: the structure $\langle S, \circ \rangle$, where S is a nonempty arbitrary set and $\circ$ a binary operation in S, is called a *semigroup* if and only if (i) S is closed under $\circ$, and (ii) $\circ$ is associative in S. The semigroup $\langle S, \circ \rangle$ is finite if S has a finite number of elements, infinite otherwise. A simple example of a finite semigroup is this: $S = \{0, 1, 2\}$ with the multiplication (concatenation) table

$\circ$	0	1	2
0	0	1	2
1	1	1	2
2	2	2	2

And a simple example of an infinite semigroup is the set $\mathbb{N}$ of positive integers with respect to addition. In other words the structure $\langle \mathbb{N}, + \rangle$ is a model of a semigroup. (For the semantic concept of a model see e.g. Vol. 2, Ch. 6, Sec. 2.)

If we interpret S as the totality of bare individuals and $\circ$ as their pairwise association, we obtain the simplest, most basic and most useful theory in scientific metaphysics: we call it *association theory*.

We shall do slightly better by adding the assumption that our semigroup, when written additively, contains the null element, and interpret the latter as the *null individual*. Roughly, the neutral element $\Box$ of S is that (fictitious) individual which, when coupled to an arbitrary member x of S, leaves x unaltered, i.e. $x \circ \Box = \Box \circ x = x$. Formally, $\Box$ is the two-sided zero of S. The new structure, $\langle S, \circ, \Box \rangle$, is called an (additive) semigroup with neutral element, or *monoid* for short. This structure is slightly richer than the corresponding semigroup, for it is more specific: now one of the elements of S, namely $\Box$, is no longer a faceless individual but is distinct from all the others – not by an intrinsic quality but by its inane behavior in company. Since $\Box$ is the nonentity or nonbeing, it is even more fictitious than any of the nondesignated elements of S. The introduction of $\square$ is justified by the fact that it will allow us to state that nothing starts from $\square$ or ends up in $\square$ – an important ontological principle indeed. (More on $\square$ in Sec. 3.1.)

1.2. *Axiomatic Foundation of Association Theory*

We postulate then that the set of individual things, when stripped of their properties and interactions – except for the property of composition – has the monoid structure. Moreover we assume that the association order makes no difference, i.e. that the concatenation $\circ$ is *commutative*: if $x, y \in S$, then $x \circ y = y \circ x$. We also assume that the association of an individual with itself leaves it unchanged, i.e. that the elements of S are *idempotent*: $x \circ x = x$. (On the other hand we do not postulate that the association operation has the cancellation property, i.e. that, for all a, b, x in S, the equality $a \circ x = b \circ x$ implies that $a = b$.) The previous assumptions entail that the result of the association of two individuals is non-null provided at least one of them is not null: for all x, y in S, if $x \neq \square$ or $y \neq \square$, then $x \circ y \neq \square$. (I.e., association does not result in annihilation.) We call this property *substance conservation*.

In sum, we assume that the set S of bare individuals is a commutative monoid of idempotents. And we assume that this structure carries over to all real things: that is, we postulate that the totality of concrete or substantial particulars, whether viewed as bare or as fully qualitied, has the monoid structure. (Caution: S is not the world, or reality, but the set of its elements. The world is one more individual, the supreme one, whereas a set is a concept. More on this in Sec. 3.1.) The preceding assumptions are compressed in the following compound

POSTULATE 1.1 Let S be a non-empty set, $\square$ a selected element of S, and $\circ$ a binary operation in S. Then the structure $\mathcal{S} = \langle S, \circ, \square \rangle$ satisfies the following conditions:

(i) $\mathcal{S}$ is a commutative monoid of idempotents;
(ii) S is the set of all substantial or concrete individuals;
(iii) the neutral element $\square$ is the null individual;
(iv) $\circ$ represents the association of individuals;
(v) the string $a_1 \circ a_2 \circ \ldots \circ a_n$, where $a_i \in S$ for $1 \leqslant i \leqslant n$, represents the individual composed of the individuals a_1 to a_n.

Remark 1 This axiom has a factual content insofar as it refers ultimately to real things and is far from being tautological. *Remark 2*

However, our postulate is hardly refutable. For, if any thing seemed to fail to satisfy our axiom, we would not say it associates. (More on this peculiarity of ontological theories in Bunge (1973a).) *Remark 3* The condition of commutativity is the only law statement involved in the above postulate. It can be rephrased thus: Association is a symmetric relation in the sense that "x associates with y" is equivalent to "y associates with x". *Remark 4* Since Postulate 1.1 states that entities can associate to form complex entities, it is inconsistent with Parmenides' block universe. *Remark 5* We have interpreted concatenation as association, whether natural or brought about by man. A narrower, operationist interpretation would view $\circ$ as the putting together of perceptible things by somebody. *Remark 6* It would be interesting to investigate even poorer structures, namely the nonassociative algebras such as groupoids (cf. Bruck, 1958). But we shall not take up this matter here. Likewise we shall not study the ontological applications of non-commutative semigroups.

Before introducing further assumptions we need some definitions. First

DEFINITION 1.1 An individual is *composite* iff it is composed of individuals other than itself and the null individual. I.e., $x \in S$ is composite iff there exist substantial individuals $y, z \in S$ such that $x = y \circ z$ and each differs from x as well as from $\square$. Otherwise the individual is *simple.*

Remark 1 According to the definition, the trivial composition $x \circ x = x$ does not count. *Remark 2* Note the difference between the composite entity $z = x \circ y$ and the set $\{x, y\}$ of its constituents or components. The latter is a concept and does not satisfy Postulate 1.1.

COROLLARY 1.1 The null individual is simple.

Proof $\square$ is simple because, for every non-null individual x, $x \circ \square \neq \square$ by Postulate 1.1 (ii).

Remark The null individual need not be the only simple one.

And now another key concept:

DEFINITION 1.2 If x and y are substantial individuals, then x is a *part* of y iff $x \circ y = y$. Symbol: $x \sqsubset y$.

Justification The relation $\sqsubset$ is a *partial order* relation, which is what the part-whole relation is supposed to be. In fact, from the previous definition and the properties of $\circ$ (Postulate 1.1) it follows that, for all $x, y, z \in S$,

(i) $\sqsubset$ is *reflexive*, i.e. $x \sqsubset x$;
(ii) $\sqsubset$ is *asymmetric*, i.e. if $x \neq y$ then: $x \sqsubset y \Rightarrow \neg(y \sqsubset x)$;
(iii) $\sqsubset$ is *transitive*: $x \sqsubset y$ & $y \sqsubset z \Rightarrow x \sqsubset z$.

Remark 1 The part-whole relation $\sqsubset$ must not be mistaken for the set inclusion relation $\subseteq$, much less for the set membership relation $\in$. Indeed whereas $\subset$ and $\in$ are relations among concepts of a certain kind (sets), $\sqsubset$ is a relation among substantial individuals. Besides, unlike $\sqsubset$, $\in$ is not transitive. (For example, if $A = \{x\}$ and $B = \{A\}$, then $x \in A$ and $A \in B$ but $x \notin B$.) *Remark* 2 We could have defined the converse $\sqsupset$ of $\sqsubset$. This is actually Whitehead's concept of "extension": *x extends over y* iff $x \circ y = x$ (Whitehead, 1919). But because here neither relation is construed in spatial terms, we might as well avoid geometric metaphors at this point. Indeed, an individual on our planet and another in a distant galaxy may be taken to associate to form a third individual, so that each component will be a part of the whole, just as much as the two components of a miscible fluid poured into a glass. *Remark 3* With the help of the previous definitions we can easily prove that

For any $x \in S$: x is simple iff, for all $y \in S$, $y \sqsubset x \Rightarrow y = x$ or $y = \Box$.

Remark 4 The condition "For every $x \in S : x \circ \Box = \Box \circ x = x$", defining the null individual, can now be read as follows: The null individual is part of every individual. We proceed to introduce the dual of this concept:

POSTULATE 1.2 There exists an individual such that every other individual is part of it. I.e., $(\exists x)[x \in S \;\&\; (y)(y \in S \Rightarrow y \sqsubset x)]$.

DEFINITION 1.3 The universal individual introduced by Postulate 1.2 is called *the world* and is denoted by $\blacksquare$.

Remark 1 Note again that the world, i.e. $\blacksquare$, is an individual not to be confused with the set S of all individuals, which is a concept not a physical object. *Remark 2* The distinguishing algebraic property of $\blacksquare$ is this: for every $x \in S$, $x \circ \blacksquare = \blacksquare \circ x = \blacksquare$. See Corollary 1.5.

DEFINITION 1.4 Let x denote an arbitrary object and $\blacksquare$ the world. Then
(i) x is *worldly* iff $x \sqsubset \blacksquare$;
(ii) x is *unworldly* iff x is not worldly (i.e. if x does not satisfy Postulate 1.1 and a fortiori x is not a part of $\blacksquare$).

Example The world is worldly but cosmology is not.

DEFINITION 1.5 Let x and y be substantial individuals. Then

(i) x and y are *detached* (*separate*) iff neither is part of the other:

$$x \lfloor y =_{df} \neg(x \sqsubset y) \;\&\; \neg(y \sqsubset x);$$

(ii) x and y *intersect* iff they are not detached:

$$x \bigcirc y =_{df} x \sqsubset y \vee y \sqsubset x.$$

The part-whole relation allows us to define yet another interesting concept that will be used throughout this work:

DEFINITION 1.6 The *composition* of an individual equals the set of its parts. I.e., let $\mathscr{C}: S \to 2^S$ be a function from individuals into sets of individuals, such that $\mathscr{C}(x) = \{y \in S \mid y \sqsubset x\}$ for any $x \in S$; then $\mathscr{C}(x)$ is called the *composition* of x.

Let us now derive a few consequences from our assumptions and definitions.

1.3. *Consequences*

First of all the trivial yet necessary

COROLLARY 1.2 The totality S of substantial individuals is partially ordered by the part-whole relation $\sqsubset$. I.e., $\langle S, \sqsubset \rangle$ is a poset.

Proof Immediate by recalling the justification of Definition 1.2.

THEOREM 1.1 The association of any two individuals is the supremum (least upper bound or l.u.b.) for them with respect to the part-whole ordering:

$$\text{If } x, y \in S \quad \text{then} \quad sup\{x, y\} = x \circ y.$$

Proof By associativity and idempotence $x \circ (x \circ y) = (x \circ x) \circ y = x \circ y$. By Definition 1.2 the last formula is the same as: $x \sqsubset (x \circ y)$. Likewise $y \circ (x \circ y) = x \circ y$, whence $y \sqsubset (x \circ y)$. Hence $x \circ y$ is an upper bound of x and y. It is also their least upper bound. In fact call z an upper bound of x and y, i.e. $x, y \sqsubset z$. Then $x \circ y \circ z = z$. Thus $x \circ y \sqsubset z$.

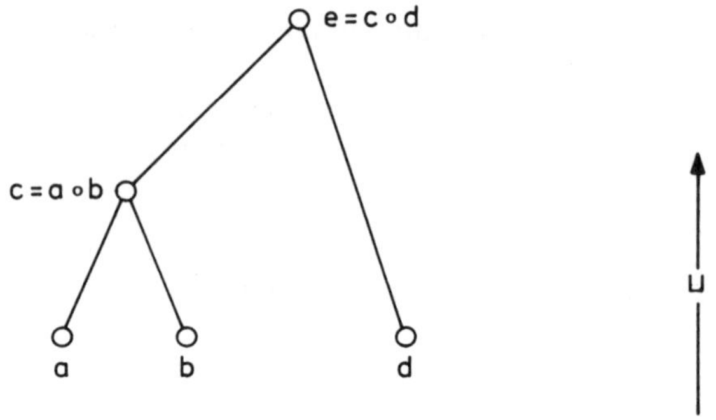

Fig. 1.1. All entities associate and they are ordered by the ⊏ relation.

The concept of supremum generalizes to arbitrary (in particular infinite) sets that of concatenation, which is defined only for pairs of individuals. We had postulated that, if x and y are substantial individuals, so is $x \circ y$ (Postulate 1.1). We now generalize that hypothesis to an arbitrary set of individuals: see Figure 1.1. That is, we assume

POSTULATE 1.3 Every set $T \subseteq S$ of substantial individuals has a supremum, which is denoted by $[T]$. I.e., for any $T \subseteq S$, there exists an individual $[T] \in S$ such that

(i) $x \sqsubset [T]$ for all $x \in T$;

(ii) if $y \in S$ is an upper bound of T, then $[T]$ precedes y: i.e., if $x \sqsubset y$ for all $x \in T$, then $sup\ T = [T] \sqsubset y$.

This axiom allows us to elucidate an important metaphysical notion:

DEFINITION 1.7 Let $T \subseteq S$ be a set of substantial individuals. Then the *aggregation* or *association* of T, or $[T]$ for short, is the supremum of T. I.e., $[T] = sup\ T$.

Remark 1 If T is a finite set then its aggregation $[T]$ is the association of all the members of T. I.e., if $T = \{x_1, x_2, \ldots, x_n\}$, then $[T] = sup\ T = x_1 \circ x_2 \circ \ldots \circ x_n$. In particular, the aggregation of a singleton is its sole occupant: if $T = \{x\}$, then $[T] = x$. *Remark 2* Because association is a finitary operation it is not possible to define the aggregation of an arbitrary (possibly infinite) set $T \subseteq S$ as the concatenation of all its members. This is why we had to employ the more general concept of *sup* or l.u.b. Let us now apply it to the entire collection of substantial individuals.

THEOREM 1.2 The world is the aggregation of all individuals:

$$▯ = [S] = \mathit{sup}\, S.$$

Proof By Postulate 1.2, ▯ exists and is the last individual, i.e. for every $x \in S$, $x \sqsubset$ ▯. But this individual fits the conditions of Definition 1.7, i.e. ▯ $= \mathit{sup}\, S$.

The preceding material is compressed into

THEOREM 1.3 The ordered quadruple $\langle S, \circ, \Box,$ ▯$\rangle$ is a sup-semilattice with least element $\Box$ and last element ▯ with respect to the part-whole relation $\sqsubset$.

Proof By Theorem 1.1 there exists a supremum for any two individuals, namely their association. Besides, $\Box$ is part of every individual, so it lies at the bottom of the net. Dually, ▯ contains every individual, so it perches on top. These conditions, together with the commutativity of association, define a sup-semilattice. See Figure 1.2.

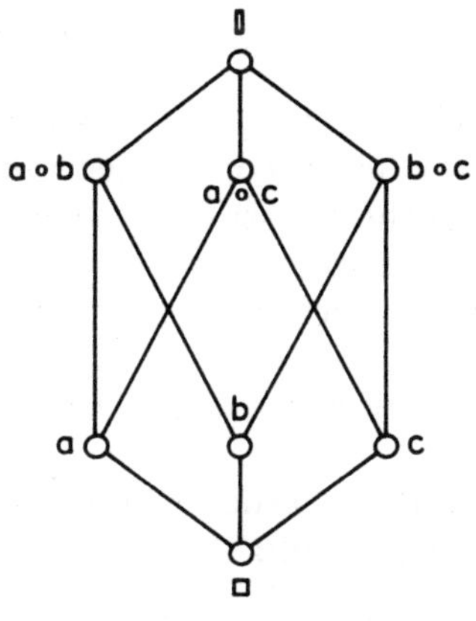

Fig. 1.2. Semilattice generated by $S = \{\Box, a, b, c,$ ▯$\}$.

THEOREM 1.4 Every substantial individual is the aggregation of its parts, i.e. of its composition. I.e., for every $x \in S$,

$$x = [\mathscr{C}(x)] = \mathit{sup}\, \{y \in S \mid y \sqsubset x\}.$$

Proof Apply Definition 1.7 to $\mathscr{C}(x)$ characterized by Definition 1.6.

COROLLARY 1.3 Every individual with a finite composition is the association of its predecessors with respect to $\sqsubset$.

Example If $\mathscr{C}(c) = \{a, b\}$, then $c = a \circ b \circ c$, and $a, b \sqsubset c$.

Now two further consequences of Postulate 1.2. First

THEOREM 1.5 There is just one world.

Proof By Postulate 1.3 every $T \subseteq S$ has a supremum and by Theorem 1.2 $[S] = \Box$. And all suprema are unique. *Alternative proof* Suppose there were a world W different from $\Box$. Then since every substantial individual is part of $\Box$, so must be W. Hence W deserves being called a subworld rather than the world.

Remark Like many statements in this chapter, this one is mathematically trivial. However, it is philosophically nontrivial in view of the fashionable possible worlds metaphysics (Ch. 4, Sec. 6.2), the many-worlds (mis)interpretation of quantum mechanics (Ch. 4, Sec. 6.6), and Popper's three worlds.

COROLLARY 1.4 Whatever is unworldly is not a substantial individual.

Proof By Postulate 1.2, for every $x \in S$, $x \sqsubset \Box$. By contraposition, if $\neg(x \sqsubset \Box)$ then $\neg(x \in S)$.

Remark Classes, such as S, every one of its subsets, and its power set, are unworldly. So is any object constructed out of classes. So is any object that fails to satisfy the clauses of Postulate 1.1. Any object other than a substantial individual (or member of S) will be called a *construct*. I.e., nothing is both a construct and a substantial individual. More on this in Sec. 3.2.

Next we prove a bunch of conservation principles. First

THEOREM 1.6 If association ensues in the null individual then the associates were null to begin with:

$$\text{For all } x, y \in S,\ x \circ y = \square \Rightarrow x = \square \ \&\ y = \square.$$

Proof Assume the antecedent, i.e. $x \circ y = \square$. Now, $x = x \circ \square = x \circ (x \circ y) = (x \circ x) \circ y = x \circ y = \square$ by associativity, idempotence, and hypothesis. Similarly for y.

THEOREM 1.7 No substantial individual associates destructively with another: If $x \neq \square$ then there is no $y \in S$ such that either $x \circ y = \square$ or $y \circ x = \square$.

Proof Assume $x \in S$ different from $\square$ and suppose there is in S an element y such that $x \circ y = \square$. Then, by Theorem 1.6 and modus tollens, $x = \square$, contrary to hypothesis.

Remark 1 In other words, except for $\square$, S has no inverses. Hence (a) S cannot possibly attain a group structure, and (b) the belief that there is an anti-thing for any given thing, such that they annihilate each other, is wrong. *Remark 2* It is sometimes said that, when particles of matter and antimatter join, they "annihilate" reciprocally. There is no such annihilation; there is rather the conversion of a pair of material entities (e.g. an electron and a positron) into a field chunk (e.g. a gamma ray photon).

COROLLARY 1.5 Provided it starts with something, association does not end up in nothingness:

$$\text{For all } x, y \in S, x \neq \square \;\&\; y \neq \square \Rightarrow x \circ y \neq \square.$$

Proof By Theorem 1.6 and contraposition.

COROLLARY 1.6 Nothing comes by association out of nought.

Proof This is the ontological interpretation of the implicit definition of the null individual, i.e. $x \circ \square = \square \circ x = x$.

COROLLARY 1.7 The world is not enlarged by associating it with any of its parts: For any $x \in S$, $x \circ ▯ = ▯ \circ x = ▯$.

Proof By the definition of world.

Remark Corollaries 1.5 and 1.6 were first stated by Epicurus, then restated by Lucretius, in the form of postulates: "Matter is indestructible" and "Matter is uncreatable". (Cf. DeWitt, 1954; Lucretius 55 B.C.)

Another axiom of Greek atomism was that every thing is either basic (simple) or composed of basics (simples). We shall adopt this hypothesis that there are basic building blocks out of which every thing, from molecule to star, from cell to ecosystem, is composed, because it has been tremendously fruitful in modern science. That is, we lay down

POSTULATE 1.4 There is a proper subset B of S such that every substantial individual is the aggregation of members of B. I.e., for every $x \in S$ other than the null individual, there is a unique subset $B_x \subset B$ such that $x = [B_x]$.

Consequently instead of using the full aggregation function $[\,]: 2^S \to S$, we need only its restriction to the power set of the simples or basics, i.e. the bijection $[\,]\|2^B: 2^B \to S$.

We shall make explicit use of Postulate 1.4 in our theory of space (Ch. 6). By the way, simples (basics) need not be spatially unextended. By Definition 1.1 a simple or basic individual is one that is not composed of any other individuals. For example, according to present day theories an electron is such a simple or basic because it is an indivisible whole even though both its electromagnetic field and its ψ-field are spatially extended. Another way of putting this is to say that electrons, though spatially extended, have no proper parts. In general,

$$\text{If} \quad x, y \in B \quad \text{and} \quad x \sqsubset y, \quad \text{then} \quad x = y.$$

So much for association theory. We shall generalize it in Sec. 2, but before that another two interesting notions have to be explored.

1.4. *Atom Aggregates*

A tacit assumption of association theory is that no two individuals are identical. In fact if a subset T of S consists of two identical individuals then, by the extensionality principle of set theory, T is a singleton. Association theory makes therefore no room for the possibility that some individuals come in multiple copies, as in the case of the letters of an alphabet, which can be duplicated *ad libitum*. Let us explore this alternative.

Assume there are simples and that every thing is either a simple or an association of simples – i.e. accept Postulate 1.4. And assume also that a single simple a may be found repeated any number of times, forming the composite entities $a \circ a$, $a \circ a \circ a$, etc. Two simples a and b will of course give rise to a greater variety: $a \circ a$, $a \circ a \circ a$, ..., $b \circ b$, $b \circ b \circ b$, ..., $a \circ b$, $a \circ a \circ b$, $a \circ a \circ b$, ..., $a \circ b \circ b$, $a \circ b \circ b \circ b$, ... Obviously even a few basic individuals, provided there are enough copies of them, will give rise to both numerosity and variety: this was a capital discovery of the ancient atomists, both Greek and Indian: *Multum ex parvo*.

To formalize the previous idea we start from a basic set A of distinct atoms or simples and give up the postulate of idempotency. Call then $A = \{a_i | i \in \mathbb{N}\}$ the set of basic elements or atoms and allow them to associate, as in $a_i \circ a_j$ and $a_j \circ a_i$, counting the association of two replicas of a given atom a_i, i.e. $a_i \circ a_i$, as a third individual distinct from a_i. Since an atom aggregate is not an atom itself, A is not closed under association. What is closed is the set A^* of all the concatenates. For example, a single atom generates infinitely many complexes. Indeed, if $A = \{a\}$, then $A^* = \{a^n | n \in \mathbb{N}\}$, where a^n abbreviates the concatenation of n identical copies of a. The set A^* is isomorphic to the set of natural numbers, so a universe formed of atoms of a single kind could still be infinitely complex. In general, if A is a non-empty set of atoms or basics, the collection A^* of concatenates of members of A is called the *free monoid generated by* A.

Example 1 The set of all conceivable chemical compounds is the free monoid generated by the set of chemical elements or atom species – just over 100. However, the set of compounds permitted by the laws of chemical binding is a finite subset of that free monoid. Thus the molecule HHHHHH, though thinkable, is not chemically possible. More on real possibility as distinct from conceptual possibility in Ch. 4.
Example 2 The minimal structure of a language with alphabet Σ is the free monoid generated by Σ: cf. Vol. 1, Definition 1.1.

We could assume, as a crude model of the world, that its furniture is a free monoid over some finite set S of simples. However, we know this to be an oversimplification: in the real world things do not come in strictly identical copies but, at most, in semi-copies. Moreover not all conceivable concatenations are really possible. However, certain subsets of S – elementary particles, atomic nuclei, atoms, molecules, cells, etc. – do come in *almost* identical copies. Hence for such things the free monoid over a finite set of atoms or simples is a sufficiently adequate (approximately true) model.

1.5. *Clustering*

The associations and dissociations of real things are brought about in specific ways – e.g. by motions or by forces –, hence their study pertains to the special sciences. However, certain features of such processes, such as the grouping of a collection of individuals into clusters, can be described a priori. So much so that this problem is the subject of a branch of pure mathematics, namely combinatorics (or combinatorial analysis). Let us take a glimpse at a couple of typical problems in combinatorics.

Problem 1 In how many ways can n individuals associate if association (concatenation) is not assumed to be commutative? That is, how many permutations of n objects (conceptual or physical) are possible? Answer: $n!$ (i.e. $1.2.3 \ldots n$). Example: the individuals a, b, c can associate thus: *abc*, *bca*, *cab*, *bac*, *acb*, *cba*.

Problem 2 Consider n objects grouped into clusters (or cells or blocks) of m individuals each (where $m \leqslant n$) without regard to order. How many such clusters are there? Answer: there are $n!/(n-m)!m!$ combinations of n individuals taken m at a time.

Problem 3 A set of n objects is partitioned into m clusters, with $m < n$. How many clusters will contain two or more individuals? Answer: at least one (Dirichlet's pigeonhole principle).

Problem 4 Consider a set of n objects. In how many ways can it be partitioned? For example, $S = \{a, b, c\}$ can be partitioned in the five following ways:

$$\{a\}, \{b\}, \{c\} \qquad \langle 3, 0, 0\rangle$$

$$\{a, b\}, \{c\}; \{b, c\}, \{a\}; \{a, c\}, \{b\} \qquad \langle 1, 1, 0\rangle$$

$$\{a, b, c\} \qquad \langle 0, 0, 1\rangle.$$

Every partition is characterized by its type $\langle m_1, m_2, \ldots, m_n\rangle$, where m_1 is the number of singles, m_2 that of couples, and so on. In every partition the number of blocks is $b = m_1 + m_2 + \ldots + m_n$, and $n = m_1 + 2m_2 + \ldots + nm_n$.

Like arithmetic, combinatorics is a theory of objects of any kind. It is therefore too general to pass for an ontological theory. (Recall the Introduction, Sec. 6.) But of course it can enter ontology, along with other mathematical theories, as an auxiliary.

1.6. *Historical Remark*

Our association theory bears a superficial resemblance to the ontology proposed by Lesniewski in 1916 (see Sobociński, 1954–55) and worked out in different directions by Tarski (1927), Leonard and Goodman (1940), Lejewski (1967), and a few others. The main differences are these: (*a*) our theory is much simpler than the others, if only because it adopts a standard mathematical formalism (namely semigroup theory) and does not eschew the language of set theory – which Lesniewski was careful to avoid; (*b*) whereas the Lesniewski theory (called *mereology*) and its offspring form the totality of contemporary nominalist ontology, our association theory constitutes only the foundation stone of a vast edifice containing items such as properties and kinds, which are anathema to the nominalist; (*c*) our association theory, and the assembly theory to follow, have been conceived not only as philosophical theories but also as theories utilizable in the foundations of science, as we shall explain in Sec. 4.

2. Assembly

2.1. *Juxtaposition and Superposition: Intuitive Idea*

The notion of association, elucidated in the previous section, is extremely general: there are many ways in which two substantial individuals can associate. We shall presently introduce two specific concepts of association: those of juxtaposition or physical sum, and of superposition or physical product. Two things placed side by side add or juxtapose, while two fluids, when mixed, superpose. We shall formalize these modes of association as so many binary operations in the set S of substantial individuals, designating juxtaposition by '$\dot{+}$' and superposition by '$\dot{\times}$'.

Example 1. Let the universe be cut into two mutually detached parts named a and b, i.e. ▯ $= a \dot{+} b$, with $a \dot{\times} b =$ □. The set $S = \{a, b,$ □, ▯$\}$ is a complete lattice, since all the members of S can join pairwise in two ways without enlarging S: look at Figure 1.3(*a*).

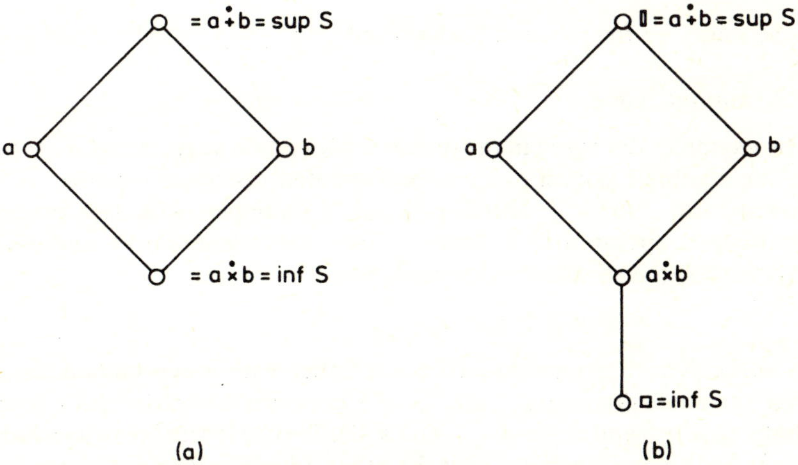

Fig. 1.3. Two examples of juxtaposition and superposition.

Example 2 Let the universe consist of two superposed things, i.e. ▯ $= a \dot{+} b$, with $a \dot{\times} b \neq$ □. The set $S = \{a, b, a \dot{\times} b,$ □, ▯$\}$, which includes the previous set, has the lattice structure too: look at Figure 1.3(*b*).

Example 3 Consider three individuals a, b, c, such that a superposes with both b and c. Suppose furthermore that the individuals are fields: look at Figure 1.4(*a*). It is seen that superposition distributes over juxtaposition.

Example 4 Juxtapose a field a and a field $b \dot{\times} c$. The result is that juxtaposition distributes over superposition. See Figure 1.4(*b*).

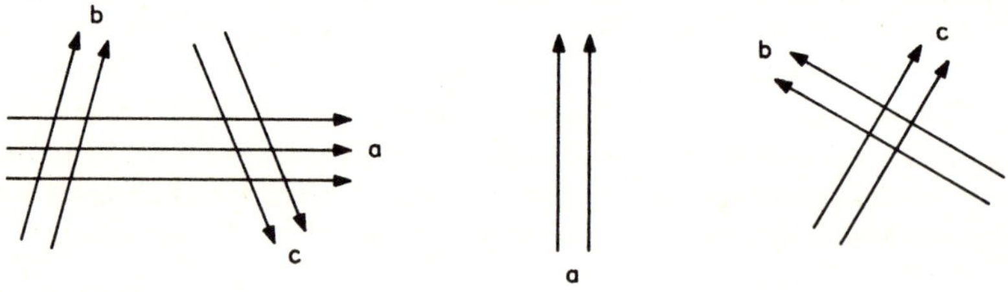

(a) (b)

Fig. 1.4. (*a*) The total entity is $a \dot{\times} (b \dot{+} c) = (a \dot{\times} b) \dot{+} (a \dot{\times} c)$. (*b*) The total entity is $a \dot{+} (b \dot{\times} c) = (a \dot{+} b) \dot{\times} (a \dot{+} c)$.

So much for the intuitive background.

2.2. *Formalization*

The examples discussed in the previous subsection suggest that we rope in the algebraic concept of a complemented distributive lattice. (Cf. Rutherford, 1965.) A lattice is a set S equipped with two binary associative operations, $\vee$ and $\wedge$, that are commutative and tied together by the so-called absorption laws

$$x \wedge (x \vee y) = x \vee (x \wedge y) = x.$$

A uniquely complemented lattice is a lattice with universal bounds □ and ▯ and such that every element x of S has a single complement x' such that $x \vee x' =$ ▯ and $x \wedge x' =$ □. And a distributive lattice is one in which every operation distributes over the other. Finally a Boolean lattice is a lattice of idempotents which is both complemented and distributive.

More explicitly: the structure $\mathscr{L} = \langle S, \vee, \wedge, ', \Box, \text{▯} \rangle$ is a *Boolean lattice* iff, for all x, y and z in S, the following equalities hold:

Associativity $x \vee (y \vee z) = (x \vee y) \vee z, \quad x \wedge (y \wedge z) = (x \wedge y) \wedge z$

Commutativity $x \vee y = y \vee x, \quad x \wedge y = y \wedge x$

Absorption $x \vee (x \wedge y) = x, \quad x \wedge (x \vee y) = x$

Distributivity $x \vee (y \wedge z) = (x \vee y) \wedge (x \vee z), x \wedge (y \vee z) = = (x \wedge y) \vee (x \wedge z)$

Least element $x \vee \Box = x, x \wedge \Box = \Box$

Last element $x \vee \text{▯} = \text{▯}, x \wedge \text{▯} = x$

Inverse $x \vee x' = \text{▯}, x \wedge x' = \Box$.

Lattice theory together with the ontological interpretation we sketched in the previous subsection constitutes the ontological theory to be called *assembly theory*. The foundation of this theory is

POSTULATE 1.5 The structure $\mathscr{L} = \langle S, \dot{+}, \dot{\times}, ', \Box, \text{▯} \rangle$, where S is a non-empty set, $\Box$ and ▯ selected elements of S, $\dot{+}$ and $\dot{\times}$ binary operations in S, and $'$ a unary operation in S, is a uniquely complemented and distributive lattice of idempotents obeying the following additional conditions:

(i) S is the set of all substantial individuals;

(ii) $\Box$ is the null individual and ▯ represents the world;

(iii) for any individuals x and y, $x \dot{+} y$ represents the juxtaposition (physical addition) of x and y, whereas $x \dot{\times} y$ represents the superposition (physical product) of x and y;

(iv) the inverse (complement) of an individual x, i.e. the individual x' such that $x' \dot{+} x = \text{▯}$ and $x \dot{\times} x' = \Box$, represents the environment or external world of x.

Remark 1 The unary operation (inversion or complementation) is defined for every individual entity. The world and the null individual are the inverses of one another. (The world, then, has no environment.) Needless to say, x' is not the opposite of x but just its complement to the world. *Remark 2* The two binary operations are defined for every pair of individuals. But whereas the juxtaposition of any two non-null entities is a third non-null individual, their superposition may be null – in which case they are detached, as we shall see in Definition 1.8. *Remark 3* At first sight superposition can result in annihilation, and wave interference would seem to illustrate this process. This is not so: two fields can superpose destructively over part of the region where they exist but not over the whole of it – otherwise there would be no conservation of energy and angular momentum. *Remark 4* When two or more individuals associate additively to form a whole they retain their identities. Consequently an aggregate, as treated by the present theory, is fully characterized by its composition. *Remark 5* By distinguishing two modes of association, i.e. juxtaposition and superposition, assembly theory allows us to perform a finer analysis of wholes than association theory did. Even so, $\dot{+}$ and $\dot{\times}$ are insufficient for scientific purposes, as things are endowed with a number of properties that determine their mode of coupling, which can be extremely complex. Just think of chemical and social bonds.

The preceding axiom will allow us to refine a few interesting notions.

2.3. *Definitions*

Since we now distinguish two modes of association we may expect roughly twice as many derived concepts as in association theory. Here are some.

DEFINITION 1.8 Let x and y be two distinct individuals. Then

(i) x and y are *detached* (or *separate*) iff their superposition is nought:

$$x \rceil y =_{df} x \dot{\times} y = \Box;$$

(ii) x and y *intersect* iff they are not detached:

$$\overset{\frown}{\underset{\smile}{x\ y}} =_{df} x \dot{\times} y \neq \Box.$$

Remark No spatiotemporal notions are involved in these definitions. On the contrary, the above notion of separateness will occur in the definitions of spatial concepts in Ch. 6.

DEFINITION 1.9 An individual $z \in S$, other than $\Box$,

(i) is *composed additively* of the individuals x and y iff $z = x \dot{+} y$;

(ii) is *composed multiplicatively* of the individuals x and y iff $z = x \dot{\times} y$.

Examples The contents of a room is composed additively of all the pieces of furniture in it. The gravitational field in a room is composed multiplicatively of the terrestrial field and the fields attached to the house.

DEFINITION 1.10 A substantial individual is *composite* (or *complex*) iff it is composed additively of individuals other than itself and the null individual. Otherwise it is *simple* (or atomic or basic).

DEFINITION 1.11 If x and y are substantial individuals, then x is a *part* of y iff x adds nothing to y:

$$x \sqsubset y =_{df} x \dotplus y = y.$$

Justification The part-whole relation $\sqsubset$ has the desired properties: reflexivity, asymmetry, and transitivity. It is then a partial order relation. If a and b are additive components of c, they are also parts of c, i.e. they precede c in the order generated by $\sqsubset$. On the other hand if a and b are multiplicative components of c, they follow the individual $c = a \dot{\times} b$.

As in Sec. 1.2, we define the composition of a substantial individual as the collection of its parts: we just copy Definition 1.6. The complexity of an individual may be defined as the numerosity of its composition, i.e.

$$\text{For any } x \in S, \text{ the } \textit{complexity} \text{ of } x =_{df} |\mathscr{C}(x)|,$$

where $|S|$ stands for the cardinality of S. However, this is a coarse measure of ontic (nonconceptual) complexity, as it accounts only for the number of components or parts not for the manner of their coupling. But since we do not yet have the concept of coupling this is all we can do for the moment.

Concerning the complexity of the world we are faced with a choice among the following hypotheses:

H1 *Finitism*: The complexity of $\square$ is finite.

H2 *Denumerable infinitism*: The complexity of $\square$ = aleph zero.

H3 *Infinitism*: The complexity of $\square$ = the power of the continuum.

However, we need not make any such choice at this point.

We now have two different concepts of aggregation:

DEFINITION 1.12 Let $T \subseteq S$ be a set of substantial individuals. Then

(i) the *additive assemblage* of T, or $[T]$ for short, is the supremum (l.u.b.) of T: $[T] = \sup T$;

(ii) the *multiplicative assemblage* of T, or (T) for short, is the infimum (g.l.b.) of T: $(T) = \inf T$.

Remark 1 Actually we have just introduced two different assemblage functions :[ ], ( ): $2^S \to S$, each of which takes a set of entities into an individual entity. By assuming Postulate 1.4 we need only the restrictions of these functions to the power set of the set of basics.

Remark 2 The two bounds, *sup* T and *inf* T, are of course defined with respect to the part-whole relation. They are such that, for all $x \in T$,

$x \sqsubset [T]$ and $(T) \sqsubset x$. Besides, if $x \sqsubset y$ for all $x \in T$, then $[T] \sqsubset y$. And, if $y \sqsubset x$ for all $x \in T$, then $y \sqsubset (T)$. *Remark 3* If $T \subset S$ is a finite set of substantial individuals then its additive aggregation is the juxtaposition of its members, whereas its multiplicative assemblage is their superposition. I.e., if $T = \{x_1, x_2, ..., x_n\}$, then

$$[T] = x_1 \dotplus x_2 \dotplus ... \dotplus x_n, \quad \text{and} \quad (T) = x_1 \dot{\times} x_2 \dot{\times} ... \dot{\times} x_n.$$

Remark 4 If T happens to contain a pair of detached entities, then its multiplicative aggregation is null, i.e. $(T) = \Box$. This holds, in particular, for any set of individuals containing the complement of one of them.

Finally we introduce a notion of decomposability. The intuitive idea is this. Let z be a substantial individual composed, no matter how, of two or more individuals. The composite individual will be said to be decomposable into x and y if these are different individuals such that $z = x \dotplus y$ and $x \wr y$. The components x and y need not be those that make up the original whole z before disintegrating into the separate entities x and y. Nor need they coexist with z: they may emerge in the act of decomposition. The exact and general notion of decomposability is this:

DEFINITION 1.13 Let x be a substantial individual. Then x is *decomposable* iff there is a set $T \subset S$ such that $[T] = x$ and $(T) = \Box$.

Remark From an operationist point of view only effective decomposition counts as an indicator of structure: if no way is known of actually splitting an entity then the latter is declared simple. Counterexample: current particle theory assigns the neutron a structure, hence a composition, but so far the neutron has withstood all efforts at decomposing it. Only the converse of the operationist thesis holds, namely: Anything effectively decomposable has a structure.

DEFINITION 1.14 A substantial individual is *uniquely decomposable* iff it is decomposable in but one way – i.e. if there is a single set $T \subset S$ such that $[T]$ equals the individual concerned and $(T) = \Box$.

Remark Here again unique decomposability is not to be mistaken for the single known way of effectively decomposing a thing. For example, so far the deuteron has been decomposed only into a neutron and a proton; but this does not exclude the possibility of decomposing it into a different set of products.

2.4. *Some Consequences*

We can retrieve a number of results obtained in association theory, since $\dotplus$ and $\circ$ are practically the same. In addition we can obtain a number of new theorems. But we shall exhibit only a few. Firstly

THEOREM 1.7 Every substantial individual is the additive aggregation of its parts. I.e., for every $x \in S$,

$$x = [\mathscr{C}(x)] = sup\{y \in S \mid y \sqsubset x\}.$$

COROLLARY 1.8 Every substantial individual with a finite composition is the juxtaposition of its parts. I.e., for any $x \in S$,

$$\text{If } \mathscr{C}(x) = T \equiv \{x_1, x_2, \ldots, x_n\} \quad \text{then}$$

$$x = [T] = x_1 \dotplus x_2 \dotplus \ldots \dotplus x_n.$$

COROLLARY 1.9 The world is the additive assemblage of all substantial individuals: $\Box = [S] = sup\, S$.

THEOREM 1.8 Every entity is part of some substantial individual: If $x \in S$ then there is at least one $y \in S$ such that $x \sqsubset y$.

Proof Take an arbitrary pair $a, b \in S$ and form the third individual $c = a \dotplus b$. This individual exists and is unique since, by hypothesis (Postulate 1.5), S forms a lattice. And, by definition of the part-whole relation, $a, b \sqsubset a \dotplus b$.

COROLLARY 1.10 Every entity is composite either properly or improperly. I.e., for every $x \in S$, $(\exists y)(y \in S \;\&\; y \sqsubset x)$.

Proof Follows from Theorem 1.8.

Remark 1 That an individual is improperly (or trivially) composite means of course that it consists just of itself. The assumption that every individual we might stumble upon might prove to be properly composite, is a powerful heuristic assumption that must not be confused with the preceding corollary. *Remark 2* Nor should the previous result be mistaken for the statement that S has a least element other than the null element and that would be part of every other individual.

We shall now use the notions of decomposability (Definition 1.13) and unique decomposability (Definition 1.14) to obtain a couple of interesting results.

THEOREM 1.9 All substantial individuals are decomposable. Decomposition includes the trivial decomposition of an individual into itself and the null entity $\square$.

Proof Let y be a part of $x \in S$ and take the environment y' of y: $y \dotplus y' = \square$ and $y' \dot{\times} y = \square$. Call $z = y' \dot{\times} x$ the superposition of y' with x. This is what completes y to form x: $y \dotplus z = x$, with $y \dot{\times} z = \square$. We have effected the decomposition of an arbitrary entity x into y and z.

THEOREM 1.10 No composite individual is uniquely decomposable. Equivalently: For all $x \in S$ other than $\square$, x is uniquely decomposable iff x is simple (i.e. decomposable only trivially: $x = x \dotplus \square$).

Proof (i) The *only if* part: let $y \sqsubset x$. Then $x = y \dotplus (y' \dot{\times} x)$ is a decomposition. In order to be unique, this must be $x = x \dotplus \square$ or $x = \square \dotplus x$. Thus either $y = x$ or $y = \square$. (ii) The *if* part: if $x = y \dotplus z$ and x is simple, then $y = \square$ or $y = x$ and also $z = \square$ or $z = x$. There are two possibilities: if $y = \square$ then $z = x$ and we have the decomposition $x = = \square \dotplus x$. And if $y = x$ and also $z = x$ then $y \dot{\times} z = x \neq \square$. But this is a contradiction, so we must have $z = \square$ and thus the decomposition $x = x \dotplus \square$.

Now the conservation law or Epicurus'

THEOREM 1.11 If $x, y \in S$ are not the null individual, then

(i) their juxtaposition does not end up in nought: $x \dotplus y \neq \square$;

(ii) provided they are not detached their superposition does not ensue in nought either: $x \dot{\times} y \neq \square$;

(iii) no being comes out of nonbeing either by juxtaposition or by superposition: $[\square] = \square$ and $(\square) = \square$.

Finally a couple of theorems concerning separation or detachment (Definition 1.8).

THEOREM 1.12 Separation is hereditary, i.e. the parts of detached entities are detached. That is, for any substantial individuals x and y,

$$x \rceil y \;\&\; u \sqsubset x \;\&\; v \sqsubset y \Rightarrow u \rceil v.$$

Proof By definition of detachment and of part,

$$x \dot{\times} y = (x \dotplus u) \dot{\times} (y \dotplus v) = \square.$$

Performing the indicated operations and recalling the distributivity of $\dotplus$ over $\dot{\times}$,

$$x \dot{\times} y = (x \dot{\times} y) \dotplus (x \dot{\times} v) \dotplus (u \dot{\times} y) \dotplus (u \dot{\times} v) = \square.$$

By Theorem 1.11 every addend must be null. In particular, $u \dot{\times} v = \square$, which amounts to $u \mathbin{\rceil} v$.

Remark This theorem shows that our detachment relation $\rceil$ for entities is analogous to the separation relation for sets (Wallace, 1941).

THEOREM 1.13 Detachment is additive. I.e., if two substantial individuals are detached from a third so is their juxtaposition: for any three entities $x, y, z \in S$,

$$x \mathbin{\rceil} z \;\&\; y \mathbin{\rceil} z \Rightarrow (x \dot{+} y) \mathbin{\rceil} z.$$

Proof Suppose the consequent is false. This assumption is equivalent to $(x \dot{+} y) \dot{\times} z \neq \square$, which expands into $(x \dot{\times} z) \dot{+} (y \dot{\times} z) \neq \square$. In turn the last statement amounts to $x \dot{\times} z \neq \square \vee y \dot{\times} z \neq \square$. But the first disjunct amounts to $\neg(x \mathbin{\rceil} z)$ and the second to $\neg(y \mathbin{\rceil} z)$, contrary to hypothesis. So, the consequent is true.

Remark In the theory of the Wallace separation spaces (Wallace 1941) additivity is postulated not demonstrated.

2.5. *Atoms and Levels*

The general notion of composition (Definition 1.6) has got to be supplemented with another, more specific notion, which will prove to be more useful than the former, namely that of composition relative to a certain set or level of entities. Thus in the case of an animal society regarded as a whole, we are interested in the set of its components not in the full set of its parts, such as the cells of the animals, even less the atomic components of their cells. That is, we want to know what the "relative" atoms of the whole are. The idea we want is this:

DEFINITION 1.15 Let $A \subset S$ be a set of entities. Then the *A-composition* (or *composition at the A level*) of a substantial individual $x \in S$ is the set of parts of x belonging to A:

$$\mathscr{C}_A(x) = \mathscr{C}(x) \cap A = \{y \in A \mid y \sqsubset x\}.$$

Example The molecular composition of a body of water is the set of its H_2O molecules. On the other hand the atomic composition of the same body is the set of H and O atoms that compose it.

We shall make ample use of this notion in Vol. 4, Ch. 7 on systems, where the composition of a system will be taken to be a certain

A-composition. We shall want the composition of a system to equal the union of the compositions of its subsystems. However, in general the composition of a complex individual is not equal to the union of the compositions of its components: see Figure 1.5. On the other hand the

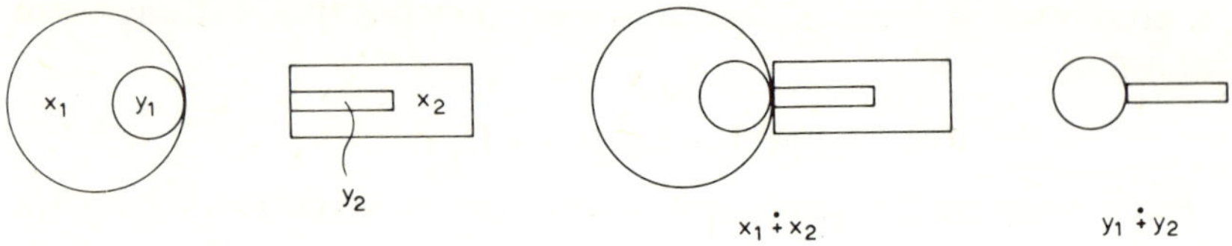

Fig. 1.5. The composite individual $x_1 \dotplus x_2$ has a part, namely $y_1 \dotplus y_2$, which is neither a part of x_1 nor a part of x_2. In general $\mathscr{C}(x_1) \cup \mathscr{C}(x_2)$ is a proper part of $\mathscr{C}(x_1 \dotplus x_2)$. Not so with the A-composition of a compound individual: see Theorem 1.14.

A-composition does have the desired property and this is one reason for preferring it over the general notion of composition. Indeed, we have

THEOREM 1.14 Let $A \subset S$ be a set of substantial individuals (level A atoms), and $x_1, x_2 \in S$ two substantial individuals. Then

$$\mathscr{C}_A(x_1 \dotplus x_2) = \mathscr{C}_A(x_1) \cup \mathscr{C}_A(x_2).$$

Proof If $x \in \mathscr{C}_A(x_1)$ then x is an A-atom (a member of A) and therefore $x \sqsubset x_1 \sqsubset x_1 \dotplus x_2$. Therefore $x \in \mathscr{C}_A(x_1 \dotplus x_2)$. Similarly $x \in \mathscr{C}_A(x_2) \Rightarrow x \in \mathscr{C}_A(x_1 \dotplus x_2)$. Thus $\mathscr{C}_A(x_1) \cup \mathscr{C}_A(x_2) \subseteq \mathscr{C}_A(x_1 \dotplus x_2)$. Now the converse implication. If $x \in \mathscr{C}_A(x_1 \dotplus x_2)$ then $x \sqsubset x_1 \dotplus x_2$, which is equivalent to $x \dot{\times} (x_1 \dotplus x_2) = x$. By distributivity it follows that $x \dot{\times} x_1 \dotplus x \dot{\times} x_2 = x$. But x is an A-atom and therefore we must have $x \dot{\times} x_1 = x$ or $x \dot{\times} x_2 = x$, i.e. $x \sqsubset x_1$ or $x \sqsubset x_2$. We have just shown that $\mathscr{C}_A(x_1 \dotplus x_2) \subseteq \mathscr{C}_A(x_1) \cup \mathscr{C}_A(x_2)$. This, conjoined with the previous result, yields the desired proof.

The entities of kind A, or A-atoms, may be simple (i.e. members of the set B of basics) or complex. To qualify as a member of a given set A, an entity must be able to associate with similar entities to form complex entities of a different kind. But we do not have yet the notion of a natural kind, which will be introduced in Ch. 4, Sec. 3.3. Therefore for the time being we shall have to make do with a presystematic or intuitive notion of an A-atom. The same holds *a fortiori* for the notion of a level. All we can say at this time is that we may split the totality S of entities into a certain number n of disjoint sets A_i, each member of which is composed – additively or multiplicatively – by entities of the next lower level A_{i-1}. The levels hypothesis may then be formulated as follows:

$$S = \bigcup_{i=1}^{n} A_i, \quad \bigcap_{i=1}^{n} A_i = \varnothing, \text{ with } A_i = \{a_i^n | n \in \mathbb{N}\}$$

and

$$a_i^n = [\alpha_{i-1}^n] \quad \text{or} \quad (\alpha_{i-1}^n), \quad \text{where } \alpha_{i-1}^n \subseteq A_{i-1}.$$

The matter of (relative) atoms and levels of organization will be taken up again in Vol. 4, Ch. 8.

2.6. *Alternative Formalizations*

Alternative formalizations of the notions of juxtaposition and superposition have been explored. One of them is in terms of ring theory, i.e. it consists in assuming that $\langle S, \dot{+}, \dot{\times}, \Box, \square \rangle$ constitutes a ring of idempotents (Bunge, 1971a, 1973a). In this construal one must abstain from interpreting the unit element of the ring as the world, for in this case the unit is not the last element. This is not a fatal flaw: the defect consists in leaving one primitive devoid of ontological interpretation. What may constitute a defect is that in ring theory $\dot{\times}$ distributes over $\dot{+}$ but not conversely. As a compensation the absorption laws, which seem idle in the ontological interpretation of lattice theory, do not occur in ring theory.

A second alternative formalization consists in a certain interpretation of Boolean algebra (Bunge, 1967b). The two formalizations, though different, are equivalent. In fact by a famous theorem of Stone's, a commutative Boolean ring $\langle S, +, \cdot, 0, 1 \rangle$ determines a Boolean algebra $\langle S, \cup, \cap, 0, 1 \rangle$ with meets and joins defined as follows:

$$x \cup y = x + y + x \cdot y, \qquad x \cap y = x \cdot y.$$

We may adopt either of these alternative formalizations in the case where the reference class is a proper subset of the total set S of substantial individuals. In this case the unit will not be interpreted as the world. The ring theoretic construal may be preferred when this condition is fulfilled and superposition is not required to be commutative, as is often the case with two parts of a machine.

Needless to say a third alternative, having none of the defects of the preceding ones, should be welcome. The same applies to every one of our ontological theories: different formalizations of the same intuitions, as well as alternative intuitions, should be explored.

2.7. *Concluding Remarks*

Our two theories of substance – namely association theory and assembly theory – systematize the intuitive ideas of a physical object and of the association of concrete individuals. We have not called any of our theories *mereology* – as we should if we had followed in Lesniewski's path – because they achieve more than an elucidation of the part-whole relation. If we were interested just in the latter concept we could adopt a simpler theory, namely that of partially ordered sets. Indeed, assuming that the structure $\langle S, \sqsubset \rangle$ is a poset and adding the obvious interpretations of S and $\sqsubset$, we would get an exact characterization of the notion of a part. And the latter would in turn allow us to define a few other concepts, notably those of detachment and composition. However, such a theory would be too poor: for one thing it would not clarify any notion of association – and this is what our theories of substance are all about.

Nor have we called our theories *calculi of individuals* – as Leonard and Goodman (1940) called their version of Lesniewski's mereology. One reason is that every first order mathematical theory is a calculus of individuals: whether it is a calculus of physical individuals is another matter. Another reason is that, unlike Lesniewski and his fellow nominalists, we draw a radical distinction between physical and conceptual individuals and reject the thesis that a single theory should be able to account for both kinds of individual. In particular our part-whole relation applies only to ontic individuals, i.e. $\sqsubset$ is defined on $S \times S$. More on this matter in the next section.

3. Entities and Sets

3.1. *The Null Individual and the World*

Both association theory and assembly theory involve the greatest fiction ($\square$) as well as the greatest reality ($\blacksquare$). The null individual has been considered by Carnap (1947), Martin (1965) and a few others. It has usually been characterized by analogy with the empty set, namely as "that thing which is a part of every thing". This characterization is not quite correct. Firstly □ is not a thing but a concept. Secondly the characterization is deficient: the null body is also the null non-body. (Besides, according to the aether theories, the aether, though not null, was supposed to be part of every thing.)

The null individual, or nonbeing, is best characterized in terms of either association or assembly, i.e. either as (i) that which, associated with any substantial individual, yields the latter, or as (ii) that which, juxtaposed to any thing leaves it unchanged and, superposed to it, annuls it. In either case □ is in S and is part of every member of S. (The claim that □ belongs to every set, even the empty set, leads to contradiction.) All these are definite properties of □. That is, according to the theories formulated in the previous sections, it is false to say that the null individual has no properties – as false as to say that it has substantial (physical, nonconceptual) properties. Which reminds one of Lucy's verdict: "For a nothing, Charlie Brown, you're really something" (Schulz, 1967).

What is true is that the null individual simply does not exist physically: it is a fiction introduced to get a smooth theory. We need this pretence not only in ontology but also in the foundations of science – which is not surprising because the two intersect. Thus the vacuum, i.e. the null field, is assigned a number of properties in field theories – such as a refractive power equal to unity. Moreover it is convenient to introduce several null individuals, one for every natural class. For example, in optics we need the void, or null body b_0, and darkness, or the null light field l_0. The reference class of a statement about a light beam l in vacuum is $\{l, b_0\}$, while a statement about a body b in darkness is about $\{l_0, b\}$, and one about the dark void concerns $\{b_0, l_0\}$ (Bunge, 1966).

We need the null entity to think about real entities but we cannot use it to build the latter. Indeed, unlike the empty set in von Neumann's reconstruction of number theory, the null individual is not the point of departure of our cosmology but only a very modest component of it, which discharges a theoretical function but has no counterpart in reality. In other words, entities are not constructible out of the nonentity. This is in sharp contrast with the central role Hegel (1812, 1816), Peirce (1892–93, 6.217) and Heidegger (1927) have assigned nonbeing. (Incidentally none of them seems to have distinguished the nonentity, or nonbeing, i.e. □, from nothingness or the empty set $\varnothing$. Yet □ and $\varnothing$ are radically different, i.e. they have different properties even though none of them is substantial or real.) Far from taking nonbeing as our point of departure, our theories of substance involve the theses of Epicurus (*apud* DeWitt 1955) and Lucretius (55 B.C.?) (*a*) that no thing emerges *ex nihilo* and (*b*) that no thing goes over into nought (Theorem 1.11). Our point of departure is the real world as known to the science of the day: wishing to understand the world we dismantle it in thought and oversimplify its ingredients as well as the manner in which they are actually assembled. We thus obtain a concept of substance, or matter, which, when enriched with properties, will ensue in the concept of a concrete thing (Ch. 3).

3.2. *Entities and Concepts*

While the null individual, a concept, *is* (identical with) □, the world (or reality) is said to be *denoted by* the symbol '▯'. In our theories of substance the world is an individual but not just one more individual: it is the entity that contains as components all other entities. But this is as far as our theories of substance go: they give no details about the structure of ▯. On the other hand they do assign definite structures (semilattice in one case, lattice in the other) to the furniture of the world, i.e. S.

We emphasize that S, the set of all entities, is not the same as the aggregate or whole $[S]$ composed of all physical objects. The difference between the concept S and the entity denoted by $[S]=$ ▯ illustrates the construct/thing dichotomy. To us the basic dichotomy in any set of objects is not that between individuals and sets but that between physical objects and conceptual ones. That is, we assume that every class O of objects is split into a class C of constructs and a class $T \subseteq S$ of substantial individuals:

$$O = C \cup T, \quad \text{with} \quad C \cap T = \varnothing.$$

Sets, relations and functions, as well as algebraic and topological structures, rank as constructs in our philosophy and, as such, they do not satisfy any theory of substance.

Whether an object is an individual or a set is of interest to ontology: if the former then it may be either physical or conceptual, if a set it can only be a concept. But whether a *conceptual* object (a construct) is an individual or a set is metaphysically and even mathematically relatively unimportant. This is for the following reasons. First, what is a (conceptual) individual in one theoretical context may become a set in another and conversely: it all depends on the fineness of our analysis. Second, some set theories draw hardly any distinction between sets and individuals. Nevertheless there is one assumption (von Neumann's D axiom) that does draw the line between the two categories by stipulating that every chain of the form ... $x \in y \in z$ has a stop. (For every nonempty set there is a member none of whose members is a member of the given set. Such a member may be regarded as an individual with respect to the given set. For example, in $A = \{\{a\}\}$ the element a is such an individual, the ascending chain being $a \in \{a\} \in A$.) Even if this axiom is not accepted, the individual/set distinction is kept in everyday mathematics. We shall keep it.

We are now in a position to clarify the difference between our ontology and Platonism with regard to individuals. (The difference with respect to properties will be discussed in Ch. 2, Sec. 5.) Suppose we agree to consider, or rather toy with, a universe composed of a finite number of atoms, or simple individuals. We start by inquiring whether these individuals are conceptual or concrete. (On the other hand Goodman, 1956, p. 16 regards this distinction to be "vague and capricious".) If they are conceptual then we let the Platonist take over, for that is his field, though where he believes in the autonomous existence of such ideas we just pretend that to be so for the sake of expediency. He should feel free to build not just the power set of the given set but also any powers of that power set if he needs them. On the other hand if the individuals are concrete then we give the Platonist a push and do our own reckoning. The upshot is this: in addition to the original atoms we get all their partial (binary, ternary, etc.) associations. And on top of these we shall have all the novel things that may ensue from such combinations, since we know from high energy physics that a couple of elementary particles may collide to produce a large number of new individuals. Thus we come up with a universe that, in numerosity, is infinitely smaller than the Platonic universe though appreciably larger than the universe of unchangeable and unreactive atoms envisaged by mechanism.

3.3. *Existence and Individuation*

Russell taught us that the existential quantifier $\exists$ cannot occur by itself but must be applied to "something described" such as Ax, where A is a unary predicate. Hence at first sight it would seem that our substantial individuals do not exist, for they have no peculiarities. However, they do share the definite property of associating with one another. That is, every member of $\langle S, \circ, \square\rangle$ is characterized by the predicate A defined by

$$Ax =_{df} (\exists y)(\exists z)(y \in S \;\&\; z \in S \;\&\; x = y \circ z \vee y = x \circ z \vee z = x \circ y).$$

Hence we may prefix (x) to this propositional schema, obtaining the proposition "All individuals associate".

Moreover we can individuate those entities with a known composition. That is, we can form definite descriptions of the form

"The member x of S with composition $\mathscr{C}(x)$".

In traditional terms: our *principium individuationis*, at this early stage, is composition. Obviously this principle fails us for simples since, with the exception of $\square$, they are devoid of peculiarities other than that they can associate. (Therefore we cannot use Quine's method of treating all singular terms as definite descriptions: Quine (1950, Sec. 37). In order to use this method we must have stock of predicates large enough to uniquely characterize every individual, and this will have to wait till next chapter.) Besides, the above principle does not even allow us to distinguish between two molecules with the same composition. But this is only a temporary difficulty, as a much finer *principium individuationis* will be adopted in Ch. 2, Sec. 3.2.

Finally a word or two about substantial existence as distinct from existence in general. The existence of simple individuals can be ascertained only by scanning the entire set S. If a belongs to S and is not the null individual we say that a exists. That is, we explicate "a exists" as "$a \neq \square \;\&\; a \in S$". This concept of existence can be called *bare substantial existence*. It suffices in the theory of substance, where the only individuating property is composition. (More on physical existence in Ch. 3, Sec. 4.4.) Note that this is a notion of ontic or physical existence to be distinguished from that of conceptual existence. In formal science to say that an individual a exists it suffices to assume or to prove that a belongs to some set that has been satisfactorily characterized. In ontology we have no use for arbitrary sets except perhaps as auxiliaries devoid of ontological import: here to exist (substantially, physically) is to have a number of substantial properties, among them that of associating with other nonconceptual objects. So, *pace* Wittgenstein (1969, Sec. 35), it makes perfectly good sense to state that there are substantial (physical) objects. Moreover ontology happens to be concerned precisely with such objects. (Recall Introduction, Sec. 2.)

4. Concluding remarks

One of the necessary conditions for an ontological theory to qualify as scientific is its relevance to science. (Cf. Introduction.) Association theory and assembly theory comply with this condition: they are relevant to any scientific or technological theory concerned with complex systems such as atoms, cells, machines, brains, or organizations. In particular any such theory employs a concept of association and therefore should welcome an elucidation of the latter. However, in the vast majority of scientific writings only an intuitive concept of association is employed. It is only when the foundations of a scientific theory are carefully laid down that $\circ$, $\dot{+}$ or $\dot{\times}$ are apt to occur explicitly, because they are needed. For example, the very statement of the hypothesis that the electric charge Q is additive involves $\dot{+}$:

If x and y are bodies, and $x \neq y$,
then $Q(x \dot{+} y) = Q(x) + Q(y)$.

Similarly with the intensity E of the superposition of two electric fields:

If x and y are fields, and $x \neq y$,
then $E(x \dot{\times} y) = E(x) + E(y)$.

If only for this reason, i.e. because it is needed, it is mandatory to include some version of assembly theory among the ontological presuppositions of any scientific theory referring to complex systems (Bunge, 1967b, 1973b, 1974d).

Although the scientific method is analytic, from time to time the scientific community is swept by holistic winds – particularly when science fails to disclose the actual composition and inner workings of certain things. At such times the black box theories come in vogue and the mechanistic approach is decried or at least shelved. A characteristic of the black box approach is that it makes little or no use of the very concept of composition: it treats every component of a system, and occasionally even the whole system, as a structureless box. No doubt this strategy is wise as long as little is known, or need be known, about the components and their mode of association. But to advertise this *pis aller* strategy as optimal (as e.g. Chew, 1966, does) is suicidal. Indeed, it amounts to obliterating the difference between simple and composite, hence between microphysical and macrophysical, and it renounces the ideal of the most complete possible account of things. Worse: the strategy ignores whatever complexity has already been disclosed. Far worse: such a drastic reduction in our cognitive goals cannot be implemented in microphysics except by borrowing from translucid box theories such as field theory and ordinary quantum mechanics (Bunge, 1964; Regge, 1967). In short, the idea of composition, and the goal of finding out the precise composition and manner of association, are not "out": at most some grapes are still sour.

So much for the concept of substance or matter. Let us now tackle the problem of characterizing the concept of a property.

CHAPTER 2

FORM

The next step in our conceptual reconstruction of the notion of a real thing is to introduce the concept of a property, trait, or character. Every real thing has a number of properties: there is no formless substance except as a useful fiction. Nor are there pure forms hovering above matter. Every concrete or substantial property, such as moving, reacting, or remembering, is the property of some thing or other – bodies, reactants, brains, or what have you. Were this not so, science would not seek to find and interrelate properties of things, much less the patterns of the associations and changes of properties – i.e. the laws.

We cannot take the notion of a property for granted because it is far from clear. In any event no discipline other than ontology attempts to clarify the notion of a property as distinct from the various specific notions of property. Logic is certainly concerned with the notion of an attribute or predicate. But, as we shall see in Sec. 1, properties should be distinguished from the attributes representing them – except of course in the context of philosophical idealism, where they are conflated.

Nor can we accept uncritically the hypothesis that things have properties, particularly since it has been challenged by nominalist philosophers. In fact the latter wish to dispense with properties, which they regard as Platonic fictions, and attempt to reduce everything to things, their names, and collections of such (see e.g. Woodger (1951) and Cocchiarella (1972)). At the very least the nominalist will want to construe every property as a class of individuals, or of n-tuples of individuals – i.e. he will adopt an ontological interpretation of the semantic doctrine called *extensionalism*, which equates predicates with their extensions. But we have seen (Volume 1, Ch. 5, and Vol. 2, Ch. 10) that extensionalism is indefensible even in mathematics, if for no other reasons than that (*a*) some of the basic mathematical predicates, notably the membership relation, are not defined as sets, and (*b*) coextensive predicates need not be cointensive or have the same meaning. Therefore extensionalists can "deplore the notion of attribute" (Quine, 1963) without being able to dispense with it, much as the Victorians deplored sex while engaging in it.

If extensionalism fails so does its ontological interpretation, namely the reduction of the world to a collection of individuals devoid of properties. For one thing this ontology is incapable of explaining what makes a given individual different from any others: it has no *principium individuationis*. For another, nominalism commits the sin of reification, or the transformation of everything into things or collections of things. (Reification, a constant of archaic thinking, is occasionally found among scientists and philosophers. Thus the vulgar materialists of the 19th century held the world to consist of things of two sorts, *Stoff* or substance and *Kraft* or force. Later on the energeticists, wishing to dematerialize the world, proclaimed that everything is a manifestation of energy. And nowadays the followers of Whitehead maintain that there are no things but just events.) For those two reasons nominalism does not tie in with science, which regards things as individuals possessing properties, and handles the notion of a property as basic – so much so that every law statement is deemed to represent an invariant relation among properties of things.

We shall propose a construal of the notion of property that avoids the extremes of nominalism (or substantialism) and Platonism (or formism), and squares with the scientific conception.

1. Property and Attribute

1.1. *Difference between Property and Attribute*

All objects have properties. If the objects are conceptual or formal, their properties will be called *formal properties*, or *attributes* or *predicates* for short. If the objects are substantial individuals, their properties will be called *substantial properties*, or *properties* for short. Because every model of a substantial individual is built with concepts, it contains attributes or predicates; and insofar as the model represents a substantial individual, some of those attributes or predicates represent substantial properties.

In the case of a conceptual object, such as a set or a theory, the words 'attribute' and 'property' are exchangeable because a conceptual object has all the properties we consistently attribute to it. But in the case of a substantial individual we must distinguish a substantial property or objective trait from the corresponding attribute(s) if any. A substantial property is a feature that some substantial individual possesses even if we are ignorant of this fact. On the other hand an attribute or predicate is a feature we assign or attribute to some object: it is a concept. A predicate may conceptualize or represent a substantial property; but then it may not or it may do so poorly, i.e. with a large margin of error. On the other hand the possessing of a property is not a matter of truth or falsity; only our knowledge of properties can be more or less true or adequate. For this reason we distinguish

$\ulcorner$Substantial individual *b possesses* property $P\urcorner$.

from

$\ulcorner$Attribute *A holds* for $b\urcorner$,

or $\ulcorner A$ is *true* of $b\urcorner$, or $\ulcorner \mathscr{V}(Ab) = 1\urcorner$,

where A is taken to represent P. Again: whereas the possessing (or the acquiring or the losing) of a property by a substantial individual is a fact often beyond our reach, our attribution of a property (via some predicate) is a cognitive act. In other words, whereas we control all predicates because we make them, we control only some properties.

Needless to say, no such difference between attributes and substantial properties is admitted by idealism: in such a philosophy the acquiring of a property coincides with its attribution. (Which is why the inadequacies of this philosophy are hardly realized in the realm of formal science.) Oddly enough naive realism, while recognizing the difference between attributes and substantial properties, asserts their one-to-one correspondence. Therefore the stand of those two poles of philosophy with respect to the theory of properties is rather close: whereas idealism asserts that the theory is the logic of predicates, naive realism is committed to the claim that the theory of substantial predicates is an ontological interpretation or model of the predicate calculus.

We proceed to approaching this matter from the vantage point of our semantics (Vols. 1 and 2).

1.2. *Attribute-Property Correspondence*

Surely we know properties only as attributes or predicates, i.e. as components of our views of things. Still, we must distinguish the object represented from its representations if we wish to account for discovery (or invention) and ignorance, truth and error. We do that by making statements of the form

$\ulcorner$Attribute *A represents* substantial property $P\urcorner$,

or $\ulcorner A \triangleq P\urcorner$ for short.

For example, in certain contexts angles represent inclinations, probabilities tendencies, densities crowdings, and so on. The attribute-property correspondence is a particular case of the knowledge-reality, or mind-world relation. This correspondence is not isomorphic because some attributes represent no substantial properties, others represent several properties, and finally some properties are represented by no attributes (because we ignore either or both) or by several predicates (often belonging to different theories of the same kind of thing).

The relation between properties and attributes may be construed as follows. Let $\mathbb{P}$ be the set of substantial properties and $\mathbb{A}$ that of attributes or predicates. Then the representation of properties by attributes is a function $\rho: \mathbb{P} \to 2^{\mathbb{A}}$ that assigns each property P in $\mathbb{P}$ a collection $\rho(P) \in 2^{\mathbb{A}}$ of attributes, so that the formula '$A \in \rho(P)$', for A in $\mathbb{A}$ and P in $\mathbb{P}$, is interpreted as *attribute A represents property P*, or $A \triangleq P$ for short. Details on the concepts of attribute (or predicate) and representation are found in Vol. 1, Ch. 3 and Vol. 2, Ch. 6.

The representation function is a correspondence between a proper subset of all the conceivable attributes (predicates) and the ill-defined set of all (known and unknown) substantial properties. That is, there are attributes with no ontic correlate. Among them we find membership in a set, the negative attributes, and the disjunctive ones. Let us discuss the three because they will show what a substantial property is not. First the matter of membership. In mathematics every member of an arbitrary (but well defined) set may be assigned the attribute of membership in that set. However, being an element of an arbitrary set does not count as a substantial property. For example, although I may decide to group my shoes together with President X's latest statement, such a common membership does not constitute an objective property shared by the two items. In short, not all sets are kinds.

Nor, *pace* a few distinguished philosophers (e.g. Bolzano (1851), Sec. 39 and Russell (1918), Secs. I and III), are "negative" properties possessed by factual items such as things or events. These have only "positive" properties (Bunge, 1974g). In the statement ⌜Neutrons are not electrically charged⌝ the negation affects the entire proposition ⌜Neutrons are electrically charged⌝: it denies the latter statement instead of attributing neutrons the negative property of not being charged, let alone that of being anticharged. We certainly need negation to understand reality and argue about it or anything else, but external reality wears only positive traits. In other words, even though a proposition of the form $\ulcorner \neg Ab \urcorner$ be true and indistinguishable from the proposition $\ulcorner (\neg A)b \urcorner$, there may not exist any real or substantial property represented by the negative attribute $\neg A$ and exemplified by the individual b. Consequently if attribute A represents substantial property P(i.e. $A \triangleq P$), then we can accept the truth condition (sufficient for qualitative attributes)

$\ulcorner \neg Ab \urcorner$ is true if and only if b fails to possess P.

but not

$\ulcorner \neg Ab \urcorner$ is true if and only if b possesses $\neg P$.

In short, negation is *de dicto* not *de re*: it is defined on the set $\mathbb{A}$ of attributes not on the set $\mathbb{P}$ of properties. And because there are no negative properties there are neither tautological nor contradictory substantial properties corresponding to tautological and contradictory attributes such as $A \vee \neg A$ and $A \mathbin{\&} \neg A$ respectively.

Likewise there are no disjunctive attributes with ontic correlates. For example, there is no substantial individual possessing the property of being heavy *or* translucid, even though the proposition ⌜Lucite front doors are heavy or translucid⌝ is true. (In other words, we employ the disjunctive attribute "heavy or translucid" but do not have it represent a property possessed by any substantial individual.) On the other hand certain properties can conjoin. For example there are things both heavy and capable of synthesizing chlorophyll. (But of course other properties are mutually incompatible, i.e. cannot conjoin.) What the expression 'conjunction of properties' signifies will be seen in Sec. 3.4.

The preceding considerations, Byzantine though they may look, are necessary for building a theory of properties and they have far reaching philosophical consequences. A first consequence of the lack of isomorphism between properties and attributes is the ruin of naive realism, both in the form of the reflection theory of knowledge and of the picture theory of language. According to these views every factually true proposition depicts a fact. But if we admit just "positive" properties then we must rule out negative facts. And if we do not admit disjunctive properties then we must write off disjunctive facts, and *a fortiori* general facts. Of course we accept negative and disjunctive propositions as long as they are sufficiently true or at least promising, but that has nothing to do with accepting negative facts and disjunctive facts. Thus "There are no elephants in your pocket." is true precisely because this proposition fails to represent a situation in which there are elephants in your pocket. (On the other hand naive realism would have us interpret that statement as "Everything is either a nonelephant or unlocated-in-your-pocket".) Last, but not least, the attribute-property gap bars any attempt to read the predicate calculus in ontological terms, i.e. to get a theory of properties without tears. Therefore we must try and build a property calculus as distinct from the predicate calculus. To work.

2. Analysis

2.1. *Property in General and Property of a Particular*

The expressions 'pure form', 'system of qualities in themselves', and even 'property' make no sense except as abbreviations or abstractions. In fact, a mathematical and semantical analysis of the concept of a property will show us shortly that every property – except of course the null property, which is a fiction – is possessed by some individual or other: there are no properties that fail to be paired to any individuals. In particular, substantial properties are properties of substantial individuals.

In other words, an attribute can only be attributed to, or predicated of, some subject or other. A snowball is white: whiteness can be predicated of snowballs and other things but it does not exist separately any more than snowballs failing to "participate" in whiteness – to lapse into Platonic jargon. In other words there are no universals in themselves but only properties that are universal in a given set of individuals, i.e. that are shared by all the members of the set. (More on universals in Sec. 5.3.) All this is rather obvious from the way properties are handled in science and predicates are analyzed in semantics. In both cases every property is represented by a function mapping individuals (or n-tuples of individuals) into statements of the form ⌜Individual (or n-tuple) x is assigned attribute A⌝. Such propositional functions, or proposition-valued functions, will be called *attributes* or *predicates*.

Example 1 The property of being able to read is representable as a function R_1 from the set H of humans into the set P_1 of propositions containing the predicate R_1:

$R_1: H \to P_1$, where $R_1(b)$, for $b \in H$, abbreviates "b can read", a member of the set P_1.

Example 2 The property of being able to read books can be represented by a function R_2 from the set of couples $\langle x, y\rangle$, with $x \in H$ and $y \in B$, where H = Humans and B = Books:

$R_2: H \times B \to P_2$, where $R_2(b, c)$, for $b \in H$ and $c \in B$, abbreviates "b can read c", a member of P_2.

Example 3 An arbitrary mathematical function f with domain D and range R can be associated the propositional function

$F: D \times R \to P_3$ such that $F(x, y) = (f(x) = y) \in P_3$ for $x \in D$ and $y \in R$.

In ontology we are concerned with properties of entities, i.e. members of the set S of substantial individuals characterized in Ch. 1. A substantial property must then be representable as a propositional function, or predicate, on a domain that somehow includes S. The function will represent the property in general, e.g. age; and its value for a particular individual, the given property of the individual concerned, e.g. its age. But sets other than S are likely to occur as well in the domain of the predicate. For example, mass is representable by a certain real valued function M on the set of quadruples ⟨body, reference frame, time, mass unit⟩. In symbols,

$$\ulcorner M: B \times F \times T \times U_M \times \mathbb{R} \to \text{Propositions containing } M \urcorner$$

represents the mass of entities $B \subset S$, whereas the value $\ulcorner M(b, f, t, u, r)\urcorner$ represents the mass $r \in \mathbb{R}$ of a particular body $b \in B$ relative to a given frame $f \in F$ at time $t \in T$, reckoned or measured in mass units $u \in U_M$ (e.g. grams or pounds).

We generalize the preceding considerations by introducing the following axiom of a methodological nature:

POSTULATE 2.1 Let S be the set of substantial individuals or some subset thereof, and let T to Z be arbitrary nonempty sets, equal to or different from S. Then

(i) any *substantial property in general* is representable as a predicate (or propositional function) of the form

$$A: S \times T \times \ldots \times Z \to \text{Propositions including } A;$$

(ii) any *individual substantial property*, or property of a particular substantial individual $s \in S$, is representable as the value of an attribute at s, i.e. as $A(s, t, \ldots, z)$, where $t \in T, \ldots, z \in Z$.

Every individual property, or property of a particular, is *dichotomic* in the sense that the individual either has it or fails to possess it. This holds not only for qualities and for quantitative properties that take on definite numerical values but also for any quantity the range of which is a family of intervals, as is the case with quantum theoretic dynamical variables. Therefore in an expression of the form ⌜$A(s, t, \ldots y) = z$⌝ it is irrelevant, for our present purposes, whether z is a single number (or an n-tuple of numbers) or a whole set of numbers (or an n-tuple of numerical intervals). In either case, whether sharp or extended, an individual or a set, z will represent a property of individual s.

Properties are then attributable to, or predicable of, individuals of some kind or other. There is no substantial property apart from entities, let alone prior to them and dwelling in a separate Realm of Forms: a form is a form (property) of a chunk of substance, a universal is a property possessed by all the substantial individuals in a given subset of S. Substrate free forms, or Platonic universals, are just as imaginary as formless substances. But at least the fiction of the bare individual makes mathematical sense while that of a pure form does not. (The barest of individuals is characterizable as the member of some set, but the simplest of forms requires a support or domain, because it is a function.)

An entity possesses certain properties but *is* not a bundle of properties. That entity b possesses properties $P_1, P_2, \ldots, P_n, \ldots$ does not entail that $b = \{P_1, P_2, \ldots, P_n, \ldots\}$. If it did we would have to define properties independently of any individuals, and condone absurd expressions such as '$\{P_1, P_2, \ldots, P_n, \ldots\}$ possesses P_1'.

Example 1 A corpuscle is not a mass-position-velocity compound but an entity possessing definite mass values, position values, and velocity values. (Such a triple is only a schema of such an entity. Recall Vol. 1, Ch. 3, Sec. 3.1.) *Example 2* Having a certain thought is a property of some animals at certain times, and as such is representable by a function on the cartesian product of the set of animals of a certain kind by the set of instants. Put another way: that thinking is a function of certain brains entails that neither thoughts nor brains of certain kinds exist separately. What do exist are brains capable of thinking. (More on the mind-body problem in Vol. 4, Ch. 10.)

The functional construal of properties solves a number of problems, among them the ancient biological conundrum: Which is prior, the organ or the function – e.g. the brain or ideation, feeling, etc.? Answer: neither, for the function is just what the organ does. Take for instance the proposition ⌜The function of organelles of kind A is to synthesize proteins of kind B⌝. This statement can be compressed into ⌜All organelles A synthesize some protein B⌝. The latter does not contain the suspect notion of biological function, which smacks of purposiveness, nor does it state that the synthesis of B is the sole function of an A. The second statement exhibits clearly the idea that the function concerned is a relation between organelles and proteins, namely the relation of synthesizing, which can be construed as a function from the set $A \times B$ of organelle-protein pairs into certain propositions. In this way the quaint idea that there could be a biological function independent of either an organ or its output vanishes.

2.2. *Intrinsic and Mutual, Primary and Secondary*

Some properties, such as radioactivity and intelligence, are inherent properties of individuals. Therefore we can sometimes represent them by unary attributes:

$$Rad\colon \text{Atoms} \to \text{Propositions involving } Rad$$

$$Int\colon \text{People} \to \text{Propositions involving } Int.$$

We call such properties *intrinsic*. Other properties, such as solubility and performance, are properties of pairs or, in general, n-tuples of substantial individuals. Accordingly we represent them by predicates of rank 2 or higher, such as

$$Sol\colon \text{Solutes} \times \text{Solvents} \to \text{Propositions involving } Sol$$

$$Per\colon \text{People} \times \text{Circumstances} \to \text{Propositions involving } Per.$$

Such properties will be called *mutual* or *relational*.

All mutual properties must be represented by predicates of rank higher than one but the converse is false. That is, some n-ary predicates, with $n > 1$, represent intrinsic properties. For example, the gross national product, which is an intrinsic property, is representable by a predicate of the form

$$GNP\colon N \times T \times U \times Q^{+} \to \text{Propositions involving } GNP,$$

where N is the family of nations, T the set of years, U the set of units of production (e.g. dollars), and Q^{+} the set of positive fractions. In the determination of the n-arity of a property, what counts is the number of sets of substantial individuals occurring in the domain of the corresponding predicate.

A mutual or relational property of an entity may or may not depend causally upon some other individual. For example velocity is a mutual property since it depends on both the moving entity and the reference frame, but it is not caused by the latter. The same holds for distance, duration, frequency, mass, temperature, electric field strength and many other properties in relativistic physics: they are frame-dependent but not causally dependent upon the reference frame, which is assumed to be passive. On the other hand certain mutual properties are environment-dependent. This is the case with the force on a body, the position and momentum distribution of a microsystem, solubility, the frequency of the cricket chirping, the performance of a student, and the rate of production of manufactured goods: in all these cases the environment exerts an influence on the entity of interest. All the phenomenal properties, such as color and perceived loudness, are mutual properties of this kind, i.e. they depend not only upon the object-in-the-environment but also upon the subject or perceiver. More on this presently.

(The way to find out whether a given property is absolute (or invariant or frame-independent) is to investigate its behavior under changes in the reference frame – e.g. under displacements or under rotations. If the corresponding attribute does not change under the given transformation then it is declared to be *invariant* under the latter (or an invariant of the entire group of transformations). This procedure establishes or refutes the hypotheses of relative invariance, i.e. absoluteness with respect to a given group of transformations. Since there is no way of finding out whether a given property is invariant with respect to an arbitrary group of transformations, there is no such thing as absolute invariance (or absoluteness). That is, *all invariance is relative.*)

Since phenomenal properties are subject-dependent they are, a fortiori, frame-dependent. (Indeed, a sentient being may be regarded as a reference frame.) The converse is not true, i.e. not every frame-dependent property is subject-dependent or subjective. For example, the frequency of an oscillation is frame-dependent but not subject-dependent. In case of doubt concerning the objective character of a property, swap subjects (observers) either experimentally or in the theoretical formulas, and see whether the values of the property change. If they do not then the property is objective on top of being relational. Outside psychology all properties, whether intrinsic or mutual, are supposed to be objective, i.e. observer-independent. This alone, and in particular the existence of invariant properties (such as electric charge, entropy, and number of entities), disqualifies any subjectivistic ontology in the eye of science. In other words, a scientific metaphysics must be just as objectivistic as science itself, i.e. thoroughly.

A scientific ontology will not discard phenomenal or secondary properties, such as color and loudness, but will attempt to account for them in terms of nonphenomenal or primary qualities. Whatever the precise account may turn out to be, it must rest on the construal of a secondary property as a mutual property at least one of whose legs is a set of sentient organisms. For example, color is wavelength as perceived by some subject, loudness felt acoustic intensity, and warmth felt heat. No sentient organism, no secondary properties. Hence color predicates are representable as certain functions on the cartesian product of the set of light signals by the set of organisms endowed with sight. The other phenomenal predicates (i.e. predicates representing secondary qualities) are similar.

The physicist assumes that there are sound waves that go unheard and light waves that are unseen. And the psychologist knows that there can be acoustic and visual sensations in the absence of concurrent physical inputs. Therefore it would seem that neither primary qualities are necessary for the corresponding secondary qualities nor conversely. However, there would be no perceptual hallucinations if there were no normal perceptions: psychologists assume tacitly that an organism deprived of every trace of a hearing system experiences no sound hallucinations, and similarly for the other kinds of sensation. Hence sound waves are necessary for hearing – normally as immediate causes, abnormally, as stimuli for past experiences (ontogenetically, phylogenetically, or both). We shall accordingly say that acoustic volume *supports* loudness, luminosity *supports* brightness, and so on, even though not every sensation is caused in an immediate fashion by a physical stimulus endowed with the corresponding supporting ability. This view goes back to Locke (1689 Bk. II, Ch. VIII).

The preceding considerations make no sense to phenomenalism, operationism, and subjectivism: according to these philosophies all properties are secondary. But of course natural science since Galilei recognizes only primary properties, and its progress has partly consisted in shoving secondary qualities to the field of psychology. Note finally that in our view secondary qualities are neither purely objective nor purely subjective, for they are possessed by the subject/object interface rather than by either component separately. This is why a secondary quality is construed as a mutual property of entities and sentient organisms.

We may summarize this subsection in the following methodological axiom, which spells out Postulate 2.1:

POSTULATE 2.2 Let $\{S_i \subseteq S \mid 1 \leqslant i \leqslant n\}$ be a family of nonempty subsets of substantial individuals not including the null individual. Further, let T to Z be arbitrary nonempty sets [of e.g. units], the same as or different from any of the S_i. Finally, let $\mathbb{R}$ be a set of numbers or of sets of numbers [such as the power set of the real line] and p a natural number. Then

(i) every *qualitative intrinsic property* (quality) of the S_i is representable by attributes of the form

$$A: S_i \times T \times \ldots \times Z \to \text{Propositions containing } A;$$

(ii) every *qualitative mutual property* of the S_i, for $1 \leqslant i \leqslant m \leqslant n$, is representable by attributes of the form

$$A: S_1 \times S_2 \times \ldots \times S_m \times T \times \ldots \times Z \to \text{Propositions containing } A;$$

(iii) every *qualitative phenomenal* (secondary) *property* is a qualitative mutual property representable as a predicate of type (ii) where $T =$ Set of sentient organisms;

(iv) every *quantitative intrinsic property* of the S_i is representable by attributes (magnitudes, quantities, variables) of type (i) where $Z = \mathbb{R}^p$;

(v) every *quantitative mutual property* of the S_i for $1 \leqslant i \leqslant m \leqslant n$ is representable by attributes (magnitudes, quantities, variables) of type (ii) where $Z = \mathbb{R}^p$;

(vi) every *quantitative phenomenal* (secondary) property is a quantitative mutual property representable by a predicate of type (v) with $T =$ Set of sentient organisms;

(vii) every *property of an individual* (or individual property) ${}^i b$ of kind S_i is representable by the value of the corresponding attribute at ${}^i b$, and every *property of an aggregation of m individuals* ${}^1 b \dotplus {}^2 b \dotplus \ldots \dotplus {}^m b$, where ${}^i b \in S_i$, is representable by the value of the corresponding n-ary attribute at $\langle {}^1 b, {}^2 b, \ldots, {}^m b \rangle$.

It might be thought that some if not all of the above distinctions are shallow or arbitrary. In particular, it might be claimed that all individual properties, whether qualitative or quantitative, can be represented by dichotomic (or presence-absence) attributes. Thus both "is married" and "is 25 years old" are dichotomic attributes in the sense that they are either true or false of a given individual. But the point is that, whereas being married is a qualitative property that admits of no degrees, age does come in degrees. What is true is that the progress of science carries with it the quantitation of a number of attributes that had previously been regarded as being inherently dichotomous.

Likewise it might be thought that, since all relations can be reduced to binary relations (Quine, 1954), the n-arity of a predicate is unimportant. This may well be so, but such a logical reduction need not be construed as mirroring anything in reality. Our planet did not cease to revolve around the Sun between Mercury and Saturn the moment ternary relations were shown to be reducible. Besides, the said reduction does not hold for the most useful of all relations, namely functions. Thus a function of two variables is not generally reducible to two functions of one variable each (i.e. two dyadic relations).

Nor does the opposite move work, viz., that of augmenting the degree of predicates on philosophical grounds. This gambit was tried by Helmholtz (1873, p. 260), who held that "each quality or property of a thing is, in reality, nothing else but its capability of exercising certain effects upon other things". Likewise Ducasse (1968) maintained that every property is a causal power to produce some effects. This view rests on the operationist confusion between a property and the way of testing for it. Surely a property that does not affect anything cannot be observed even indirectly and therefore can hardly be said to be possessed by the entity of interest. But if scientific theory construes a given property as an n-ary predicate, our metaphysics has no right to "interpret" it as an $(n+1)$-ary property.

3. Theory

3.1. *Unarization and Dichotomization*

Thus far we have made only a few rather informal remarks on properties and have analyzed the way they are representable by predicates. In the present section we shall make definite ontological assumptions embodied in a general theory of properties and their scopes, i.e. kinds. Given the variety of properties (intrinsic and mutual, qualitative and quantitative, primary and secondary) it may well be wondered whether the task is feasible. Fortunately we can introduce a remarkable uniformity, and thus pave the way for the search for structure, by using two devices. One is to replace every mutual property by a bunch of intrinsic properties, the other is to concentrate on individual intrinsic properties. The end result is in each case a set of unary dichotomic predicates every one of which is either true or false of a given substantial individual. A couple of examples will show how to proceed in general.

Consider the property of falling. In ordinary knowledge this is represented by a unary predicate. In science, falling is analyzed as a binary predicate F such that 'Fxy' is interpreted as "x falls on y". We can now freeze the second argument, i.e. take it for granted that whatever falls does so on a fixed body b such as our planet. That is, we can form the *pseudo-unary* predicate F_b such that

$$F_b x = x \text{ falls on } b.$$

If now we change the value of the parameter, say to $c \neq b$, we obtain another predicate, namely F_c, such that $F_c y = y$ falls on c. In this way *the single binary predicate F is replaced by an infinite set of unary predicates* F_z. (No return then to the original innocence of ordinary knowledge.) There is of course no economy at all in the unarization procedure: this is just a trick allowing one to speak, though at greater length, of any property of an individual *as if* it were intrinsic. Therefore it has nothing to do with either Bradley's attempt to eliminate relations in favor of monadic predicates or "internal relations" (Bradley, 1893). Nor is it related to the even more quaint claim that a binary relation can be construed as composed of three separate parts: an outgoing arrow, an incoming arrow, and the glue between them (Harary, 1971).

The unarization procedure can be generalized to cover not only mutual properties of n-tuples of individuals but all kinds of complex properties. Consider the statement

> ⌜Body a pulls body b, at time t, relative to reference frame f, with force p in force units u⌝.,

or ⌜$P(a, b, t, f, u, p)$⌝ for short. The 6th rank quantitative predicate P represents the property of pulling, and any of its values represents the individual property of a given thing pulling another with a fixed force relative to some frame at some instant. By focusing on the agent a we manufacture one predicate P_{btfup} for each circumstance $\langle b, t, f, u, p\rangle$, such that

$$P_{btfup}(a) = a \text{ pulls } [b \text{ at time } t \text{ relative to frame } f \text{ with force } p \text{ in units } u].$$

Every one of these predicates may be taken to represent an intrinsic property and moreover an individual one, for it is either possessed or not possessed by the corresponding substantial individual, in this case a. Since the sets of instants, frames, force units, and force values are infinite, the number of such artificial unary predicates is infinite as well. This is a high price to pay and for this reason the transaction is not performed in science. But it is the price to pay if we want a general theory of properties and kinds. Let us then proceed to make

DEFINITION 2.1 Let $S_1, S_2, \ldots, S_n \subseteq S$, with $n > 1$, be n nonempty sets of substantial individuals and let $T_1, T_2, \ldots, T_m$, with $m \geqslant 0$, be m arbitrary nonempty sets. Furthermore let

$$A: S_1 \times S_2 \times \ldots \times S_n \times T_1 \times T_2 \times \ldots \times T_m \to \text{Propositions}$$

involving A

be an attribute representing an n-ary substantial mutual property of such individuals. Then for each $1 \leqslant i \leqslant n$ and for each $(n+m-1)$-tuple

$$\langle x_1, x_2, \ldots, x_{i-1}, x_{i+1}, \ldots, x_n, t_1, t_2, \ldots, t_m \rangle$$
$$\in S_1 \times S_2 \times \ldots S_{i-1} \times S_{i+1} \times \ldots \times S_n \times T_1 \times T_2 \times \ldots \times T_m,$$

the *i-th unary attributes associated to A* are the functions

$$Ax_1x_2 \ldots x_{i-1}x_{i+1} \ldots x_nt_1t_2 \ldots t_m(x) =$$
$$=_{df} A(x_1, x_2, \ldots, x_{i-1}, x, x_{i+1}, t_1, t_2, \ldots, t_m).$$

We shall henceforth take this unarization for granted throughout the remainder of this chapter. And we shall also take it for granted that, when dealing with a property, we actually handle an individual property, i.e. a property of some substantial individual. These two conventions allow us to introduce

DEFINITION 2.2 P is a *substantial property* (or a member of the set $\mathbb{P}$ of substantial properties) iff some substantial individuals possess P:

$$P \in \mathbb{P} =_{df} (\exists x)(x \in S \ \& \ x \text{ possesses } P).$$

In what follows we shall investigate the structure of the set $\mathbb{P}$ of substantial properties. That such an investigation is called for is obvious from the result of Sec. 1, namely that although all properties may be representable by attributes, not all attributes represent substantial properties.

3.2. *Basic Assumptions and Conventions*

We proceed to formulate a handful of assumptions and conventions that will allow us to lay bare the structure of the set of all substantial (intrinsic and dichotomic) properties. (The formal treatment from Definition 2.5 on follows closely Bunge and Sangalli (1977).)

The collection of properties of an individual, and that of a set of individuals, deserve special names that will be used frequently in the sequel:

DEFINITION 2.3 Let $T \subseteq S$ be a nonempty set of substantial individuals and $\mathbb{P}$ the set of (unarized dichotomic) substantial properties. Then

(i) the set of (unarized dichotomic) *properties of individual* $x \in T$ is called

$$p(x) = \{P \in \mathbb{P} \mid x \text{ possesses } P\};$$

(ii) the set of (unarized dichotomic) substantial *properties of the T*'s is called

$$p(T) = \{P \in \mathbb{P} \mid \text{For all } x \in T, x \text{ possesses } P\}.$$

Consider for a moment the set $p(x)$ of (unarized dichotomic) properties of an entity x. Since every one of these properties is taken at a given time and relative to a given reference frame, when reckoning with all the possible instants and frames we end up with a nondenumerable infinity of (potential) properties of x, no matter how modest this individual may be. (Recall Sec. 3.1.) On the other hand it seems reasonable (and comforting) to assume that the set of substantial properties in general, though extremely numerous, is finite. Thus one thinks of an animal as having different weights, metabolic rates, and ages at different times; yet all these are just different values of only three general properties – weight, metabolic rate, and age at different times and relative to different frames. In any case we stick our neck out and assume

POSTULATE 2.3 The set of general substantial properties is finite, and that of substantial individual (unarized and dichotomized) properties, i.e. $\mathbb{P}$, is nondenumerable.

Next we accept Aristotle's ontological axiom that every individual either possesses a given property (in a given respect at some time) or not:

POSTULATE 2.4 For every $x \in S$ and every $P \in \mathbb{P}$,

$$x \text{ possesses } P \vee \neg x \text{ possesses } P.$$

This axiom must not be mistaken for the excluded middle principle, which concerns predicates. Hence a mathematical intuitionist can accept the former. By logic, whether ordinary or intuitionistic, it follows immediately that no individual has a given property and fails to possess it. That is, there are no inherently contradictory things: contradiction proper is always *de dicto* never *de re*. What there can be is of course mutually opposing properties, such as excitation and inhibition, but these do not illustrate contradiction. Also, by adopting Postulate 2.4 or its corollary, one does not rule out change, in particular the acquisition or loss of a property, because every property occurring in the postulate is taken at a given time under given circumstances. Which leads us to

DEFINITION 2.4 Two properties are *incompatible* over a set $T \subseteq S$ of substantial individuals iff possessing one of them precludes having the other: If $P_1, P_2 \in \mathbb{P}$, then

(i) P_1 is *incompatible* with P_2 over $T =_{df} (x)(x \in T \Rightarrow (x \text{ possesses } P_1 \Rightarrow \neg x \text{ possesses } P_2))$;

(ii) P_1 and P_2 are mutually *compatible* over T iff they are not incompatible over T.

Note that two properties may be compatible yet not both actually possessed by a given entity. For instance honesty and wealth are compatible even though rarely concomitant. (Compatibilities are dispositional properties.)

An interesting case of property incompatibility is that of microsystems. According to the quantum theory a microsystem does not have both a sharp spatial localization and a precise velocity value. (What are compatible, moreover concomitant, are the probability distributions of position and of momentum.) In this theory several properties (the dynamical variables) are represented by operators, some of which fail to commute pairwise. If A_1 and A_2 are operators representing mutually incompatible properties, then $A_1A_2 \neq A_2A_1$ and conversely. When this is the case then, whenever A_1 takes on a precise (eigen)value, its partner A_2 exhibits a whole range of values and conversely. That is, noncommuting variables do not have joint sharp values: they have only distributions of values, each with a given probability. Operationism notwithstanding this – insofar as quantum mechanics is true – is an objective trait of nature not a feature of measurement. In other words the incompatibility between properties represented by noncommuting operators is verified in experiment not caused by the latter (Bunge, 1967b).

We adopt next a *principium individuationis*:

POSTULATE 2.5 No two substantial individuals have exactly the same properties:

$$\text{For all} \quad x, y \in S, \quad \text{if} \quad x \neq y \quad \text{then} \quad p(x) \neq p(y).$$

By contraposition it follows that, if two entities have exactly the same properties, then they are one:

COROLLARY 2.1 For all $x, y \in S$,

$$\text{if} \quad p(x) = p(y), \quad \text{then} \quad x = y.$$

Either of these hypotheses may be called *Leibniz' law* (Leibniz, third and fourth letter to Clarke, in Alexander, Ed., 1956). They take identity seriously without mistaking it for mere similarity: the slightest difference between two entities – such as a difference in relative position with respect to a third entity – results in difference. It is only in the realm of constructs that there can be several identical copies of one and the same object, as with the cases of $a + a$ and $A \times A$. But even here one may argue that there is a single object taken repeatedly.

It is usually maintained that the concept of identity occurring in Leibniz' law is that of *contingent identity* as opposed to formal identity. Thus a living brain is said to be contingently identical with the interior of the skull of a living animal, and gene duplication contingently identical with the uncoiling of the double helix. I do not understand the alleged difference between contingent and formal identity: since both concepts have exactly the same properties they must be identical and must fall under the theory of identity.

Because the relation of identity is reflexive, we retrieve a famous statement sometimes regarded as a basic principle of both logic and ontology, at other times decried as absurd or, even worse, as trivial, namely

COROLLARY 2.2 Every entity is identical to itself (i.e. self identical).

Some philosophers have held that this statement is incompatible with change. Nothing of the sort. If an entity changes then it becomes a different entity – or an entity in a different state – even though we may go on calling it by the same name. All that Corollary 2.2 asserts is that every entity preserves whatever "identity" it may have – until it loses it and acquires another. (Caution: in the last sentence 'identity' signifies "collection of peculiarities".)

Note that identity, therefore also difference, are relations, hence concepts not facts. Even though it is true that a dog is not the same as a star, there is no connection or coupling between the two – i.e. "$\neq$" mirrors no such relationship or link. In other words both identity and difference are *de dicto*. This is one of the crucial places at which naive realism (i.e. the fact-thought isomorphism thesis) fails.

One last caution concerning identity: this concept must not – *pace* Strawson (1959) – be mistaken for the pragmatic concept of *identification*. One may by mistake identify two things that are actually different; or one may "identify" an object *qua* member of a certain class, thereby failing to use the concept of identity. For this reason we cannot adopt the operationist claim that "the relation of identity or distinctness has no other meaning than that two objects have been identified or distinguished" (Yessenin-Volpin, 1970, p. 7). Whether or not an identity statement has been put to the test, it should be meaningful to begin with. (For the relation between meaning and testability see Vol. 2, Ch. 7, Sec. 5.1.)

Finally we reach a central convention in our theory:

DEFINITION 2.5 The *scope* of a substantial property is the collection of entities possessing it. In other words, the scope $\mathcal{S}$ is the function $\mathcal{S}: \mathbb{P} \to 2^S$ from the set of all substantial properties to the set of all the subsets of substantial individuals, such that "$x \in \mathcal{S}(P)$", for $x \in S$, is interpreted as "*Individual x possesses property P*".

For example, the scope of the mass property is the set of all bodies, and that of the social mobility property is the set of all human societies.

A first, surely modest use of $\mathscr{S}$ is in defining the fiction called *null property*: N is a null property iff $\mathscr{S}(N)=\varnothing$.

Note the difference between the scope of a property and the extension of a predicate or attribute (cf. Vol. 2, Ch. 9, Sec. 1). For one thing $\mathscr{S}$ is defined on properties not on predicates; for another $\mathscr{S}$ is not defined for negation and disjunction.

Note also that some of the ideas handled in the preceding can be reformulated in terms of scopes. For example the Definition 2.4(i) of property incompatibility becomes

P_1 and P_2 are *mutually incompatible* over $T =_{df} \mathscr{S}(P_1) \cap \mathscr{S}(P_2) \cap T = \varnothing$.

Given a property other than a null property, there is always a nonempty set of individuals that possess it since, by Definition 2.2, a substantial property is one possessed by at least one substantial individual. See Figure 2.1. But the converse is false: an arbitrary collection of objects need not share a given property. When they do, and no object outside the collection has the property of interest, the set is called a class:

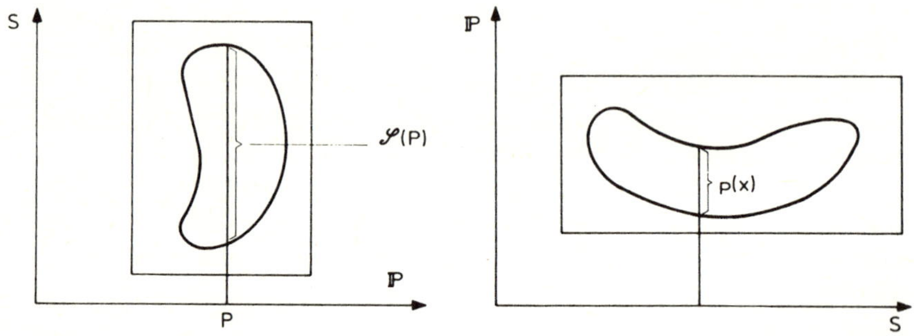

Fig. 2.1. Two complementary concepts: the scope of a property and the set of properties of an individual.

DEFINITION 2.6 A subset X of S is called a *class* (of substantial individuals) iff there exists a substantial property $P \in \mathbb{P}$ such that $X = \mathscr{S}(P) \in 2^S$.

Since all substantial individuals can associate (Ch. 1), S forms a class – in fact the largest of all classes studied by ontology, i.e. the universe of discourse of the latter.

Our next assumption is

POSTULATE 2.6 The intersection of any two classes is a class: For any two compatible substantial properties $P, Q \in \mathbb{P}$, there is at least one other property $R \in \mathbb{P}$ such that $\mathscr{S}(R) = \mathscr{S}(P) \cap \mathscr{S}(Q)$.

Note that the third property determined by any two given properties need not be unique. In fact there might be a fourth property with the same scope as R. Secondly, if P and Q are mutually incompatible over some domain of individuals, then they determine an empty class. Thirdly, the above assumption is not necessary for attributes or predicates because, if A and B are attributes with overlapping domains, then their conjunction A & B exists on their overlap and it determines a set, namely the extension of the composite predicate A & B. But we still do not know what the conjunction of properties might be, so the above postulate is nontrivial. Fourthly, we are not asserting that the union of two classes is a class, for this would be definitely false in the particular case when the classes are natural kinds: think of the set (not the class) equal to the union of the class of clouds and the class of flies.

So much for properties in general. Let us now study a particular class of properties, namely laws.

3.3. *Laws as Properties*

So far as Postulate 2.6 is concerned, all the classes could be pairwise disjoint. Our final assumption will deny this possibility, i.e. it will assert that some pairs of classes overlap and, moreover, that some are contained in others. In other words, we shall assume that all entities satisfy some laws – or, what amounts to the same, that every substantial property is lawfully related to some other substantial property. To formulate this idea with precision we need

DEFINITION 2.7 If $P, Q \in \mathbb{P}$ are substantial properties, then the statement $\ulcorner \mathscr{S}(P) \subseteq \mathscr{S}(Q) \urcorner$ or its converse or, equivalently,

> $(x)Lx$ with $Lx = \ulcorner x$ possesses $P \Rightarrow x$ possesses $Q \urcorner$ or its converse

is called a *law statement* relating P to Q.

If the scope of a property is included in that of another, the two properties are said to be *lawfully related.* A property not so related to any other property is said to be *stray* or *lawless.*

This notion of a law will be considerably refined in Ch. 3, Sec. 2.3. Note that we have not defined laws, or objective patterns, but law statements, or conceptual reconstructions of objective patterns. The relation between the two categories is an instance of the property-attribute relation discussed in Sec. 1.2.

It might be objected that there are single property laws, such as the law of inertia, which can be abbreviated to "$V = \text{const.}$" This would be mistaken, for the preceding formula is a poor rendering of the correct statement, which involves further variables. Firstly there is the property of being a free body (predicate B). Secondly, there is a tacit reference to a frame (predicate F) as well as to time (T). Indeed, the complete formulation is: "If x is a free body (B) and y a reference frame (F) and t an instant of time (T), then the velocity of x relative to y is constant for all t in T". The function V represents then a mutual property (of body and frame) and the last statement makes no sense without the antecedent, which contains the predicates B and F. True, time (T) looks external to things – common to all hence the property of none. But it actually summarizes the state of the reference frame. (See Ch. 5, Sec. 3.2.) In fact t can be replaced by the phase of some process occurring in a frame $y \in F$. (For example t could be the angular position of the pointer of a chronometer attached to y, or the mass of water flowing in a clepsydra.) This renders it a property of entities. In any event the initial simplicity of the law of inertia was deceptive.

We can now formulate the hypothesis that there are no stray or lawless properties:

POSTULATE 2.7 Every substantial property is lawfully related to some other substantial property. I.e., if $P \in \mathbb{P}$, then there exists a $Q \in \mathbb{P}$ such that either $\mathcal{S}(P) \subseteq \mathcal{S}(Q)$ or $\mathcal{S}(Q) \subseteq \mathcal{S}(P)$.

This ontological principle underlies all science and all technology. (Recall Introduction, Sec. 7, principle *M8*.)

Note that, since laws interrelate substantial properties and the latter are properties, *laws themselves are properties of entities*. So much so that they can be expressed by formulas of the basic form $\ulcorner(x)Lx\urcorner$. Thus it is a property of a free body that it satisfies the law of inertia. Like every other property, it is represented by a predicate analyzable in the manner prescribed by Postulate 2.2. However, laws are properties *lato sensu*, not in the strict sense we have been considering heretofore. For, if every law were to belong to $\mathbb{P}$, then by Postulate 2.7 it would occur in some other law – and so on *ad infinitum*. But this would contradict the Postulate 2.3 that $\mathbb{P}$ is finite. Hence when claiming that laws are properties we use the term 'property' in a broader sense than when we state that laws relate properties. More in a while.

Postulate 2.7 is our present version of the *principle of lawfulness*. (For a discussion of the latter see Bunge (1959).) It will play a decisive role in Ch. 3, Sec. 3.3, in helping us define the notion of a natural kind as distinct from a mere class. And it provides the following propertyhood

CRITERION 2.1 P is a property iff P occurs in at least one law.

By this criterion predicates such as "non-smoker", "fast or thinking", and "green before AD 2000 and blue thereafter" (the infamous "grue") do not represent any substantial properties. This is of course in agreement with Sec. 1. Moreover we are now in a better position to distinguish an arbitrary function from one representing a substantial property:

CRITERION 2.2 A function F represents a property of substantial individuals of class $T \subseteq S$ iff

(i) T^n, with $n \geqslant 1$, occurs in the domain of F;

(ii) F occurs in at least one law statement referring to T.

By this criterion "differentiable" (in the mathematical sense) does not represent a substantial property because its domain is the set of functions. On the other hand "differentiable function" can be made to represent properties that happen to be smooth.

Finally we introduce a useful convention:

DEFINITION 2.8 The totality of laws possessed by substantial individuals is denoted by $\mathbb{L}$.

Remark 1 The ontological concept of a law can be illustrated. On the other hand it is not ostensive: it is impossible to point to a law – as different from a sentence expressing a law statement conceptualizing a law. Furthermore nobody could possibly exhibit or even mention the set of all laws. (At most one could try to list the known law statements in a given field of research – or rather a subset of standard law statements with neglect of their innumerable instances.) Thus $\mathbb{L}$ is a nonconstructive set and is, if anything, even more metaphysical than its members. It is one of those ontological notions used, though not elucidated, in the everyday metascientific discourse of scientists. *Remark 2* Since laws are properties in a broad sense, we can form the totality $\mathbb{P} \cup \mathbb{L}$ of properties *lato sensu*, where $\mathbb{P}$ is the set of properties *sensu stricto* (those occurring in the members of $\mathbb{L}$). *Remark 3* Because laws are properties, they can be assigned scopes – provided we agree to introduce a new scope function defined on $\mathbb{P} \cup \mathbb{L}$, not just on $\mathbb{P}$. The scope of a law would then be the set of entities possessing ("obeying") the law. This set is the same as the reference class of the law statement representing the objective law – not however the same as its extension, which is the set of entities for which the law statement holds exactly. *Remark 4* Another consequence of construing laws as properties (in a broad sense) is that the scope of a law may include or be included in that of some other property, in particular that of another law. Hence the set $\mathbb{L}$ of laws, far from being amorphous, is partially ordered in respect of breadth of scope. However, in the technical developments to follow we shall keep to the narrow construal of 'property'.

So much for our main assumptions and definitions. Let us now proceed to make the most of them.

3.4. *Precedence and Conjunction of Properties*

The concept of a law introduced by Definition 2.7 and occurring in Postulate 2.7 suggests the formation of the following preordering relation:

DEFINITION 2.9 If P and Q are substantial properties (i.e. members of $\mathbb{P}$) then *P precedes Q* iff P is more common than Q, i.e.

$$P \preccurlyeq Q =_{df} \mathscr{S}(Q) \subseteq \mathscr{S}(P).$$

Since in turn inclusion is defined in terms of implication, the previous definition is equivalent to

DEFINITION 2.10 If P and Q are substantial properties, then *P precedes Q* iff P is necessary for Q, i.e.

$$P \preccurlyeq Q =_{df} (x)(x \in S \Rightarrow (x \text{ possesses } Q \Rightarrow x \text{ possesses } P)).$$

In other words, "precedes", "is more common than", and "is necessary for" are coextensive. Equivalently: "follows", "is rarer than", and "is sufficient for" are coextensive. Example: the property of thinking follows that of being alive, which in turn follows that of containing genetic material.

Warning: the preceding definitions should not be read in terms of attributes or predicates, if only for the following reason. Let A and B be two attributes with the same referents and the same rank, and assume that B is necessary for A, i.e. $A \Rightarrow B$. Then by Craig's interpolation theorem there exists at least one other attribute C with the same referents and the same rank that mediates between A and B, i.e. such that $A \Rightarrow C \;\&\; C \Rightarrow B$. This, which is a theorem in logic, does not guarantee the existence of a substantial property represented by the interpolated predicate. Such an existence hypothesis, if made, would give us a different theory of properties.

The precedence relation is reflexive and transitive, i.e. it is a preorder relation. In fact it is not antisymmetric, i.e. $P \leqslant Q$ and $Q \leqslant P$ does not imply $P = Q$. Indeed different properties may have the same scope – just as two attributes can have the same extension without being identical. In order to obtain an order and thus uncover richer structures we must shift attention from properties to classes (or scopes of properties). To begin with we make

DEFINITION 2.11 If P and Q are any two substantial properties, then they are *concomitant* iff they have the same scope:

$$P \sim Q =_{df} \mathscr{S}(P) = \mathscr{S}(Q).$$

In other words, two properties are concomitant if the corresponding attributes are equivalent. Or also:

$$P \sim Q =_{df} (x)(x \in S \Rightarrow (x \text{ possesses } P \Leftrightarrow x \text{ possesses } Q)).$$

The concomitance of properties is what Hume called 'constant conjunction of properties' (Hume, 1739–40). In our ontology, by virtue of Definition 2.7, *the concomitance of properties is not coincidental but lawful*; or, if preferred, it is nomically necessary rather than accidental.

The relation $\sim$ of concomitance is important because it is an equivalence relation on $\mathbb{P}$, i.e. it splits $\mathbb{P}$ into disjoint sets (equivalence classes), namely of those properties that happen to come together. To investigate the structure of the resulting sets we need some further notation:

DEFINITION 2.12 Let $\mathbb{P}$ be the collection of substantial properties and $\sim$ the relation of property concomitance defined on $\mathbb{P}$. Then for any $P \in \mathbb{P}$:

(i) the *totality of concomitants* of P: $[P] =_{df} \{Q \in \mathbb{P} \mid Q \sim P\}$;

(ii) the *family of sets of concomitants*: $[\mathbb{P}] =_{df} \{[P] \mid P \in \mathbb{P}\}$;

(iii) the *scope of sets of concomitants*: $[\mathscr{S}]: [\mathbb{P}] \rightarrow \mathscr{P}(S)$.

Clearly, for any substantial property P, $[\mathscr{S}]([P]) = \mathscr{S}(P)$ is the class of entities that possess P and all the other properties concomitant with P.

The preorder relation $\preccurlyeq$ on $\mathbb{P}$ induces a partial order relation $[\preccurlyeq]$ on $[\mathbb{P}]$ according to

DEFINITION 2.13 Let $[\mathbb{P}]$ be the collection of all equivalence classes of properties. Then, for any two $[P], [Q] \in [\mathbb{P}]$,

$$[P][\preccurlyeq][Q] =_{df} \mathscr{S}[Q] \subseteq \mathscr{S}[P].$$

This new precedence relation has all the formal properties of the inclusion relation. It enables us to compare not properties but their scopes – thus obliterating any difference among properties with the same scope. Hence $\langle [\mathbb{P}], [\preccurlyeq] \rangle$ is a partially ordered set. Actually $[\mathbb{P}]$ has a richer structure, as shown by

THEOREM 2.1 The family of sets of concomitant properties has the sup-semilattice structure, where the supremum of $[P]$ and $[Q]$ is defined as

$$[P] \sqcup [Q] =_{df} \{R \in \mathbb{P} \mid \mathscr{S}(R) = \mathscr{S}(P) \cap \mathscr{S}(Q)\}.$$

Proof It is easily verified that $[P] \sqcup [Q]$ is a set of concomitants and the least (in the sense of $[\preccurlyeq]$) upper bound of $\{[P], [Q]\}$, i.e. $[P], [Q][\preccurlyeq]([P] \sqcup [Q])$.

COROLLARY 2.1 If a substantial property R belongs to the supremum of two sets $[P]$ and $[Q]$ of concomitant properties, i.e. $R \in [P] \sqcup [Q]$, then an entity possesses R just in case it possesses both P and Q: $(x)(x \in S \Rightarrow (x \text{ possesses } R \Leftrightarrow x \text{ possesses } P \,\&\, x \text{ possesses } Q))$.

We shall shorten "x possesses P & x possesses Q" to "x possesses $P \wedge Q$", where the complex property $P \wedge Q$ is characterized, albeit not defined, by

$$\mathscr{S}(P \wedge Q) = \mathscr{S}(P) \cap \mathscr{S}(Q).$$

$P \wedge Q$ will be called the *conjunction* of P and Q, or the *complex property* composed of P and Q. This is not a definition proper because equality of scope (which is all we are requiring) is necessary but not sufficient for identity of properties. (Caution: property conjunction should not be mistaken for property concomitance or "constant conjunction".)

Properties then may be simple (or basic) or complex (or derivative). For example in contemporary physics mass, charge, spin and strangeness are regarded as basic (simple) properties in the sense that they are not reducible to any other properties. On the other hand weight is a derivative (complex) property, as it consists in an entity's having a mass *and* being placed in a gravitational field. That is, weight is the conjunction of two properties.

Only a scientific theory can say whether a given property is basic or derivative. But such a dichotomy holds only for the proper subset $\mathbb{T}$ of $\mathbb{P}$ represented in the theory, so the most one can venture is to say that a certain property P is simple, or else complex, *in* $\mathbb{T}$. Alternative theories dealing with the same properties may decide differently. For example energy and momentum, which are simple in certain theories, are treated as complex in others. Therefore we cannot pass final judgment on the degree of complexity of a property on the basis of an analysis of the corresponding attributes. Nevertheless such indications, ambiguous and fickle as they are, are the only ones we have.

It seems reasonable to assume that, even though we often find it difficult if not impossible to ascertain whether a given property is basic, reality has no such problem and builds complex properties out of simple ones – or rather complex entities out of simpler ones. It is therefore worth while to state that assumption in exact terms. But to this end we first have to introduce an exact concept of a basic property. And this in turn calls for the explicit introduction of the concept of a universal property, i.e. one possessed by all entities. More precisely, we form the equivalence class of all universal properties and call it

$$[U] =_{df} \{P \in \mathbb{P} \mid \text{For all } x \in S : x \text{ possesses } P\}.$$

Clearly, $[U]$ precedes every other set of concomitant properties in the sup semilattice $\langle [\mathbb{P}], \sqcup \rangle$, hence it is the smallest element of it, i.e. its zero. In other words $[U]$ is the root of the tree of (sets of concomitant) properties: see Figure 2.2. Note that there is no top element formed by the properties that no entity possesses because, by Definition 2.2, a substantial property is possessed by at least one entity.

We are now ready to elucidate the notion of a basic (or generating) property. We stipulate it to be a property occurring as a conjunct in some other (complex) property. More exactly, we adopt

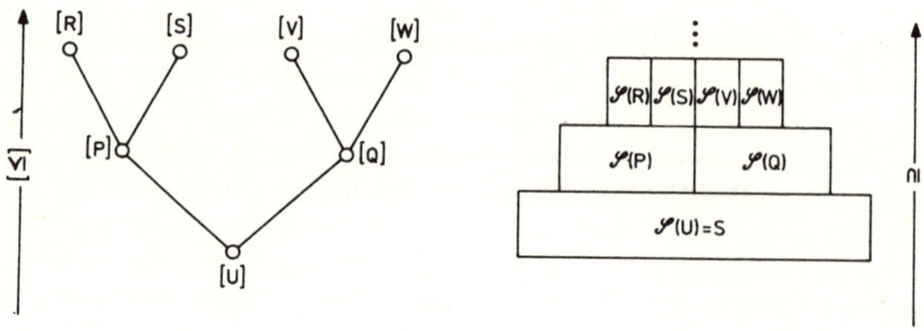

Fig. 2.2. The tree of (sets of concomitant) properties and the pyramid of classes of entities with the same properties. The more basic a property the more common it is.

DEFINITION 2.14 A subset $\mathbb{B}$ of $\mathbb{P}$ is called a set of *independent generators* of $\mathbb{P}$ iff

(i) for every $P \in \mathbb{P}$ there exist $B_1, B_2, ..., B_n \in \mathbb{B}$, $n \geqslant 1$, such that $P = B_1 \wedge B_2 \wedge \ldots \wedge B_n$;

(ii) for all $n > 1$ and all $B_1, B_2, ..., B_n \in \mathbb{B}$, if $B_1 = B_2 \wedge \ldots \wedge B_n$ then either $B_i = B_1$ or $B_i = U$ (a universal property) for $i = 2, 3, ..., n$.

Every set $\mathbb{B}$ of independent generators splits into a set of universal properties and another of *atomic* properties, i.e. properties which are basic (or generating) but not universal:

$$\mathbb{B} \subseteq At(\mathbb{P}) \cup [U].$$

If there are such atomic properties then this is a trait of reality not a mathematical "fact". And if we do assume their existence then we must suppose that they are infinitely many. In fact a finitely generated semi-lattice is finite but, by Postulate 2.3, $\mathbb{P}$ is infinite. Therefore the assumption that there are basic properties that determine all others reads:

POSTULATE 2.8 The set of atoms of $\mathbb{P}$ is infinite and it generates $\mathbb{P}$.

An immediate consequence is that each property is of either of the following types:

(i) *basic* properties (i.e. members of some $\mathbb{B} \subseteq At(\mathbb{P}) \cup [U]$);

(ii) properties that are *finite conjunctions* of two or more basic properties;

(iii) properties that are *infinite conjunctions* of basic properties (in the sense that their scopes are infinite intersections).

Infinite conjunctions of properties are not ghostly or even exceptional. Thus the trajectory of a particle relative to some reference frame may be construed as the conjunction of its (nondenumerably many) successive positions.

The epistemological consequence of Postulate 2.8 is this. For any class $T \subset S$ of entities, in order to know all the properties in $p(T)$ it suffices to know the atomic properties of the T's, i.e. the atoms of $\mathbb{P}$ belonging to $p(T)$.

We close this section with an algebraic note showing that the collection of substantial individuals, far from being an amorphous set, has a definite structure. Consider the family $\mathbb{K}$ of all classes of substantial individuals. Since by Postulate 2.6 the intersection of any two classes is a class, $\mathbb{K}$ has the semilattice structure. In other words, we have proved

THEOREM 2.2 The system $\langle \mathbb{K}, \cap \rangle$, where

$$\mathbb{K} = \{\mathscr{S}(P) \in 2^S \mid P \in \mathbb{P}\}$$

is the family of all classes of substantial individuals, is an inf-semilattice.

We shall have much more to say on the algebra of classes of substantial individuals in Ch. 3, Sec. 3.

3.5. *Similarity*

Knowing that two given concrete objects are different is not enough: we should know what their degree of difference is – or, which amounts to the same, in which respects they are the same. To be the same in some respects is to be similar. Now, similarity can be weak or strong, superficial or deep, according as the overlap in properties is slight or large. Most assertions of identity with reference to concrete objects are actually statements of strong similarity (or equality) not of strict identity. For example the statement that all the atoms of a given kind in a certain state are identical means that they are the same barring certain extrinsic differences – e.g. in position relative to other things.

We must therefore investigate the concept of degree of similarity and its dual, degree of dissimilarity. We start with the convention that the similarity between two things is the collection of their shared properties. More precisely, we make

DEFINITION 2.15 Let $\sigma: S \times S \rightarrow 2^{\mathbb{P}}$ be the function such that $\sigma(x, y) = p(x) \cap p(y)$ for any $x, y \in S$. Then

(i) $\sigma(x, y)$ is called the *similarity* between x and y, and

(ii) two entities are said to be *similar* ($\sim$) iff their similarity is not nought:

$$\text{If } x, y \in S \quad \text{then} \quad x \sim y =_{df} \sigma(x, y) \neq \varnothing.$$

Note that $\sim$ is a similarity relation, not an equivalence relation, because it is reflexive and symmetric but not transitive. (One entity may resemble another and the latter a third while the former does not resemble the third. Just think of face similarity.)

By Postulate 1.1 in Ch. 1, Sec. 1.2, all substantial individuals have the property of associating. (Besides, we know from physics that all entities have energy, are capable of moving, etc.) Hence

THEOREM 2.3 All entities are similar [or equal in some respect]: If $x, y \in S$, then $x \sim y$.

In other words, no matter how dissimilar two things may be in most respects, they share some property or other – in fact a number of properties. This is one of the reasons that all things can be studied scientifically (with the help of the scientific method) – the other reason being that they are all studied by beings equipped with similar neurosystems. Caution: this is not to conclude, with mechanism and animism alike, that all things have the same properties, only in varying degrees – as Whitehead (1929) and Teilhard de Chardin (1964) believed.

If, for the sake of expediency rather than truth and depth, we assign all properties the same weight, then we can define a quantitative concept of degree of similarity provided we restrict ourselves to the subset of the known properties of a thing (to avoid division by infinity):

DEFINITION 2.16 The *degree of similarity* between two entities $x, y \in S$ relative to a finite subset $\mathbb{B}$ of $\mathbb{P}$ [such as, e.g., the genetic similarity or the cultural similarity of two human populations] is

$$s(x, y) = \frac{|\sigma(x, y) \cap \mathbb{B}|}{|(p(x) \cup p(y)) \cap \mathbb{B}|}.$$

This is none other than the simple matching coefficient used in numerical taxonomy to estimate the distance between biological species (Sneath and Sokal, 1973). Because $\mathbb{B}$ is a finite set, $s(x, y)$ can be zero or it can be equal to unity. For example, the degree of genetic similarity between any two cells of a single organism equals 1. But by taking $\mathbb{B}$ to be a big enough set of properties we obtain a smaller value of s.

The above concepts of similarity can be generalized to whole sets of entities, namely as follows. Let $x_1, x_2, \ldots, x_n$ be the members of a finite set T of entities. Then the *similarity among the members of* T is defined as

$$\sigma(x_1, x_2, \ldots, x_n) = \bigcap_{x_j \in T} p(x_j)$$

and the corresponding degree of similarity as

$$s(x_1, x_2, \ldots, x_n) = \frac{|\sigma(x_1, x_2, \ldots, x_n)|}{\left|\bigcup_{x_j \in T} p(x_j)\right|}.$$

However, we shall not employ the last two concepts.

The notion of dissimilarity is mathematically more interesting than its dual. We define it thus:

DEFINITION 2.17 Let $\delta: S \times S \to 2^{\mathbb{P}}$ be the function such that

$$\delta(x, y) = p(x) \mathbin{\Delta} p(y) \quad \text{for any} \quad x, y \in S,$$

where 'Δ' designates the symmetric or Boolean difference. $\delta(x, y)$ is called the *dissimilarity* between x and y.

By Definitions 2.15 and 2.17, the relation between δ and σ is

$$\delta(x, y) = p(x) \cup p(y) - \sigma(x, y).$$

The function δ is a set valued distance function, hence $\delta(x, y)$ should be a good measure of the "distance" between x and y in the space of properties. More precisely, we have

THEOREM 2.4 The dissimilarity function δ is a set valued distance function in a one dimensional space: for all $x, y, z \in S$,

(i) $\varnothing \subseteq \delta(x, y) \subset \mathbb{P}$;
(ii) δ is symmetric: $\delta(x, y) = \delta(y, x)$;
(iii) $\delta(x, y) = \varnothing$ iff $x = y$;
(iv) $\delta(x, y) \mathbin{\Delta} \delta(y, z) = \delta(x, z)$.

Proof The first part is obvious from the definition. (ii) is a consequence of the symmetry of Δ. (iii) follows from $p(x) \mathbin{\Delta} p(x) = \varnothing$. Conversely, by

Postulate 2.5 if $p(x)\,\Delta p(y) = \varnothing$ then $x = y$. (iv) follows upon replacing the LHS of (iv) by the definiens:

$$(p(x)\,\Delta\,p(y))\,\Delta\,(p(y)\,\Delta\,p(z)) = p(x)\,\Delta\,p(y)\,\Delta\,p(y)\,\Delta\,p(z) =$$
$$= p(x)\,\Delta\,p(z).$$

By taking the cardinality of the dissimilarity and normalizing it to unity we obtain a quantitative concept of dissimilarity:

DEFINITION 2.18 Take a finite subset of the set of all the properties of two entities $x, y \in S$. Then their *degree of dissimilarity* is

$$d(x, y) = \frac{|\delta(x, y)|}{|p(x) \cup p(y)|}.$$

The values of d, just as those of the degree of similarity s, are bounded by 0 and 1. It can be shown that d is a metric, hence a good measure of the distance between entities in respect of their properties. (Needless to say, there is no relation, other than a mathematical similarity, between this distance and a spatial distance.) That is, one can prove

THEOREM 2.5 The structure $\langle S, d\rangle$, where d is the degree of dissimilarity, is a metric space.

Extreme difference is sometimes called *opposition*. This concept is elucidated by

DEFINITION 2.19 Two entities are *opposite* to one another iff their similarity is nil.

But, according to Theorem 2.3, all things share some properties. Hence we have

THEOREM 2.6 There are no opposite entities.

Opposition, when it exists, is relative to a proper subset of properties. Moreover only "positive" properties should count: the mere absence of a property is not to be equated with the opposite of that property. For instance dark is not the opposite of illuminated, as the former is nothing but the absence of illumination. Real opposition, when it exists, is active and consists in having contrary effects. Moreover real opposition does not exist between utterly different entities. For example two animals of the same species may fight over the same pool of water, or the same prey, or the same female, precisely because they have similar drives.

This point deserves a few more examples. In an electric circuit the self-inductance initially opposes the action of the impressed electromotive force, so that the two aspects are mutually opposed in their effects on the overall current. Likewise if a substance A promotes the synthesis of a protein while A' inhibits it, then A and A' may rightfully be regarded as opposites with regard to protein synthesis no matter how much they resemble in other respects – and they do since both can act on the given protein. Likewise in a mammal the sympathetic and the parasympathetic systems oppose one another in the sense that they check each other: when one of them excites a given organ the other inhibits the latter. But even this opposition is not complete, as there is also some cooperation between the two systems. And in any case this interaction ensues in homoeostasis or equilibrium not in qualitative change (such as the breakdown of the main system) – whence it does not illustrate dialectics.

In sum, although *some* pairs of things do have parts and features that are mutually opposed in *limited* respects, no thing is composed of absolute opposites. Consequently the thesis of dialectical ontologies, that *every* thing is a synthesis or unity of opposites (in what respects?) is, if intelligible at all, false. Moreover the assumption that there are absolute opposites contradicts the materialist thesis, incorporated into our system, that all things – no matter how different – share some properties precisely by virtue of being entities. Besides, the thesis that all things are unities of opposites entails the infinite divisibility of every single entity into two mutually opposed entities, since every part should in turn be a synthesis of opposites – and there is no empirical evidence for this consequence. The most we can assert is that there are relative opposites, i.e. things with features that oppose one another in their effects. But this timid thesis can hardly serve as a cornerstone of a philosophical cosmology. (More on dialectics in Bunge, 1975.)

The upshot of this subsection is this. Although by Postulate 2.5 every entity is unique in some respect(s), all entities are similar in some other respect(s). Shall we then say that all entities are alike and aren't? Not at all: 'alike' (or 'similar') is a syncategorematic term or, if preferred, a vague notion that should be qualified by the expression 'in the respect(s) P'. The complete expression 'x and y are alike (similar) in the respect(s) P' is no longer vague, hence it does not lead us to contradiction. It is moreover compatible with 'x and y are dissimilar in the respect(s) Q, where $Q \neq P$'. In other words 'similar' is not an attribute word on the same footing as 'alive': it is an incomplete or partial attribute word – though not more so than 'blue' or 'heavy'. Hence the need for replacing wherever possible the relation $\sim$ of similarity by the function σ or by the function d.

3.6. *Indiscernibility*

We hold that there are no two identical entities – yet it is common experience that some things are indiscernible or indistinguishable. There is no contradiction here, as two different concepts of difference are involved: the *ontological* concept of (objective) difference on the one hand, and the *epistemological* (or pragmatic or psychological) concept of (subjective) differentiability, or discernibility, or ability of somebody to distinguish empirically, e.g. by observation. Two entities may be indistinguishable to a given subject equipped with certain means of observation but may become discernible to the same subject using improved equipment. (The concept of resolving power of a microscope or of a telescope presupposes the difference between objective difference and perceived difference.)

We must therefore distinguish between factual difference and empirical difference or discernibility. This will require roping in a notion that, strictly speaking, we have no right to use systematically since it will be introduced in Volume 4, Ch. 10, namely that of a cognitive subject. Therefore we shall not number the definitions that follow.

DEFINITION. Let $\mathbb{P}_s \subset \mathbb{P}$ be a proper subset of the totality of substantial properties, and s a cognitive subject in a given state and equipped with certain cognitive means (instruments, theories, etc.). Then for $x, y \in S$, x and y are *indiscernible* (indistinguishable) *to* s iff x and y are the same in every respect $P \in \mathbb{P}_s$ known to s. Notation: $x \sim_s y$.

The terms 'indiscernibility' and 'indistinguishability' have been misused at several critical points. *Example 1* Leibniz himself stated his principle (our Postulate 2.5) in an equivocal manner, namely by writing that "il n'y a point dans la nature deux êtres réels absolus indiscernables", and calling this the *principium identitatis indiscernibilius.* As Russell (1900) saw, all Leibniz intended to affirm is "that any two substances [things] differ as to their predicates [properties]". Leibniz cannot have intended to state that if two things are indiscernible then they are identical, for he was well aware of the marvels of microscopy.

Example 2 The logical principle of substitutivity is sometimes called the *principle of the indiscernibility of identicals* – another obvious mistake. Consider: the principle holds that if b has P, and if $b = c$, then P is true of c as well. It has become fashionable to claim that this principle fails in the so-called "intensional contexts" and thus doubts are cast on the truth of Leibniz' ontological principle (our Postulate 2.5). For example, it is clear that "Smith doubts that $x = y$" does not entail "Smith doubts that $x = x$". The rejoinder is that, if Smith doubts that $x = y$, then he will be well advised to abstain from applying it the principle of substitutivity. And, if a principle is not applied, it runs no risks. Anyway this has nothing to do with indiscernibility. *Example 3* In microphysics one assumes that particles of the same kind and in the same state (i.e. with no intrinsic differences) count as equal or equivalent and so can be exchanged both in fact and in the calculations. This assumption is usually cast with the term 'indistinguishable' replacing 'equal' or 'equivalent' – as if the particles could care about our ability to discern among them. The fact is that we do distinguish among them by their extrinsic (spatiotemporal) properties and so are able to count them. See Figure 2.3. In other words, the truth of the matter is that the "elementary particles" are distinct and often distinguishable in practice, but (a) they can be counted as equal or equivalent and (b) when constituting certain wholes they surrender part of their individuality. (See Ch. 5, Sec. 4.2 for the partial loss of "identity" of things when they become components of a system.) In short, Postulate 2.5 is confirmed by microphysics. Plato (*Phaedo* 74a–75c), Leibniz (1704, Ch. XXVII), and Bolzano (1851, Sec. 50) were right in assuming it.

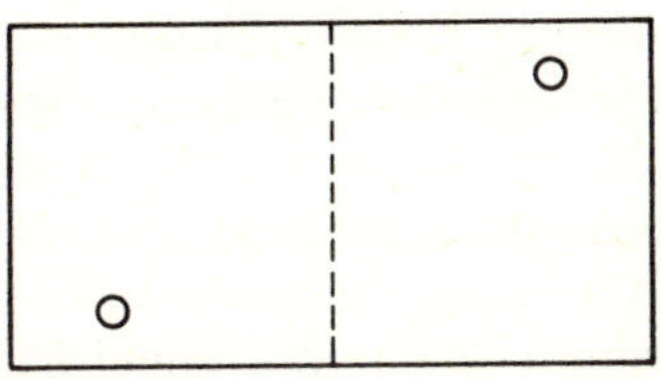

Fig. 2.3. Two electrons in a box: exchangeable if in the same state, yet distinct and distinguishable (discernible). I.e. their exchange has no effect on the states of the system as a whole. But it can happen precisely because the particles are distinct.

Indiscernibility is a special kind of similarity (see Sec. 3.5). Hence, unlike identity, indiscernibility is not transitive. Therefore it does not generate equivalence classes. Instead, indiscernibility generates indiscernibility spaces, also called tolerance spaces (Zeeman, 1962; Schreider, 1975). These spaces have interesting topological properties and occur in the study of perception. We shall take a quick look at them, starting with

DEFINITION A *tolerance space* $\langle T, \sim_s \rangle$ is a set $T \subset S$ of entities together with an indistinguishability (indiscernibility) relation $\sim_s$ on T.

Example Let T be the collection of entities in the visual field of a subject s (in a given state) and let $\sim_s$ be the visual acuity tolerance for the same subject, i.e. (extensionally) the collection of pairs of entities that s cannot distinguish visually from one another. The structure $\langle T, \sim_s \rangle$ is a tolerance space.

The topology of a tolerance space is determined by the collection of neighborhoods defined by

DEFINITION The *vicinity* of an entity $x \in S$ is the collection of entities indistinguishable from x:

$$N(x) = \{y \in S | y \sim_s x\}$$

Any given set of entities can be covered by such neighborhoods. (Clearly no topology generated in this way is Hausdorff. Hence it has little if any interest in the study of physical space.) What holds for individual entities holds for aggregates of such since these, too, are entities. That is, we could define tolerance spaces for sets of aggregates of entities. But we cannot define indistinguishability between sets of things, except of course conceptual indistinguishability which is the same as identity. Hence most of the results on tolerance spaces obtained by Zeeman (1962) have no application to reality. A case in point is the theorem according to which if $\langle T, \sim_s \rangle$ is an indistinguishability space so is $\langle 2^T, \sim_s \rangle$. In fact we do not perceive or fail to perceive the elements of 2^T, for they are sets.

4. Properties of properties

4.1. *Identity and Difference of Properties*

How can property identity be characterized, i.e. under what conditions can two properties be said to be the same? We shall begin by examining, and eventually rejecting, two seemingly natural solutions to this problem – only to end up by declaring the problem itself to be ill conceived.

Suppose properties were conceived of in a nominalistic fashion, i.e. as collections of individuals, or of n-tuples of such (recall the Introduction). Then identical properties would be those possessed by exactly the same individuals (cf. Wilson, 1955). That is, every equivalence class $[P]$ of concomitant properties (Definition 2.11) would be just a singleton. But this proposal is not viable, for it obliterates the difference between properties with equal scope, hence between coextensive though not cointensive attributes, such as "fluid" and "viscous", or "living" and "mortal". Hence here as elsewhere nominalism is a failure.

A second possibility that suggests itself is to adopt a paraphrase of Leibniz' law of individual identity (Postulate 2.5), namely this: "Two properties are the same just in case they have the same (second order) properties". Unfortunately we do not know exactly what a second order property might be. We know only what a second order *attribute* or predicate is – namely one that applies to first order predicates and obeys some system of second order predicate logic. Therefore the logical criterion of attribute identity does not help us out.

Our theory of properties concerns only properties of individuals, hence it does not contain a definition, let alone a criterion, of property identity. This is just as well, for nobody needs such a definition and such a criterion. In science, as in everyday life, one uses properties to individuate entities, not the other way around. Therefore one characterizes the identity (or the difference) of things, namely by Postulate 2.5.

For the above reasons the discussions on property identity in the current philosophical literature are wrong headed hence sterile. Moreover they are sloppy because they are not conducted in the context of a full fledged theory of properties. Let us therefore turn to some genuine problems, and in the first place to the problem of the identity of attributes.

The latter problem is genuine and manageable, if only because we manufacture attributes ourselves. Roughly, the criterion of attribute identity is this: "Two attributes are the same iff they are the same (propositional) function and are assigned the same interpretation". Function identity is not enough because one and the same function may be assigned any number of inequivalent interpretations. To obtain a full characterization of an attribute representing a given substantial property we must append the semantic assumption (or "correspondence rule") telling us what property of which entities the given predicate represents.

A more precise characterization of attribute identity is obtained with the help of our theory of meaning (Vol. 2, Ch. 7). In the latter two predicates are the same just in case they have the same sense and the same reference class. If the predicate happens to belong to a definite closed context, such as a scientific theory, then its sense is defined as the union of its principal ideal and its principal filter. And in any case the reference class of a predicate equals the union of the sets occurring in its domain. Therefore two predicates are the same just in case (*a*) the unions of their respective principal ideals and filters coincide, and (*b*) their domains are the same. In sum, we have

CRITERION 2.3 Two attributes (predicates) are the same iff they have the same sense and the same referents.

Needless to say, two predicates are different just in case they are not the same. Moreover their precise difference can be computed with the help of Definition 7.13 in Vol. 2, Ch. 7.

So much for the pseudoproblem of the identity of properties.

4.2. *Property Weight*

Aristotle saw that not all properties are equally weighty and classed them into essences and accidents. He erred of course in reducing essences to the four basic elements, in regarding them as absolute and unchanging, and in believing change (in particular motion) to be just as accidental as location. Galilei discarded the Aristotelian essences for being unknowable and, by investigating the modest but knowable *accidentia* and their lawful interrelations, changed the face of science. (See Shea, 1972, pp. 70–72.)

Although modern science rejects the essence/accident dichotomy it does not brush aside the very distinction between essential and accidental properties. Far from levelling properties it recognizes that some are necessary for others (Sec. 3.3); moreover, it assumes that there are, as it were, degrees of necessity. So much so that, whenever one sets up a theoretical model of a concrete entity, he tries to seize on its salient properties discarding all others as inessential – at least until further notice. Thus the ontological hypothesis that properties have different weights underlies the strategy of model building.

Where theories are scarce but data plentiful, statistical measures of property weight can be used. The most obvious candidate is of course the coefficient of linear correlation. With its help the following measure of relative property weight can be defined. (Cf. Blalock, 1961.) Let P, Q, and R be properties such that R depends (statistically) upon P and Q. Then

$$P \text{ is } \textit{weightier} \text{ than } Q \text{ for } R \text{ iff } |r_{PR}| > |r_{QR}|,$$

where r_{AB} is the linear correlation coefficient between A and B.

As soon as precise functional dependences among variables can be assumed, a more precise and stable measure of relative property weight can be established. Thus if F is a differentiable function of n variables x_i, where $1 \leqslant i \leqslant n$, each of these "independent" variables contributes its own share to F. An obvious measure of the contributions of x_i to F (or of the weight of x_i relative to F) is

$$w(x_i, F) = \int_{D_i} dx_i \left| \frac{\partial F}{\partial x_i} \right|$$

where D_i is the domain of x_i.

Regardless of the precise mode of computation of property weights, they fall under

DEFINITION 2.15 Any function $w: \mathbb{P} \times \mathbb{P} \to [0, 1]$ such that, for all P, Q, $R \in p[T]$, where $T \subset S$,

(i) $w(P, P) = 1$ for all P;

(ii) if P precedes (is necessary for) Q, then $w(P, Q) \geqslant w(Q, P)$;

(iii) if P and Q are incompatible [by Definition 2.4], then $w(P, Q) = w(Q, P) = 0$;

(iv) if P and Q are concomitant [by Definition 2.11], then $w(P, Q) = w(Q, P) \neq 0$;

(v) for any given $P \neq Q$, $\sum_{Q \in \mathbb{P}} w(P, Q) = 1$;

(vi) $\ulcorner w(P, R) > w(Q, R) \urcorner$ is interpreted as P being weightier than Q for R,

is called a *property weight*, and its value $w(P, Q) \in [0, 1]$ the *weight* of P relative to Q.

What about essential properties? The traditional answers are:

(i) *nominalism and conventionalism*: there are no essential properties;

(ii) *essentialism*: every entity has some properties which are essential, all others being accidental.

The former doctrine is at odds with the practice of science, where one does not assign all properties the same weight but regards some of them as basic and the others as derived. (For example, some chemical properties are steric, i.e. shape-dependent. And some biological properties depend upon the chemical composition.) As for classical essentialism, it contains a grain of truth – namely the recognition that not all properties are removable tags – but it fails in conceiving of accidents as unimportant and lawless. And the latest version of essentialism, namely modal essentialism (cf. Teller, 1975), is useless because it revolves around the vague notion of necessity elucidated in modal logic – the metaphysical inanity of which will be shown in Ch. 4.

Our own brand of essentialism, which may be called *nomological essentialism*, boils down to the following points:

(i) An essential property is, by definition, one that, far from being stray, participates in some law. (On the other hand an accidental property is a nonessential property.) Now, by Postulate 2.7, all properties participate in some laws. Hence *all properties are essential* – which is to say lawful. In other words there are no accidental properties.

(ii) Although there are no accidental properties, there are *attributes* accidental in a given context. Thus the price of gold is accidental from a chemical point of view, not so from a financial point of view.

(iii) Every law of an entity is a property of it (in the extended sense clarified in Sec. 3.3), hence it is an essential property of the entity.

(iv) The real divide is not the nonexisting distinction between *essentia* and *accidentia* but that between *basic* and *derived* properties. This distinction is inherent in the very notion of property precedence: P is *basic to* Q iff $P \preccurlyeq Q$, or $\mathscr{S}(Q) \subseteq \mathscr{S}(P)$. But if P is basic to Q then P is also basic to the law $\ulcorner \mathscr{S}(Q) \subseteq \mathscr{S}(P) \urcorner$.

Nomological essentialism has the following advantages over its rivals: (*a*) it is consistent with a full fledged theory of properties; (*b*) it fits in with the practice of science; (*c*) it discourages the popular view that the essence of a thing is a part of it, namely its unchanging inner core – which core proves in most cases to be imaginary; and (*d*) it puts an end to the medieval dispute (revived by existentialism) concerning whether essence is prior to existence or conversely.

4.3. *Resultants and Emergents*

Let us return to a consideration of properties in general, such as having red blood cells, rather than individual properties, such as having so many red blood cells at a given moment. Some properties may be called *bulk* or *global* properties because they characterize an entity as a whole. Bulk properties are of two kinds: resultant and emergent (Lewes, 1879). Energy is a resultant or hereditary property in the sense that it is possessed by every part of a thing. On the other hand being stable, being alive, having a certain structure, and undergoing a social revolution, are emergent or nonhereditary properties because they are not possessed by every component of the whole: see Figure 2.4. Accordingly we are justified in proposing

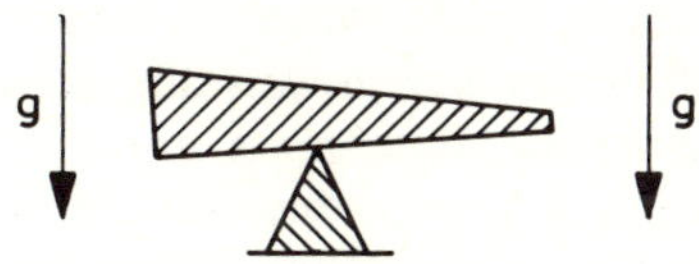

Fig. 2.4. A lever supported by a wedge in equilibrium in a gravitational field. Being in equilibrium is a property of the system as a whole.

DEFINITION 2.16 Let $P \in p(x)$ be a property of an entity $x \in S$ with composition $\mathscr{C}(x) \supset \{x\}$. Then P is a *resultant* or *hereditary property* of x iff P is a property of some components $y \in \mathscr{C}(x)$ of x other than x; otherwise P is an *emergent* or *gestalt property* of x. That is,

(i) P is a *resultant* or *hereditary* property of $x =_{df} P \in p(x)$ & $(\exists y)(y \in \mathscr{C}(x)$ & $y \neq x$ & $P \in p(y))$;

(ii) P is an *emergent* or *gestalt property* of $x =_{df} P \in p(x)$ & $(y)\neg(y \in \mathscr{C}(x)$ & $y \neq x$ & $P \in p(y))$.

Mechanism and individualism assume all properties to be hereditary or resultant, hence explainable by reduction, as is the case with the total charge on a body or the total consumption of goods in a society. We follow instead the tradition of the Greek atomists in recognizing the existence of emergent properties. But, unlike holism, we regard emergents as rooted in the properties of the components, hence as explainable in terms of the latter, though not by reduction to them. For example, temperature is explainable as average kinetic energy of the molecules, but this does not exemplify reduction because averages, though *computed* from individual values alone, are collective properties. Were this not so it would make sense to attribute a temperature value to an individual molecule and a social structure to a person – but it does not. Likewise the cohesiveness of a society can be explained in terms of the participation of its members in a number of social cells but it is societal property not an individual one. These remarks can be generalized into

POSTULATE 2.9 Some of the substantial properties of all composite things are emergent, and all of the emergent properties of a whole are preceded by properties of some of its parts:

(i) $(x)(x \in S \ \& \ \mathscr{C}(x) \supset \{x\} \Rightarrow (\exists P)(P \in p(x) \ \& \ P \text{ is emergent}))$

(ii) $(x)(x \in S \ \& \ \mathscr{C}(x) \supset \{x\} \Rightarrow (P)((P \in p(x) \ \& \ P \text{ is emergent}) \Rightarrow (\exists y)(\exists Q)(y \in \mathscr{C}(x) \ \& \ Q \neq P \ \& \ Q \in p(y) \ \& \ Q \leqslant P)))$.

Whereas the first clause should warm the hearts of holists, the second should cool it. In fact, whereas (i) asserts that some bulk properties are emergent, (ii) suggests that emergents are analyzable – though not reducible. In other words, whereas some bulk properties (the hereditary or resultant ones) are eliminable (definable) in favor of microproperties, others are not – yet in either case they are analyzable or explainable. Let us emphasize that, whereas (ontological) reducibility entails (epistemological) analyzability, the converse is false. There is epistemic novelty in the formation of attributes representing emergent (ontological) novelty. Nor does the explanation of emergence involve the elimination of ontological novelty: a mountain is not explained away when explained as composed of atoms. Explained emergence is still emergence.

4.4. *Properties of Properties*

Every predicate has a number of properties – second order predicates. The *n*-arity of a predicate is such a second order predicate. Properties such as this one belong in the very definition of the predicate. A second order property of a different kind is that exemplified by the predicate "mechanical predicate". By this one means a predicate, such as the mechanical stress tensor, occurring exclusively in mechanical theories. Such properties are then dependent upon the mode of division of intellectual work. In either case a property of a predicate is a construct not a property of a concrete object.

What about substantial properties: does every such property have properties, and do the latter represent features of a concrete individual? Consider a couple of examples. First: the weight of an organism has the property that it varies in time. But this is just a way of saying that organisms have a *variable weight*, and this is a property of theirs not a second order property. Second: the specificity peculiar to enzymes is (believed to be) a steric property, i.e. one deriving from the shape of the enzyme molecule. But this is just to say that molecule *shape precedes specificity* – and this is a law concerning enzymes, i.e. a first order property.

We make bold and generalize: While every predicate has some (second order) predicates, there is no such thing as a second order substantial property. Whatever we predicate of substantial properties is a predicate – a construct or artifact. (Of course some such predicates are attributed truly and others not, but this is a different story.) This is just as well, for otherwise the symbol '$p(x)$' could not designate the set of *all* properties of x.

5. Status of Properties

5.1. *The Reality of Properties*

We have tacitly regarded all substantial properties as real, though not as autonomously real, or real in themselves, i.e. apart from the individuals possessing them. More precisely, we have implicitly employed

DEFINITION 2.17 A property P is *real* $=_{df}$ There is at least one individual $x \in S$, other than the null individual, that possesses P (equivalently: $\mathscr{S}(P) \neq \varnothing$).

This definition applies not only to intrinsic properties (represented by unary predicates) but also to mutual properties (represented by n-ary predicates). Thus to say that a certain relation R is real amounts to saying that there are R-related entities or substantial individuals. (Cf. Bolzano, 1837, Sec. 80.5.) We do not assign intrinsic properties a greater ontological weight, or a higher degree of reality, than mutual properties. Thus a mutual property like gravitational interaction (or a certain parent-child mutual action) is just as real as an intrinsic property such as composition (whether chemical or social).

Surely the above view is not the only possible one. Essentially the following doctrines concerning the reality of properties have been put forth:

(ia) *Properties, whether intrinsic or mutual, are real, nay supremely real, and individuals only exemplify them.* (This generalizes Plato's original doctrine of forms, all of which were unary.) This view is untenable, as a property is null unless it is possessed by some individual. Consider in particular a mutual property represented by a binary predicate R. The extension of R is the set of ordered pairs $\langle x, y\rangle$ such that x bears R to y. If the extension is empty, i.e. if R has no examples, then R happens to be the null relation – the one that fails to hold.

(ib) *Whereas intrinsic properties are real, mutual properties are not.* This thesis seems to find support in ancient logic and ancient grammar, with their insistence that all predicates are unary. In turn, this view may spring from a spontaneous tendency to deal with things separately rather than in their mutual relations. It was vehemently defended by Bradley (1893) and still has defenders among those philosophers who regard, say, "South of the Canadian border" as a "spurious property". The thesis has recently been revived in the proposal to split a relation, or arc of a graph, into two demiarcs and a junction or glue – without of course any attempt at characterizing mathematically either of these alleged components (Harary, 1971). Needless to say, the thesis is indefensible and, if adopted, would kill science, which consists largely in an endeavour to disclose connections. And it would render metaphysics quite uninteresting, as it would force us to conceive of reality not as an aggregate of systems but as a mere collection of disconnected individuals. This borders on

(ii) *All properties, whether intrinsic or mutual, are unreal: only individuals are real.* This is of course the nominalist thesis and the exact dual of the generalized Platonist thesis (ia). According to it a property, if intrinsic, is identical with a collection of individuals and, if mutual, is nothing but a collection of ordered n-tuples. This view is muddled: the examples of a relation, i.e. the individual n-tuples, are characterized by their being related. The relata constitute the extension or graph of the relation, not the relation itself. (More against the extensionalist identification of predicates with sets in Vol. 1, Ch. 4, Sec. 1.2, and Vol. 2, Ch. 10, Sec. 1.2.)

(iii) *Neither properties nor individuals are independently real.* Some individuals and some properties constitute things, their states and their changes of state – the only realities. In the fact that *thing b pulls thing c* what is real is neither of the three items separately but the fact as a whole. Only in pure mathematics does one find (or rather create) individuals bereft of all relations, namely the members of structureless sets. And again only in pure mathematics can one maintain that either the individuals or the relations (in particular the functions) constitute the basic objects out of which everything else can be built. Thus set theory takes the view that individuals and collections of such are basic, everything else being reducible to them; and category theory starts on the other hand from the functions which the individuals happen to satisfy. Neither point of view carries over to ontology: here we must regard individuals as well as their properties as so many abstractions. The real thing is the substantial individual with all its intrinsic and mutual properties. Everything else is fiction. Needless to say we champion this third position and do it on the strength of its being inherent in science. Indeed, science has no use for either propertiless individuals or properties that happen not to be properties of substantial individuals.

What holds in general for relations holds in particular for polarities, or relations among polar opposites such as source and sink. A pole is a constituent of a polarity, which in turn is an even degree relation such as a binary interaction between particles differing only in the sign of their electric charge. By definition then poles come in pairs or, generally, in 2 n-tuples: there are no monopoles. (The expression 'electrical monopole', applied to an electrically charged body, is a misnomer.) Therefore it is mistaken to claim that the existence of any one pole renders its partner possible (Stiehler, 1967, pp. 15, 17). However, we shall not set much store by polarity – a pitfall of archaic metaphysics – since most polarities are imaginary: in fact most of them are just not pairs of opposites and others arise from reifying negation. Thus quantity-quality, one-many, subject-object, appearance-reality, and mind-body are not pairs of opposites but of differents. Others, like being-nothingness, identity-difference and the like are genuine pairs of opposites but have no ontic status. Indeed nonbeing does not oppose being except logically, hence they cannot fight except metaphorically. Likewise the opposition between identity and difference is conceptual not a "struggle of opposites". In both cases the confrontation is strictly logical: A "versus" non-A. Which brings us to the important difference between two kinds of relation: binding and nonbinding.

The relation of being greater (or older or richer or worse) than is "external" to its relata in the sense that it does not alter them and does nothing to keep them together or even apart. Similarly with all the spatiotemporal relations such as those of contiguity, betweenness, precedence, and simultaneity: they do not affect their relata. In general, all comparative (order) relations and all equivalence relations are of this kind, i.e. nonbinding. Not so the hydrogen bond or a relation of economic bondage or of cultural influence: these "make a difference" to the individuals so related: they are in a way "internal" to them. This distinction, ignored by Hume and his followers, was pointed out by Peirce (c. 1909, 6.318), who wrote of "existential relations" or relationships in contrast to mere relations. It was also a favorite with Whitehead (1929) and Woodger (1929) who called them 'organic relations' although they also hold among parts of a machine.

Binding relations, i.e. those that "make a difference" to the relata, may be characterized as follows. Two entities $x, y \in S$ are *bound* (or *connected* or *coupled*) iff some changes in x are accompanied (or preceded or followed) by some changes in y. This must be considered a preliminary elucidation, since we do not yet have any exact concept of change. (See Ch. 5, Sec. 4.1.) Nevertheless the previous characterization should suffice to draw the attention to the two main types of substantial mutual property and to avoid some mistakes.

One mistake stemming from overlooking the above distinction concerns the relation between being connected and being a part. Clearly any two things with a common part are connected. But this is a very special kind of connection: overlapping is sufficient but not necessary for connection. Hence, *pace* Leonard and Goodman (1940), and Feibleman and Friend (1945), things need not overlap in order to be coupled. For example, the feudal baron and his serf are connected without having any common part. Another mistake worth warning against is the belief that, since logic does not discriminate between binding relations and nonbinding relations, it ought to be augmented with ontic or physical predicates (Lewis, 1946, pp. 218 ff.). This is not necessary: we may take off the rack any of the n-ary predicates found ready made in logic (and in mathematics) and endow it with the appropriate ontological interpretation. In general it is not advisable to try and supplement logic with ontology: these two disciplines have very different goals and methods, and a single logic should underly all the varieties of ontology.

5.2. *A Critique of Platonism*

The Platonic doctrine of properties is, in a nutshell, the following: (*a*) form is self-existent, ideal, and external to matter, and (*b*) form precedes matter and can take on individuality, or be realized (exemplified) in particulars. Thus a white thing is said to "participate" in the universal Whiteness – or, as logic teachers are still fond of saying, a white individual *instantiates* the predicate Whiteness. (Aristotle corrected this doctrine by denying the autonomy of forms and holding that matter develops form but, to him, the formless substance was real not just a useful fiction as it is for us.) This theory of forms lingers on in a number of contemporary scholars. While only a few (notably Castañeda, 1974) are avowed Platonists most of them would be surprised to learn that they have adopted the least reputable of all of Plato's teachings. Here are a few examples of Platonism picked at random in recent publications.

(*a*) The biologist Paul Weiss (1963) invokes a super-ordinating "principle" hovering above the molecules that make up a cell and regulates it.

(*b*) E. W. Sinnot (1963, p. 194), another biologist, writes that "Form consists of *relations* among particles, or orderly patterns in them. (...) It is a category of being very different from matter, for it is not the nature of the material particles themselves that is involved, but rather how they are related to one another. Form may disappear and appear, as order yields to randomness or comes again, but matter (in its widest sense, as matter-energy) is conservative, moving toward uniformity and maximum entropy".

(*c*) Nobel laureate Werner Heisenberg (1969, pp. 324–325) states: "'In the beginning was Symmetry' – this is surely truer than the Democritean thesis 'In the beginning was the particle'. The elementary particles embody the symmetries, they are the simplest representations of the latter but they are just a consequence of the symmetries".

(*d*) The philosopher U. J. Jensen (1972) has claimed that mental phenomena are not particulars but universals, hence unmeasurable.

A related view is that properties, far from being separate from things, constitute the latter: a thing would be a bundle of qualities – hence a noun could be decomposed into a pack of adjectives. This view, a thoroughly monistic doctrine of form, has been held by the following contemporary thinkers:

(*a*) Russell (1940 p. 98) wrote: "Common sense regards a 'thing' as having qualities, but not as defined by them; it is defined by spatio-temporal position. I wish to suggest that, wherever there is, for common sense, a 'thing' having the quality *C*, we should say, instead, that *C* itself exists in that place, and that the 'thing' is to be replaced by the collection of qualities existing in the place in question. Thus '*C*' becomes a name, not a predicate".

(*b*) A. I. Ujomov (1965, p. 17) states: "The thing is a system of qualities". And the parts of a thing are not parts of space but "parts of a system of qualities" (p. 18).

(*c*) G. Falk (1966, I, p. 77): "A body moving with velocity v is, according to dynamics, nothing but a certain momentum p and a certain amount of energy $\varepsilon = \varepsilon(p)$ that are transported through space with the velocity v". (Disregard the heresy of identifying a function with an arbitrary value of it.)

(*d*) H. Hiż (1971) too has held that any entity can be identified with the set of its properties and moreover that "an individual may be construed as a set of properties of an individual" – pardon the circularity.

Both the Platonic doctrine of forms and the view that things are bundles of qualities are muddled and therefore neither can be formulated exactly. Organic forms are forms *of* organisms; particle and field symmetries are symmetries of properties *of* things (e.g. of their hamiltonians or of their basic laws); a kind of mental phenomena is a class of *individual* mental phenomena; a thing is *characterized* by its properties but not identified with or defined by them, if only because some of its properties are shared by other things; and a set of properties may be used to sketch or model a thing but it does not *replace* the latter. In sum let us heed Ockham's injunction: *Do not detach forms from particulars* – e.g. speak not of Motion but of moving entities. (Or, if you do employ the concept of motion, construe it as the set of moving things.) Likewise let us speak not of Life but of living things, not of Mind but of sensing and thinking things, and so on. Firstly because there is not a shred of empirical evidence for the hypothesis that forms are detachable from their carriers. Secondly because every property is conceptualized as a function defined on some set of entities. (Recall Postulate 2.1 in Sec. 2.1.)

5.3. *The Problem of Universals*

The problem of universals boils down to the questions: What are universals and how do they exist if at all? This problem looks now rather different from what it looked like in the Middle Ages. Indeed we have come to realize that, before rushing to propose an answer such as that universals exist *ante rem* (Platonism), *in re* (Aristotelianism), or *post rem* (nominalism), we should clarify the question itself. To do this we shall define the concept of a universal or rather two radically different concepts of universal in tune with the methodological dualism we adopted in Ch. 1, consisting in splitting every set of objects into entities (S) and constructs (C). Indeed we shall distinguish the *substantial universals*, or widespread properties of entities, from the *conceptual universals* – among them the predicates representing the former. This distinction is introduced explicitly by

DEFINITION 2.18 Let $A \in \mathbb{A}$ be an attribute, $P \in \mathbb{P}$ a property, and T a set. Then

(i) A is a (*conceptual*) *universal* in the set $T \subset C$ of constructs iff the extension of A equals T:

$$A \text{ is } \textit{universal} \text{ in } T =_{df} \mathscr{E}(A) = T,$$

(where the extension function is defined in Vol. 2, Ch. 9 of this *Treatise*);

(ii) P is a (*substantial*) *universal* in the set $T \subset S$ of entities iff the scope of P equals T:

$$P \text{ is } \textit{universal} \text{ in } T =_{df} \mathscr{S}(P) = T.$$

These notions of a universal are then relativized to a set of objects. And since nothing is said about the cardinality of this set, the above definition does not allow one to distinguish a "genuine" universal such as womanhood from an idiosyncrasy such as being the 20th president of México. One might of course try to improve on this situation by introducing a quantitative concept of degree of universality. A suitable candidate for finite sets would be the function characterized by

DEFINITION 2.19 Let $P \in \mathbb{P}$ be a substantial property and $T \in \mathscr{P}(S)$ a nonempty finite set of substantial individuals. Then the *degree of universality* of P in T is the value of the function

$$u\colon \mathbb{P} \times \mathscr{P}(S) \to [0, 1] \text{ such that } u(P, T) = \frac{|\mathscr{S}(P)|}{|T|},$$

where $|A|$ designates the numerosity of A.

This concept would allow one to overcome the primitive dichotomy of the universal and the particular. Still, this more refined concept does by itself not solve the ontological and epistemological problem of universals. To begin with the so called *formal properties of things*, such as number and shape, appear to be full fledged conceptual universals inherent in concrete entities and thus defy the substantial/conceptual dichotomy assumed in Definition 2.18. Thus numerosity is a property of any collection whatever the nature of its components. True, but then a set is a concept. Surely dogs are normally four legged, this being a substantial property of theirs. The corresponding mathematical property is the cardinality of the set of legs of a dog. In other words four leggedness is a substantial property to be distinguished from the mathematical property of the set of legs of a quadruped. The relation between the predicate and the corresponding property is that of representation: the former represents the latter in the case we are considering. Likewise when we say that the history of a person is uninterrupted we do not assign the person a formal property but choose the mathematical property of continuity (a conceptual universal) to represent the smoothness (or absence of gaps) in a sequence of states of an entity. To sum up: not things but our models of them have mathematical properties, and this because we conceptualize substantial properties as functions. This mode of representation is so deeply ingrained in our habits of thought that we often mistake the deputy for his constituency.

In the medieval dispute about universals only conceptual universals seem to have been considered by both the "realists" (Platonists) and the nominalists. In that context we would have sided with Ockham (ca. 1320 I, C.xvi): "nothing is universal except by signification, by being a sign of several things". The conceptual universal "humanity" is not in the individual Socrates but in our idea of him. Of course Socrates is human: he has a property that is universal in the class of humans. This is what the Aristotelians replied: humanness, as a substantial universal, "inheres" in every human being. (In our terminology: the scope of the substantial property of humanness is mankind.) In sum substantial universals (properties and in particular laws) are *in re*. On the other hand conceptual universals (predicates) are *entia rationis*: they are *post rem* if they happen to represent substantial universals preexisting knowledge, and *ante rem* if they anticipate experience or action.

Our theories of substance (Ch. 1) and form (present chapter) allow us to see a further aspect of the problem of universals in a different light, namely the particular/universal dichotomy. First of all this dichotomy, when legitimate, is conceptual not real: the world contains neither bare particulars nor pure forms. Substance and form, individual and universal, are distinct aspects of our conceptual analysis and theoretical modelling of things and facts. Lacking an independent existence, they are not mutually reducible. That is, universals are not collections of particulars (nominalist reduction) and particulars are not bundles of universals ("realist" reduction). The nominalism-"realism" issue, though *de facto* not dead, is finished *de jure*: neither contender was completely right because each focused on just one aspect of the situation while ignoring the other. And each side is done justice by our representing every substantial property by a predicate, and construing the latter as a function from individuals to statements.

Moreover the individual/universal dichotomy, valid and so useful in first order logic, may collapse elsewhere. *Example 1* Sets are neither individuals nor universals – unless they occur as members of a collection of sets, in which case they become individuals, without however the family being a universal. (A nonempty set is constituted by individuals but this does not make it into a universal, much less an individual.) However, some sets are determined or defined by universals (predicates) via the scope function or the principle of abstraction. *Example 2* Some "universals", such as "the set contained in every set", are instantiated by a single object (in this case the empty set). Why call them 'universals' if they are unique? *Example 3* One and the same object can now be regarded as an individual, now as a set (or as a concrete collection). There is nothing final about being an individual. *Example 4* To say that b is an individual is saying too little: what are the properties of b? And saying that P is a property is just as little: what individuals does P apply to, i.e. what is its scope?

The upshot of the preceding discussion is this: Platonism is not a viable metaphysics any more than nominalism is. It might be contended that, though unsuitable as an ontology of substantial individuals, Platonism is the correct philosophy of mathematics because "mathematics has no need for non-classes, like cows or molecules; all mathematical objects and relations can be formulated in terms of classes alone" (Mendelson, 1963, p. 160). This, it is sometimes claimed, is why the von Neumann–Bernays–Gödel version of set theory deals only in classes and sets. Yet (*a*) the basic predicate of this theory, namely the membership relation, is not construed as a class, and (*b*) the individuals of the theory are, precisely, sets and classes (as pointed out by Bernays, 1937). Moreover one needs properties even in pure mathematics as soon as one becomes interested in particular sets, i.e. in characterizing them rather than in handling them indiscriminately (i.e. in general) as set theory does. Just think of characterizing, say, the interior of a circle of a given radius without using any predicate. Finally, even if mathematicians one day succeed in producing a mathematics using only classes, metaphysics is likely to continue using the notions of individual and of property if only because both are essential to science: scientific research involves finding out the properties of individuals. Not even the most sophisticated reconstruction of a scientific theory could eliminate properties in favor of classes if only because distinct properties can have the same scope. Nor could it succeed in eliminating individuals, because properties are properties of individuals or of sets of such.

6. Concluding remarks

To summarize. There is but one world, denoted $\Box$, and it is composed of entities or substantial individuals, i.e. members of the set S introduced at the outset (Ch. 1). Every entity is either simple or composite. And every entity has a certain number of properties in addition to its composition. For example a lump of copper, i.e. an entity composed of Cu atoms, is malleable, shiny, melts at 1083 °C, is a good conductor of electricity, is rather scarce, its ore is coveted by all industrial powers, and so on. These are systemic properties of the nonhereditary or emergent kind; on the other hand having a mass and being capable of moving are hereditary properties of a lump of copper, i.e. properties shared by its atomic components.

A substantial property may also be called a *physical* property, in the largest possible sense of 'physical', and in contrast to a conceptual item such as an attribute or predicate. Thus being feathery must be distinguished from the attribute(s) or concept(s) representing featherhood or denying it (e.g. 'featherless'). Firstly because every property is representable by at least one attribute. Secondly because some attributes (in particular the negative and the disjunctive ones) represent no substantial property at all.

Further, there are no properties or forms fluttering over substantial individuals: there are only propertied entities, such as feathery animals. And properties, far from coming either in isolation or in amorphous droves, form clusters and drag one another. That is, every property is either necessary or sufficient for some other properties of the same entity. I.e., there are laws. And these laws are complex properties of entities or, rather, of entire kinds. So much so that they determine the natural kinds. But this is a subject for the next chapter.

CHAPTER 3

THING

So far we have dealt largely with fictions: entities deprived of most of their properties, and forms without a definite stuff. However, we warned that these are fictions and anticipated that they would enable us to construct the notion of a real thing as a fully qualitied individual. This is indeed what a concrete or material object, such as a radio wave or a person or a society is: namely an entity endowed with all its properties, both intrinsic and mutual, permanent and transient.

In the present chapter we shall deepen our study of the concept of a substantial individual and shall investigate in particular the following problems. One of them concerns the characterization of the states of a thing. We shall use the results of this investigation in our treatment of the problem of change in general, to be tackled in Ch. 5. Another question to be dealt with in this chapter is this: What groups different things into natural kinds or species?

Surely the objects we are concerned with "exist in space and time", as the obsolete formula has it. (We shall argue in Ch. 6 that concrete things constitute space and time.) However, we shall still disregard all of the spatiotemporal characteristics of things, characteristics which will be taken up in Ch. 6. Moreover we shall not particularize any properties: that is, our concept of a concrete object, or thing, will apply to a community just as well as to an electron. Finally we shall defer to Vol. 4, Ch. 7 the study of the particular things known as systems. In fact our definition of the system concept will employ the notion of a thing we proceed to elucidate.

1. Thing and Model Thing

1.1. *Thing*: *Definition*

We stipulate that a thing is an entity or substantial individual (Ch. 1) endowed with all its (substantial) properties (Ch. 2). As may be recalled (Definition 2.3 in Ch. 2, Sec. 3.2) the totality of properties of a substantial individual $x \in S$ was called

$$p(x) = \{P \in \mathbb{P} | x \text{ possesses } P\},$$

where $\mathbb{P}$ is the collection of substantial (nonconceptual) unarized properties.

It may also be remembered that, although we identify or spot any concrete individual by its properties, it is impossible to define an entity as the set of its properties, if only because the expression 'Entity $p(x)$ possesses property P which is a member of $p(x)$' makes no sense. A final caution: although usually a proper subset of $p(x)$ will suffice to *distinguish* x from other entities, nothing short of the totality $p(x)$ of properties of x will *constitute* and *individuate* it, i.e. render it ontically distinct from every other entity. In fact by Postulate 2.5 and its corollary in Ch. 2, Sec. 3.2, what makes a thing what it is, i.e. a distinct individual, is the totality of its properties: different individuals fail to share some of their properties.

The preceding considerations suggest introducing

DEFINITION 3.1 Let $x \in S$ be a substantial individual and call $p(x) \subset \mathbb{P}$ the collection of its (unarized) properties. Then the individual together with its properties is called the *thing* (or *concrete object*) X:

$$X =_{df} \langle x, p(x) \rangle.$$

When referring to a fully qualitied entity, i.e. a thing, we shall often write '$X \in \Theta$' or simply '$x \in \Theta$', where Θ is characterized by

DEFINITION 3.2 The *totality of things* is called Θ:

$$\Theta =_{df} \{\langle x, p(x) \rangle | x \in S \;\&\; p(x) \subset \mathbb{P}\}.$$

This concept of a thing synthesizes the notions of substance and of form studied in Chs. 1 and 2 respectively. We might of course have proceeded the other way around, i.e. starting from things and analyzing them into bare individuals and bunches of properties. But this alternative would have violated the norm that enjoins us to build out of simple units and it would have broken the line of progress we intend to follow.

It might be objected that, since an ordered pair is often defined as a set of sets, namely thus: $\langle a, b \rangle = \{\{a\}, \{a, b\}\}$, our definition of a thing does not retrieve the idea that a thing is a particular. However, it is well known that an ordered pair need not be analyzed and defined as a set but may be characterized as an individual satisfying the single axiom:

$$\langle a, b\rangle = \langle a', b'\rangle \text{ if and only if } a = a' \text{ and } b = b'.$$

Besides, there need be no similitude between a definition and the thing defined. After all, Definition 3.1 characterizes a concept of a thing rather than being the "real definition" of a thing. More on this in Sec. 1.3. And now we turn to some assumptions and a few more definitions.

1.2. *Assumptions*

Our first assumption will be that there are things and indeed infinitely many and moreover a whole continuum of them:

POSTULATE 3.1 The totality Θ of things is an uncountable set.

Remark 1 From a mathematical point of view it would have been more natural to postulate the cardinality of the set S of bare individuals introduced in Ch. 1. That of Θ would have followed because of the one to one correspondence between the two sets. But since bare individuals are fictitious it seemed preferable – from a metaphysical point of view – to assume as little as possible about them. *Remark 2* That an existence hypothesis must be either affirmed or denied explicitly at this point, is shown by the existence of subjective idealism, according to which there are only mental states – or, according to a recent version, just propositions (Fitch, 1971). *Remark 3* Another reason for having to postulate the existence of things is that, if we want to prove anything about existents, we must posit them. We cannot prove the existence of concrete things any more than we can prove the existence of deities or of disembodied minds. What can be proved is that, unless there were things, other items – such as acting on them and investigating them – would be impossible. *Remark 4* That the hypothesis $\ulcorner \Theta \neq \varnothing \urcorner$ is true, is suggested by the fact that all of the factual sciences adopt it tacitly: far from doubting the existence of all things, they take some of them for granted, hypothesize others, and investigate them all. *Remark 5* That the infinity of things is nondenumerable is suggested by the existence of physical continua, such as gravitational fields, all parts of which are fields themselves. Even if the world were to consist of a single particle surrounded by its own gravitational field, it would contain a nondenumerable infinity of things, namely all the parts of the field, which – according to our present knowledge – is infinitely divisible. But of course such a world would contain no detachable parts. *Remark 6* The quantization of a field does not alter the situation: not the field itself but the set of its quanta is denumerable. And even in the absence of quanta a residual or background field remains, namely the one corresponding to the vacuum state. (The latter is not nought.) *Remark 7* The postulated existence of a nondenumerable infinity of things is compatible with the controversial hypothesis that they are organized in a finite number of systems – the only chance for the universe to be spatially finite. But we are not committing ourselves on this point which belongs to Ch. 6 anyhow. In the hypothetical example mentioned in Remark 5 the system is of course the single particle accompanied by its gravitational field.

Having defined the concept of a thing (Definition 3.1) we now define that of the juxtaposition or physical addition of things:

DEFINITION 3.3 Let $X = \langle x, p(x)\rangle$ and $Y = \langle y, p(y)\rangle$ be two things. Then the *juxtaposition* of X and Y is the third thing

$$Z = X \dotplus Y = \langle x \dotplus y, p(x \dotplus y)\rangle.$$

Note that we are not assuming that the properties of the whole are all and only those of the parts, i.e. that $p(x \dotplus y) = p(x) \cup p(y)$. This assumption would be false if only because the compound $x \dotplus y$, for $x \neq y$, has the property of being composed of x and y, which its parts fail to have.

Next we posit that the association theory expounded in Ch. 1, Sec. 1 carries over to things, and we use the symbol '$\dotplus$' to denote the association or juxtaposition of things. More precisely, we lay down

POSTULATE 3.2 Let Θ be the totality of things, $\Box$ a distinguished element of Θ, and $\dotplus$ a binary associative operation in Θ. Then the following holds by stipulation:

(i) $\langle \Theta, \dotplus, \Box\rangle$ is a commutative monoid of idempotents – for any x, y, z in Θ, $x \dotplus y = y \dotplus x$, $x \dotplus (y \dotplus z) = (x \dotplus y) \dotplus z$, $x \dotplus x = x$, and $x \dotplus \Box = \Box \dotplus x = x$;

(ii) $\dotplus$ represents the association or juxtaposition of things;

(iii) the string $x_1 \dotplus x_2 \dotplus \ldots \dotplus x_n \in \Theta$, where $n > 1$, represents the concatenate of things x_1 to x_n;

(iv) the neutral element $\square$ of the monoid is the null thing or nonentity – i.e. not a thing proper.

We can retrieve for things everything we defined or proved for bare individuals (or things deprived of their properties other than the property of associating and the properties, such as composition, derived from associability). In particular we can make

DEFINITION 3.4 Let $x, y \in \Theta$ be things. Then x is a *part* of y iff $x \dotplus y = y$. Symbol of the part-whole relation: $\sqsubset$.

Our next assumption is

POSTULATE 3.3 There exists a thing such that every other thing is part of the former. That thing is unique and we call it *the world* or ▯.

Since the world is a thing, and indeed the biggest of all, $x \dotplus ▯ = ▯ \dotplus x = ▯$.

Because the juxtaposition $x \dotplus y$ includes x and y as parts (by Definition 3.4), the former is the supremum of x and y with respect to $\sqsubset$, i.e., $sup\ \{x, y\} = x \dotplus y$. This remark allows one to prove the analogue for things of Theorem 1.4, namely

THEOREM 3.1 The totality of things has the semilattice structure. More precisely, $\langle \Theta, \dotplus, \square, ▯ \rangle$ is a semilattice with least element $\square$ and last element ▯.

The generalization of the notion of concatenation or juxtaposition to an arbitrary set $T \subseteq S$ is that of *sup T*. But the existence of this element has got to be assumed:

POSTULATE 3.4 For any subset T of Θ, its supremum $[T]$ exists.

This hypothesis allows us to make

DEFINITION 3.5 The *aggregation* of a set $T \subset \Theta$ of things, or $[T]$ for short, is the supremum of T with respect to the part-whole relation, i.e. the thing such that

(i) $x \sqsubset [T]$ for all $x \in T$;

(ii) if $y \in T$ is an upper bound of T, then $[T]$ precedes y: i.e. if $x \sqsubset y$ for all $x \in T$, then l.u.b. $T \equiv sup\ T = [T] \sqsubset y$.

Remark Just as Postulate 3.2 contained the statement that every finite association of things is a thing (clause (iii)), so the previous definition tells us, among other things, that every association or juxtaposition, whether finite or infinite, yields a thing. This is no ontological triviality: magicians claim to be able to create things out of nothing and to annihilate other things. And some emergentist philosophers believe that certain types of aggregation ensue in objects transcending the realm of things.

The previous assumptions and definitions entail

COROLLARY 3.1 Every part of a thing is a thing:

$$\text{If } x \sqsubset y \text{ and } y \in \Theta, \text{ then } x \in \Theta.$$

Proof The part-whole relation has been defined for things only – not e.g. for concepts – so the assertion that $x \sqsubset y$ is a tacit recognition that both x and y are things.

Remark Nor is this result self evident. Think of the claim that the constituents of ordinary self existents are not things existing autonomously but are the outcome of certain cognitive acts (such as measurements) performed by experimenters with the help of things of a special kind (pieces of apparatus).

Finally we state a consequence of the previous Corollary 3.1 and Postulate 3.3:

COROLLARY 3.2 Every part of the world is a thing:

$$\text{For every } x\text{: If } x \sqsubset \square \quad \text{then} \quad x \in \Theta.$$

It follows by contraposition that whatever is not a thing fails to be part of the world. But this matter borders on the next subsection.

We close this subsection with two notes of caution. The first is that the world $\square = [\Theta]$ does not have all the properties of each of its parts: for example, it does not spin, it does not interact with other things, it is not alive, and it has no mind. Indeed, by using Definitions 3.1 to 3.3 we find that

$$\square = \langle [S], p([S]) \rangle$$

where, in accordance with the warning following Definition 3.3,

$$p([S]) \neq \bigcup_{x \in S} p(x).$$

The second note of caution is that the statement that two or more things form an aggregate satisfying Postulate 3.2, is a factual statement, i.e. one concerning matters of fact. However, this does not render it falsifiable. If any of the clauses of the postulate – e.g. commutativity – should fail, we would conclude not that the postulate has been refuted but that the things of interest have not formed an aggregate. (Equivalently: the resulting thing does not satisfy our axiomatic definition of an aggregate.) For example, if a catalyzer, such as an enzyme, acts on two different reactants, it may happen that it binds faster with one of them than with the other, so that the order in which the three things join does make a difference. But of course this order is temporal, and so has nothing to do with associativity.

1.3. *Thing and Construct*

In the previous chapters we have distinguished objects of two kinds: substantial or concrete entities, and constructs. It is now time to sharpen somewhat this distinction – a matter that will be taken up again in Sec. 4.3. To this end we shall start by outlining the concept of a construct, leaving its detailed study to the sciences of constructs, namely logic, semantics, and mathematics.

We feign that there are constructs, i.e. creations of the human mind to be distinguished not only from things (e.g. words) but also from individual brain processes. (Only, we do not assume that constructs exist independently of brain processes.) We distinguish four basic kinds of construct: concepts, propositions, contexts, and theories. Concepts, such as the notion of a thing, are the building blocks of propositions – such as 'All things change' – which are in turn the constituents of contexts – such as the set of all propositions concerning dogs – and of theories.

Basically a concept is either an individual (such as "3" in elementary arithmetic), or a set (such as the set of integers), or a relation (such as "$<$"). Whether either category is reducible to the other will not concern us here. The most interesting relations are the functions, which are either many-one or one-one relations. We distinguish two kinds of function: propositional and nonpropositional. A propositional function, or predicate, or attribute, is a function that maps some domain of individuals (or n-tuples of such) into some set of propositions. (Recall Ch. 2, Sec. 2.1.) A nonpropositional function takes values in a set other than a set of propositions. (However, every nonpropositional function may be paired off to a propositional function. For example the sine function, which maps reals into reals, is a nonpropositional function that can be assigned the propositional function

$$\text{Sin}: R \times R \to P, \quad \text{where}$$
$$\text{Sin}(x, y) = (\sin x = y) \quad \text{for} \quad x, y \in \mathbb{R},$$

P being the set of all propositions containing the predicate Sin.) Finally, a context is a set of propositions sharing a reference class, and a theory is a context closed under the operation of deduction. In sum we have the following classing:

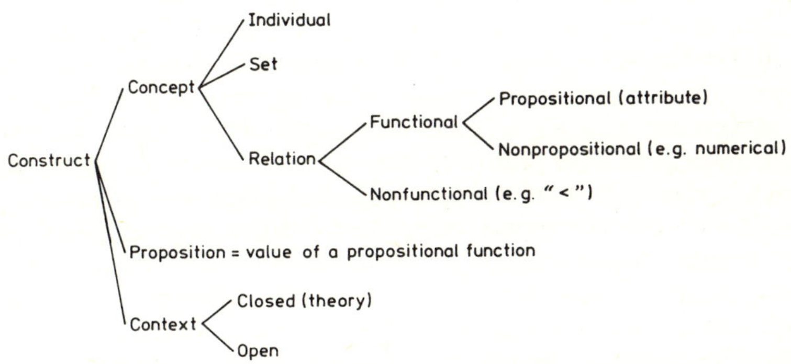

Constructs do not have all the properties of things. For example sets add and intersect but do not aggregate, do not move around, have no energy and no causal efficacy, etc. Constructs, even those representing things or substantial properties, have a conceptual structure not a material one. In particular, predicates and propositions have semantical properties, such as meaning, which is a nonphysical property. We summarize and extrapolate:

POSTULATE 3.4 Every object is either a thing or a construct, no object is neither and none is both.

Equivalently: Every set O of objects splits into two disjoint sets, one of things, the other of constructs. In the simplest nontrivial case, the set O of objects contains but one thing ϑ and one construct, namely $\{\vartheta\}$. That is, $\Theta = \{\vartheta\}$, $C = \{\{\vartheta\}\}$, $O = \{\vartheta, \{\vartheta\}\}$. Clearly, $\Theta \cap C = \varnothing$.

Substantial individuals devoid of properties, and substantial properties apart from the individual things possessing them, rank as constructs. Likewise a change in properties apart from every changing thing, and a change in an individual apart from a change in its properties, are fictions, i.e. members of some C.

Postulate 3.4 is an axiom of *methodological dualism*. It does not commit us to metaphysical dualism: we are not claiming that there are two kinds of thing, the *res extensa* and the *res cogitans*, or things proper and ideas. We take it that constructs, whether useful or idle, scientific or mythical, are fictions not entities. Hence they are not part of the real world even when they take part in our representations of the latter.

Moreover the preceding postulate entails that each category of objects, Θ and C, has its own peculiarities but does not preclude the possibility that they share some properties. For example some constructs concatenate just like things, and many things are just as thinkable as constructs. However, these two properties are mutual not intrinsic. Intrinsic properties are either of things or of constructs. On the other hand a mutual (e.g. binary) property can link things with constructs. An example of a mutual property of this kind is that of representation, such as it occurs in the statement ⌜Proposition p represents thing b⌝.

The totality of objects has no property other than that of being the union of the class of things and that of constructs. Therefore ontology cannot be construed as the theory of the arbitrary object, or of all objects: there is nothing to be said about O beyond what has just been said. This does not preclude the possibility of making long discourses about the ability to think of all objects, for when so doing we employ binary predicates such as "representing" or "thinking of", which are not subject to the above restriction.

Another obvious consequence of the preceding considerations is that concrete objects (things) have no intrinsic conceptual properties, in particular no mathematical features. This last statement goes against the grain of objective idealism, from Plato through Hegel to Husserl, according to which all objects, in particular material things, have ideal features such as shape and number. What is true is that some of our *ideas* about the world, when detached from their factual reference, can be dealt with by mathematics. (For example, by analysis and abstraction we can extract the constructs "two" and "sphere" from the proposition "That iron sphere is composed of two halves".) In particular, mathematics helps us to study the (mathematical) form of substantial properties. In short, not the world but some of our ideas about the world are mathematical.

Any construct that violates Postulate 3.4 will be declared *metaphysically ill formed*. The attribution of conceptual properties to things, and the attribution of substantial properties to constructs, are in the category of metaphysically ill formed statements. Examples of such metaphysical misfits are ⌜A body is a point set⌝ and ⌜The empty set is grey⌝. Likewise the thesis that a thing is identical with the set of its properties (i.e. $\vartheta = p(\vartheta)$) is metaphysically ill formed because it entails attributing substantial properties to a set, as shown by the following example. Let $p(\vartheta) = \{\text{Drunkard, Cantankerous, ...}\}$. Identifying this obnoxious individual with his properties leads to attributing drunkenness, cantankerousness, etc., to the set {Drunkard, Cantankerous, ...}. It could be argued that some metaphysically ill formed statements, such as ⌜Santa Claus has a white beard⌝ and ⌜Number three is perching on a branch⌝ are meaningful in certain contexts. But they do not belong in the body of factual knowledge and it is a matter of indifference whether or not they can be assigned truth values in certain contexts, such as fairy tales – e.g. possible worlds metaphysics.

Let us now turn to a special kind of construct, namely the conceptual sketch or model of a thing.

1.4. *Model Thing*

Theoretical science and ontology handle not concrete things but concepts of such, in particular conceptual schemata sometimes called *model things*. Our construal of a thing as a substantial individual together with the set of all its properties (Definition 3.1) is of course such a model thing – albeit a rather poor one. A richer characterization of a thing is given by a set equipped with specified relations, such as functions or operations. For example, if the thing represented is a force field, then the set will be a portion of a geometrical manifold M – e.g. a region in Euclidean three space – together with a tensor field F on M. Briefly, $\langle M, F\rangle \triangleq$ field. We shall adopt this mode of representation by making

DEFINITION 3.6 Let $X = \langle x, p(x)\rangle$ be a thing of class $T \subseteq \Theta$. A *functional schema* X_m of X is a certain nonempty set M together with a finite sequence $\mathbb{F}$ of nonpropositional functions on M, each of which represents a property of T's. Briefly,

$$X_m =_{df} \langle M, \mathbb{F}\rangle, \text{ where}$$
$$\mathbb{F} = \langle F_i \mid F_i \text{ is a function on } M \ \&\ 1 \leqslant i \leqslant n < \infty\rangle.$$

The base set M will be denumerable or nondenumerable, as the case may be. It may or may not be thought of as mapped on a subset of physical spacetime. (In systems theory M is usually taken to be a set of time instants.) As for the functions F_i in $\mathbb{F}$, every one of them refers to the individual $x \in T$ of interest and is thus evaluated at a fixed x in T even though it may be a function of whatever other variables may be the case. The finiteness of the set of components of $\mathbb{F}$ agrees with the part of Postulate 2.3 specifying that there are finitely many general properties (such as length or longevity). And it does not contradict the second part of that postulate, according to which $p(x)$ is nondenumerable for each $x \in S$. Indeed, recall that a single continuous general property, such as age, gives rise to infinitely many individual properties (such as successive ages) as the property in question takes its values. It might be objected that restricting the number of components of $\mathbb{F}$ is artificial since any function, even the constant function, can be decomposed into infinitely many functions, as is the case with Fourier series expansions. But such mathematical tricks have no ontological significance: they are purely conceptual.

Example 1 The simplest functional model of a corpuscle with variable mass is the classical mass point. Here $M = F \times T$, where F is the set of reference frames and $T = \mathbb{R}$ the real line, every point of which is interpreted as an instant of time. And $\mathbb{F} = \langle \mu, \pi, \varphi \rangle$ is a triple of functions on $M = F \times T$, such that $\mu(f, t)$ represents the mass, $\pi(f, t)$ the position, and $\varphi(f, t)$ the force acting on the corpuscle, relative to frame $f \in F$, at time $t \in T$. This functional schema of a particle is consistent with a number of alternative equations of motion (law statements), i.e. it can be shared by a number of different theories hinging on the same model object. *Example 2* A simple electrical model of the brain consists of a finite region of a four manifold (namely spacetime) together with a single function on it representing the electric potential. And a possible chemical model of the same thing is constituted by the same set together with n functions with domain on it, each of which represents the concentration of a given chemical at a given point in space and time.

Finally we assume that the set of functional schemata is nonempty and can grow indefinitely. That is, we lay down the following methodological

POSTULATE 3.4 Any thing can be modelled as a functional schema: For any $X \in \Theta$ there is at least one $X_m = \langle M, \mathbb{F} \rangle \in C$ such that $X_m \triangleq X$.

Remark 1 The failure to distinguish the thing represented from its model is not just a form of mental derangement: it is also at the root of black magic and subjectivism. The idealist who does not distinguish a thing from any of its models cannot account for the multiplicity of schemata of one and the same thing. Consequently he cannot understand the history of theoretical science, which consists partly in the replacement of some schemata by others. *Remark 2* Our characterization of a functional schema is consistent with current scientific practice. However, two notes of caution are in order. Firstly, more sophisticated modes of representation are conceivable, e.g. with the help of the concept of a category (cf. Padulo and Arbib, 1974). Secondly, sometimes a thing is *defined* as a certain relational structure, i.e. it is identified with one of its model objects. This is mistaken, for only (some) constructs can be defined: things can only be represented (or misrepresented), and occasionally also manipulated. But the mistake is harmless if we are aware of it. *Remark 3* Nominalists are bound to dislike our construal of a sketch or schema of an individual thing as a structured set: they may demand "individualism" for both the world and our conceptual representations of it. Thus Goodman (1956, p. 16): "Nominalism for me consists specifically in the refusal to recognize classes"; it "requires only that whatever is admitted as an entity at all be construed as an individual" (ibid., p. 17). (In particular it does not require the concrete/conceptual distinction (ibid., p. 16).) That demand is not met by science, so we can ignore it. *Remark 4* The set M, if infinite, may represent an actual infinity such as a gravitational field or some other continuum. Such an infinity will be looked at askance by Aristotelians and empiricists alike. On the other hand it will warm the hearts of Leibnizians. Thus Bolzano's *Paradoxien des Unendlichen* (1851) carried as a motto the following quotation from Leibniz: "Je suis tellement pour l'infini actuel, qu'au lieu d'admettre que la nature l'abhorre, comme l'on dit vulgairement, je tiens qu'elle l'affecte partout, pour mieux marquer les perfections de son auteur".

The various functional schemata of a given thing need not be equivalent: for example they may exhibit different amounts of structure. Thus a set whose elements are related pairwise is more structured, integrated, or tightly knit, than one whose elements have only unary properties. Moreover an amorphous set does not become structured by increasing the number of properties of its elements even though it does become more complex. (I.e. the more structured the more complex but not conversely.) This suggests measuring the amount of structure of a relational structure by the number and rank of the predicates included in it. A translation of this idea into our terminology is as follows:

DEFINITION 3.7 Let $X_m = \langle M, \mathbb{F} \rangle$ be a functional schema of a thing X and call the *rank* of $F_i \in \mathbb{F}$ the number of independent variables or argument places of F_i. Then the *amount of structure* of thing X exhibited by its functional schema X_m equals

$$\alpha(X_m) = \sum_k n_k (r_k - 1),$$

where n_k is the number of functions of rank r_k in $\mathbb{F}$.

Now, two relational structures, such as $X_m = \langle M, \mathbb{F} \rangle$ and $X'_m = \langle M', \mathbb{F}' \rangle$, are said to be *similar* iff $\mathbb{F}$ and $\mathbb{F}'$ have the same number of components and their corresponding components are of the same rank (Tarski 1954). In our terminology: two model things X_m and X'_m are *similar* iff their amounts of structure are the same, i.e. if $\alpha(X_m) = \alpha(X'_m)$. This notion suggests attributing the modeled things themselves the property of being similar:

DEFINITION 3.8 Let X and X' be two things modeled as $X_m = \langle M, \mathbb{F} \rangle$ and $X'_m = \langle M', \mathbb{F}' \rangle$ respectively. Then X and X' are *formally* (morphologically) *similar under the same modelings* iff their respective functional schemata X_m and X'_m are similar (i.e. if they exhibit the same amount of structure).

Remark 1 In particular, things described with the help of the same mathematical formalism are modeled similarly and can thus be declared to be similar – under the given modeling! (See Bunge, 1973a, Ch. 5.) However, since morphological similarity as defined above is not an intrinsic property of things but rather a property of pairs thing-model thing, we must be careful not to mistake the above concept for that of objective similarity introduced in Ch. 2, Sec. 3.5. Whereas the latter is an ontological concept the former is an ontological-epistemological one. *Remark 2* It goes without saying that things belonging to the same natural kinds – e.g. electrons, neurons, peasant societies – are represented by the same functional schemata. In particular, indiscernibles are representable by the same model things. In other words, for theoretical purposes we may treat indiscernibles *as if* they were identical – which of course they are not (cf. Ch. 2, Sec. 3.6).

So much for a preliminary characterization of model things, or sketches of things. We proceed to refine the notion of a functional schema of a thing, by taking a closer look at the n-component function $\mathbb{F}$. This will enable us to elucidate the notion of a state of a thing, a notion which will be crucial to our investigation of change in general.

2. State

2.1. *Centrality of the State Concept*

Every thing is – at a given time associated with a given reference frame – in some state or other. This is a hypothesis about the furniture of the world and, since it does not specify either the kind of thing or the kind of state, it is ontological. It is not a self evident assumption. For one thing it is rarely if ever formulated explicitly. For another the world might be constituted in such a way that the assumption be false. Moreover according to the classicist interpretations of the quantum theory the latter fails to assign definite states to its referents. But this is mistaken: the truth is that the theory does not assign its referents classical states such as sharp position and velocity values, but quantum states such as position and velocity probability distributions.

The ontological hypothesis that every thing is in some state or other underlies all science, has invaded philosophy, and has spilled over to ordinary knowledge, to the point that statesmen speak of the state of the nation. Each exact scientific theory is of course in a position to give a precise characterization of the concept of a state. But it will be concerned with a particular kind of state, such as the dynamical or the chemical or the physiological or the economic state of its referents, not with the generic concept of state of a thing. To elucidate the latter concept is the business of exact ontology, in particular of systems theory.

Surprisingly enough, systems theory has so far failed to give an exact and general enough analysis of the concept of the state of a system. To begin with the usual accounts (e.g. Zadeh and Desoer, 1963; Mesarović and Takahara, 1975; Padulo and Arbib, 1974) do not apply either to continuous systems such as fields or to quantum-mechanical systems. Moreover the concept of a reference frame has no place in those accounts, probably because it is not needed in automata theory, electrical network theory, and a few other theories. Yet the concept is central to many other theories. (Recall Example 1 in Sec. 1.4.) For example, it occurs in the explanation of the working of an electric motor – not to speak of the underlying theories of mechanics and electrodynamics. The notion of a reference frame is in fact so central in physics that the states of any real physical system are relative to some frame or other: just think of the state of motion of a body, or of the state of an electromagnetic field.

A second concept one misses in systems theory is that of a law. To be sure a general theory of things should neither presuppose nor contain any specific (e.g. thermodynamical or chemical or biological) law, for otherwise it would not be so general. (Equivalently, the systems it describes could not be realized in a number of alternative ways, in particular employing materials of different kinds – hence satisfying different laws.) However, the general concept of a law – the philosophic concept as distinct from a particular law statement such as Ohm's – ought to play some role in systems theory if only for a mathematical reason. Indeed, but for the laws – which place restrictions upon the ranges of the components of the function $\mathbb{F}$ occurring in any functional schema of a thing – we should accept the usual characterization of the collections of possible states (or state spaces) as vector spaces or even as inner product or metric spaces. The existence of laws ruins this characterization. In fact if a law restricts the range of a variable, as it usually does, then it is no longer true that the product of a variable by an arbitrary scalar belongs to the same space. In sum, not even the most general scientific theories – namely those that qualify also as metaphysical theories – have given satisfactory characterizations of the general concept of a state.

Nor have metaphysicians exactified the notion of state – nor, a fortiori, that of change of state. Not even the process metaphysics – such as those of Hegel, Bergson, Whitehead, and their numerous followers – have taken it upon themselves to clarify the notions of state and of change of state – although in science, at least, a process is a sequence of different states. Consequently those philosophers have bequeathed us obscure philosophies of change. Much the same holds for the fashionable possible worlds metaphysics: here the concept of state of a possible world is central, yet it remains undefined. Likewise the various systems of inductive logic employing the concepts of state and of state description fail to analyze them in a manner consistent with the concepts of state occurring in science.

Such a lack of analyses is deplorable if only because the concept of a state, hence the notions of a state description and of a change of state, pose interesting philosophical problems. For example, since every state of a thing is determined by a bunch of state variables for the thing, and since the choice of the latter is partly determined by the state of the art, what right do we have in assigning a thing an objective state portrayed by a state description? Another example: the so-called identity theory (or rather hypothesis) is usually formulated as stating the identity of mind and brain. Would the hypothesis not gain in precision by construing it as stating that mental states are states of the nervous system (or rather of subsystems thereof) and therefore mental events changes in the states of neuron assemblies? These and other philosophical problems call for, even before they are properly stated, an elucidation of the notion of a state of a thing. Let us therefore turn to this task.

2.2. *State Function*

Recall the notion of a functional schema $X_m = \langle M, \mathbb{F} \rangle$ representing a thing X (Definition 3.6). It consists of a set M and a list $\mathbb{F} = \langle F_i \mid 1 \leq i \leq n \rangle$ of n functions with unspecified domain M and equally unspecified codomains V_i. Each component $F_i : M \to V_i$ of $\mathbb{F}$ is supposed to represent a general property, or property of things of the kind to which X belongs. And each value of F_i for a particular entity at a point $x \in M$ is supposed to represent an individual property, or property of X. In Example 1 of Sec. 1.4 the function $\mu : F \times T \to \mathbb{R}^+$ represents the mass (a general property), and its value $\mu(f, t)$ represents the mass of the given corpuscle, relative to frame f, at time t. Unlike the F_i, which are property-representing "variables" (or rather functions), f and t are not "variables". They are arbitrary members of certain sets (F and T respectively) and they are not possessed by any thing in particular: on the contrary they are rather public in the sense that they can be "used" by a number of things.

The components F_i of the list $\mathbb{F}$ of functions in a functional schema are usually called *state variables* because their values contribute to characterizing or identifying the states the thing of interest can be in. We shall call them *state functions* because this is what they are. And we offer the following preliminary characterization: A function is a *state function* for a thing of a given kind if it represents a property possessed by the thing. Whether this representation is faithful (true) is immaterial to the function's qualifying as a state function. What is decisive is that the function refers to the thing and can be interpreted as representing or conceptualizing the intended property.

A few more examples should give a better feel for the concept of a state function.

Example 1 Qualitative global property: social structure. Let $F_i \triangleq$ social structure in ith respect (occupation, or income, or number of school years, or political power, or whatever). Call Σ the set of all social systems (communities or organizations) and $T \subseteq \mathbb{R}$ the set of instants. Further, call $\mathscr{C}(\sigma)$ the (atomic) composition of system $\sigma \in \Sigma$ at the person level, i.e. the set of individuals composing σ. Define on $\mathscr{C}(\sigma)$ a social equivalence relation $\sim_i$ as that which induces the partition of $\mathscr{C}(\sigma)$ into groups of people that are homogeneous in the ith respect (roughly the same occupation, or the same income, etc.). Call $\mathscr{C}(\sigma)/\sim_i$ such a partition. Then a social structure is a list $\mathbb{F}$ of n state variables of the form

$$F_i: \Sigma \times T \to \{\mathscr{C}(\sigma)/\sim_i \mid \sigma \in \Sigma\}.$$

Example 2 Population, a conspicuous biological and sociological variable, may be conceptualized as a state function

$$F: \Sigma \times T \to \mathbb{N}$$

from pairs ⟨community σ of organisms of a kind, instant t⟩ to the natural numbers $\mathbb{N}$. Each value $F(\sigma, t) = n$, for $n \in \mathbb{N}$, represents an individual property of σ, whereas the function F itself represents a general property, or a property of all the members of the set Σ.

Example 3 Quantitative stochastic property. Let $F \triangleq$ Momentum probability distribution. Here $M = \{q\} \times \{f\} \times \mathbb{R}^3 \times T$ and $V = \mathbb{R}$, where $q \in Q$ is a quantum-mechanical entity of kind Q, $f \in F$ a reference frame of kind F, and $\mathbb{R}$ the real line. The function $F: M \to V$ satisfies a certain law statement and '$F(q, f, p, t)\, dp$' is interpreted as "The probability that entity q, relative to frame f, at time t, possesses a momentum comprised between p and $p + dp$". (In rather common notation, $F = |\varphi|^2$, where φ is the Fourier transform of the state function ψ representing the position distribution.)

We are now ready for a general characterization of our concept:

DEFINITION 3.9. Let X be a thing modeled by a functional schema $X_m = \langle M, \mathbb{F}\rangle$, and suppose that each component of the function

$$\mathbb{F} = \langle F_1, F_2, \ldots, F_n\rangle: M \to V_1 \times V_2 \times \ldots \times V_n$$

represents a property of X. Then F_i, for $1 \leqslant i \leqslant n$, is called the ith *state function* for X, $\mathbb{F}$ the *total state function* for X, and its value

$$\mathbb{F}(m) = \langle F_1, F_2, \ldots, F_n \rangle(m) = \langle F_1(m), F_2(m), \ldots, F_n(m) \rangle$$

for $m \in M$ is said to represent the *state of* X *at* m – all this *in the representation* X_m. Furthermore, if all the V_i for $1 \leqslant i \leqslant n$ happen to be vector spaces, then $\mathbb{F}$ is called a *state vector* for X in the representation X_m.

Notice the cautious expression 'in the representation X_m'. The reason is that there is no such thing as the absolute state function for a given thing: indeed there are as many state functions as functional schemata of the thing can be conceived, i.e. any number of them. (For example, whereas lagrangian theories employ generalized coordinates and velocities as the basic state functions, hamiltonian theories choose generalized coordinates and momenta.) And even one and the same representation is compatible with infinitely many choices of systems of units, every one of which may ensue in a different state function.

The sole test for the adequacy of a choice of state functions is the adequacy (factual truth) of the theory as a whole, in particular that of its key formulas, which are those interrelating the various components of the total state function – namely the law statements and constraint formulas of the theory. Even so there may be alternative though basically equivalent formulations of one and the same theory. (For example, most field theories can be formulated using either field strengths or field potentials, and the latter are mutually equivalent modulo certain arbitrary constants or even functions.) In the case of equivalent theories there are no preference criteria other than those of computational convenience, or else heuristic power, or even sheer beauty or fashion.

To put it negatively: The choice of state functions is not uniquely determined by empirical data but depends partly on our available knowledge, as well as upon our abilities, goals, and even inclinations. This consideration will play an important role in any talk of states and state spaces, of which more in Sec. 2.4. But lest we have given the impression that the choice of state functions is totally arbitrary and a matter of sheer taste, let use hasten to add that whatever set of state functions be chosen, they are supposed to abide by certain law statements – and this is not a matter of convention. More on this in the next section.

Finally a point of mathematical detail. Typically the state functions for a given thing are defined on a single domain M. Think of the mass density, stress tensor, velocity field, entropy density, and the other "variables" concerning a fluid. Likewise the set of dynamical variables ("observables") of a quantum mechanical system are all operators on the Hilbert space associated with the system. However, it might happen that a given set of state functions for things of a given kind fail to be defined on the same domain. In this case a harmless trick will allow us to assign them all the same domain and thus comply with a tacit condition of Definition 3.9. Thus if

$$F_1: A \to B \quad \text{and} \quad F_2: C \to D, \quad \text{where} \quad A \neq C \quad \text{and} \quad B \neq D,$$

we can adopt the new state functions

$$\left.\begin{array}{ll} F_1': A \times C \to V_1 & \text{such that} \quad F_1'(a, c) = F_1(a) \\ F_2': A \times C \to V_2 & \text{such that} \quad F_2'(a, c) = F_2(c) \end{array}\right\} \quad \text{for all} \quad a \in A, c \in C.$$

But, as we noted before, it is unlikely that we shall ever have to make use of this artifice.

2.3. *Law Statements as Restrictions on State Functions*

In Ch. 2, Sec. 3.3, we gave an exact but not very illuminating definition of a law. Roughly, we stated that if P and Q are two properties of entities of a kind, then any inclusion relation between the scope of P and Q counts as a law. We are now in a position to offer a deeper analysis of the law concept, namely as a condition on certain state functions for some thing. Such conditions may take a number of forms, depending not only on the things themselves but also on the state of our knowledge. All of the following are conspicuous simple forms:

Example 1 Range of $F = V$, where F is a state function and V some well defined set. (Think of the relativistic restriction on the velocity values.)

Example 2 $\partial F/\partial t \geqslant 0$, where $t \in T$, and $T \subseteq \mathbb{R}$ occurs in the domain of F.

Example 3 Kinematical model of a thing (usually called dynamical system):

$$\frac{d\mathbb{F}}{dt} = \mathbb{G}(\mathbb{F}, t), \text{ with } \mathbb{G} \text{ a specific function.}$$

Example 4 Lagrangian model of a system:

$$\int_{t_1}^{t_2} dt\, F(q, \dot{q}, t) = \text{extremal, with } t_1,\ t_2 \text{ two selected elements of } T \subseteq \mathbb{R}.$$

Example 5 Field theoretic model of a system:

$$F_2(x, y) = \int_V du\, dv\, F_1(u - x, v - y), \text{ with } F_1, F_2\colon E^3 \times E^3 \to \mathbb{C}.$$

Example 6 Another model of the same genus:

$$\nabla^2 F_1 = F_2, \quad \text{with} \quad F_1, F_2\colon E^3 \to \mathbb{R}.$$

The preceding considerations suggest adopting the following characterization:

DEFINITION 3.10 Let $X_m = \langle M, \mathbb{F}\rangle$ be a functional schema for a thing X. Any restriction on the possible values of the components of $\mathbb{F}$ and any relation among two or more such components is called a *law statement* iff (i) it belongs in a consistent theory about the X's and (ii) it has been confirmed empirically to a satisfactory degree.

(It might be objected that, since we know whether a function is a state function only once we have satisfied ourselves that it occurs in some law statement, the above definition is circular. Not so, because we have not *defined* state functions in terms of law statements. All we have asserted is that lawfulness is a test or criterion of state functionality.)

If a law statement concerns a certain thing x, we call it $L(x)$, and we may call $\mathbb{L}(x) = p(x) \cap \mathbb{L}$ the totality of laws, or rather law statements, for x. Likewise we call $L(T)$ a law possessed by every thing in the set T, and $\mathbb{L}(T) = p(T) \cap \mathbb{L}$ the totality of laws "obeyed" by everything in T. This is not a mere matter of notation for, as we saw in Ch. 2, Sec. 3.3, laws are properties. And, being properties, they are representable as functions. Indeed, a law statement may be construed as the value of a certain function – a *law function* – with domain the class T of things concerned, and codomain the set of law statements of the form $L(x)$. In short,

$$\text{If } T \subset \Theta, \quad \text{then} \quad L: T \to L(T).$$

For example, Ohm's law for a battery-resistor circuit x (an individual of a certain class C) can be written

$$L: C \to L(C), \quad \text{where} \quad L(x) = \ulcorner e(x) = R(x) \cdot i(x) \urcorner$$
$$\text{for every} \quad x \in C,$$

where in turn e, R and i are real valued functions on C representing the electromotive force, the resistance, and the current intensity respectively. This manner of writing shows clearly that the laws are properties of things and some of them interrelate properties of things.

The same ideas can be represented in an alternative way that lends itself more easily to generalization, namely as follows. Consider again the set C of battery *cum* resistor circuits. As a state vector for things of this kind we may choose either the voltage-current pair or the voltage-charge pair. We pick the former:

$$\mathbb{F} = \langle e, i \rangle : C \to \mathbb{R} \times \mathbb{R} \quad \text{where } \mathbb{R} \text{ is the set of real numbers.}$$

This function assigns to each thing $x \in C$ the pair of real values $\langle e(x), i(x) \rangle$, in such a way that $e(x) = R(x) \cdot i(x)$, where $R(x)$ is a real number. We define now the first and second projections, G and H, of $\mathbb{F}$, as follows:

$$G: \mathbb{R} \times \mathbb{R} \to \mathbb{R}, \qquad H: \mathbb{R} \times \mathbb{R} \to \mathbb{R}$$

such that

$$\langle e(x), i(x) \rangle \overset{G}{\mapsto} e(x)$$

$$\langle e(x), i(x) \rangle \overset{H}{\mapsto} R(x) \cdot i(x)$$

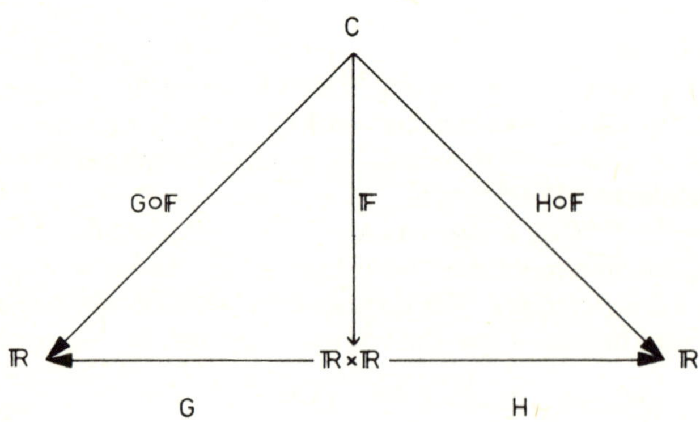

Since

$$(G \circ \mathbb{F})(x) = e(x) \quad \text{and} \quad (H \circ \mathbb{F})(x) = R(x) \cdot i(x),$$

Ohm's law now reads

$$G \circ \mathbb{F} = H \circ \mathbb{F},$$

i.e. as a statement of the identity of two function compositions.

The previous treatment carries over to any other situation where a fixed thing is being considered and the domain of the state function $\mathbb{F}$ is taken to be an arbitrary set M, such as the collection of all instants, or a portion of a certain manifold. We could multiply the examples, analyzing more and more complex cases, but this would not lead us very far. It is far more profitable to look into general frameworks for scientific theories, such as lagrangian and hamiltonian dynamics (for which see Bunge, 1967b) – but we must get on with the matter of states.

2.4. *State Space*: *Preliminaries*

Every theoretical model of a thing is concerned with representing the really possible (i.e. lawful) states, and perhaps also the really possible (lawful) changes of state, of the thing. A few typical examples should give us a feel for this matter.

Example 1 Let N be a set of neurons, or of nerve fibres, and suppose that each such unit can be in either of two states: *on* (firing) or *off* (not firing). That is, we can introduce a state function $F: N \to \{0, 1\}$ representing the neuronal activity, such that

$$\text{For each } n \in N, F(n) = \begin{cases} 1 & \text{iff } n \text{ is on} \\ 0 & \text{iff } n \text{ is off.} \end{cases}$$

The functional schema of the neural system is then $\langle N, F \rangle$ with N and F as described above. And the corresponding state space for each neuron $n \in N$ is $S(n) = \{0, 1\}$. If the neurons are regarded as mutually independent, the state space of the aggregate $[N]$ of neurons (i.e. the system of interest) is $S([N]) = \{0, 1\}^{|N|}$. Thus for a system composed of 3 neurons, the state space has 2^3 elements: $S = \{\langle 0, 0, 0\rangle, \langle 1, 0, 0\rangle, \langle 0, 1, 0\rangle, \langle 0, 0, 1\rangle, \langle 1, 1, 0\rangle, \langle 1, 0, 1\rangle, \langle 0, 1, 1\rangle, \langle 1, 1, 1\rangle\}$. *Example 2* In the genetics of populations, three state functions are often employed: the size N of a population, the probability p of some particular gene, and the latter's adaptive value v. Hence for a system composed of two interacting populations, A and B, the state space is the region of $\mathbb{R}^6$ spanned by $\langle N_A(t), N_B(t), p_A(t), p_B(t), v_A(t), v_B(t)\rangle$ in the course of time. *Example 3* In the elementary theory of the ideal gas, the state function is the triple pressure-volume-temperature functions. The corresponding state space is a cube contained in $(\mathbb{R}^+)^3$. *Example 4* In chemical kinetics the instantaneous state of a chemical system is described by the values of the partial concentrations of both reactants and products. Therefore the state space of the system is inside $(\mathbb{R}^+)^n$, where n is the number of system components (reactants, catalyzers, and products). *Example 5* In elementary electrostatics the state function is $\mathbb{F} = \langle \rho, \varphi \rangle$, where ρ represents the electric density and φ is the electric potential. Hence the local state of the given field is the value of $\mathbb{F}$ at $x \in E^3$, where E^3 is Euclidean three space. And the entire state space is the set of ordered couples $\{\langle \rho(x), \varphi(x)\rangle \in \mathbb{R}^2 | x \in V \subseteq E^3\}$, where V is the spatial region occupied by the field. *Example 6* In quantum mechanics the state of a system is represented by a one dimensional subspace (or ray) of the Hilbert space assigned to the system. Since a thing of this kind is typically not attributed a pointlike location but assumed instead to be spread over some spatial region $V \subseteq E^3$ (with a definite probability distribution), the state of the thing is the set of all values its state vector takes in V.

Before attempting to jump to any generalizations let us emphasize a point of method made in Sec. 2.2. We may certainly assume that, whether we know it or not, each thing – in particular each isolated thing – is in a definite state relative to some reference frame and at each instant (or else point in spacetime). Yet our *representation* of such a state will depend upon the state function chosen to represent the thing – which choice depends in turn upon the state of our knowledge as well as upon our goals. What holds for each single state holds a fortiori for the entire state space for a thing. That is, far from being something out there, like physical space, a state space for a thing stands with one leg on the thing, another on a reference frame, and a third on the theoretician (or modeller). To persuade oneself that this is so it is enough to take a new look at Example 5 above, where a reference frame at rest relative to the field source was assumed. If the system is now considered relative to a moving frame – moving, that is, with respect to the field source – a four vector current density will have to replace the single charge density, and a four vector potential will take the place of the single scalar potential. (Alternatively – and this is where the scientist's freedom comes in – the four potential can be replaced by an antisymmetric tensor representing the electric and magnetic components of the field relative to the chosen frame.)

Having emphasized the conventional ingredient of every representation of states, let us now stress that it has an objective basis as well. This can be seen from the existence of two different concepts of state space, one of which is more realistic than the other. Suppose we have settled on a certain total state function $\mathbb{F}$. If we form the cartesian product of the codomains of the various components of $\mathbb{F}$ (in tune with Definition 3.8), we obtain the codomain V of $\mathbb{F}$ itself, a set that will be called the *conceivable state space* for the thing represented. This is precisely what we did in the examples that sparked off this section.

However, a state function may not take values in its entire codomain but may be restricted to a subset of the latter, and this by virtue of some law. (Recall Sec. 2.3.) Therefore for every component of $\mathbb{F}$ our concern should be the range of it rather than its codomain. For example, the total population of organisms of a given kind in a given territory is constrained not only by the carrying capacity of the latter but also by the birth and death rates, as well as by additional factors such as sunshine and rainfall. Again, although the range of the speed function for a body is the entire real interval $[0, c)$, where c is the speed of light in a void, an electron traveling in a transparent medium will not come close to the upper bound c, for it is subject to further laws. In general: Only those values of the components of the total state function that are compatible with the laws will be really (not just conceptually) possible. In other words, because the laws impose restrictions upon the state functions and their values, hence upon the state spaces, only certain subsets of the latter are accessible to the thing represented. We shall call the accessible part of the state space the *lawful state space* of the thing in the given representation and relative to a given frame. (It is sometimes called the space of physical states: cf. Hirsch and Smale, 1974.) To say that a thing *behaves lawfully* amounts then to saying that the point representing its (instantaneous) state does not wander beyond the bounds of the state space chosen for the thing (See Figure 3.1).

2.5. *Definition of a State Space*

The preceding remarks can be summarized in terms of the concepts introduced in Sec. 2.3:

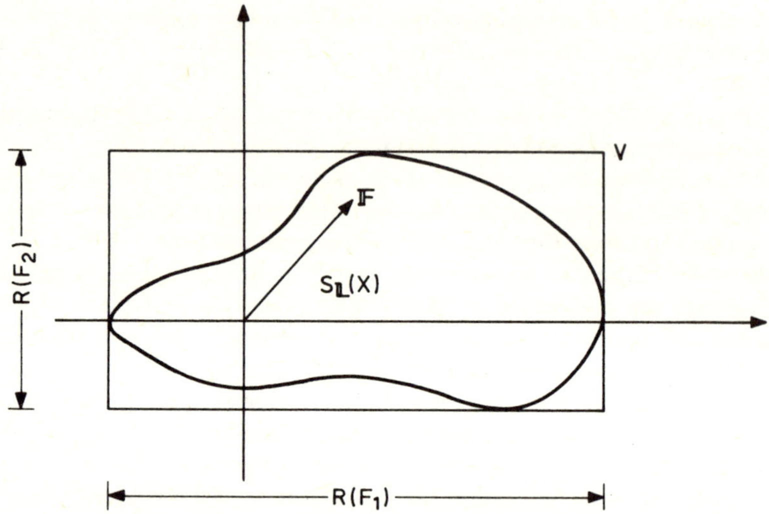

Fig. 3.1. The lawful state space $S_{\mathbb{L}}(X)$ for a thing X is a subset of the cartesian product of the codomains of the components of the state function (in this case only two).

DEFINITION 3.11 Let $X_m = \langle M, \mathbb{F}\rangle$ be a functional model for a thing X, where $\mathbb{F}: \langle F_1, F_2, ..., F_n\rangle: M \rightarrow V_1 \times V_2 \times ... \times V_n$ is the state function, and call $\mathbb{L}(X)$ the set of all law statements of X. Then the subset of the codomain $V = V_1 \times V_2 \times ... \times V_n$ of $\mathbb{F}$ restricted by the conditions (law statements) in $\mathbb{L}(X)$ is called the *lawful state space* of X in the representation X_m, or $S_{\mathbb{L}}(X)$ for short:

$$S_{\mathbb{L}}(X) = \{\langle x_1, x_2, ..., x_n\rangle \in V_1 \times V_2 \times ... \times V_n \mid \mathbb{F} \text{ satisfies jointly every member of } \mathbb{L}(X)\},$$

and every point of $S_{\mathbb{L}}(X)$ is called a *lawful* (or *really possible*) *state* of X in the representation X_m.

Clearly, the lawful state space is included in the corresponding conceivable state space: $S_{\mathbb{L}}(X) \subseteq S(X)$ for every $X \in \Theta$ and in any fixed representation X_m of X. For example, if $\mathbb{F} = \langle F_1, F_2\rangle$ takes values in $\mathbb{R}^2$, subject to the restriction (law statement) $\ulcorner F_2 = kF_1 \urcorner$, where k is a real number, then $S_{\mathbb{L}}$ is a one dimensional subspace of $S = \mathbb{R}^2$, i.e. a curve on the plane.

Example In Example 5 (electrostatics), of Sec. 2.4 the conceivable state space was

$$S = \{\langle \rho(x), \varphi(x)\rangle \in \mathbb{R}^2 | x \in V \subseteq E^3\}.$$

Since the two components of the state function $\mathbb{F} = \langle \rho, \varphi\rangle$ are tied together by the law statement: $\ulcorner \nabla^2\varphi(x) = 4\pi\rho(x)\urcorner$, the lawful state space of the thing is the subspace of S defined by

$$S_{\mathbb{L}} = \{\langle \rho(x), \varphi(x)\rangle \in \mathbb{R}^2 | x \in V \subseteq \mathbb{R}^3 \ \& \ \nabla^2\varphi(x) = 4\pi\rho(x)\}.$$

Because most of the state functions occurring in scientific theories are real valued, and because $\mathbb{R}$ and any of its cartesian powers are vector spaces, the majority of the conceivable state spaces are vector spaces. However, because of the restrictions imposed by the law statements, the lawful state spaces, though usually included in vector spaces, need not be vector spaces themselves – let alone metric spaces.

We close this subsection with a handful of assorted remarks.

Remark 1 If we pretend that the collection of states of a thing is finite, then we are dealing with a *finite state model* of the thing, otherwise with an *infinite state model* of it. Actually there are no finite state things but only model things restricted to a finite number of states, such as automata. The finite state models are reasonable only when our interest is restricted to global properties. In such cases one can use the notion of *capacity* of a thing, defined as $I = \log_2 n$, where n is the number of states of the thing. *Remark 2* The aggregate (or noninteracting composition) of two finite state things can be represented by a state space with $n_1 \cdot n_2$ points, where n_1 and n_2 are the number of states of components 1 and 2. Accordingly the capacity of such an aggregate equals the sum of the partial capacities:

$$I(1 \dot{+} 2) = \log_2(n_1 \cdot n_2) = \log_2 n_1 + \log_2 n_2 = I_1 + I_2.$$

But this case, central to linear system theory (cf. Zadeh and Desoer 1963), is too particular to constitute a general basis for ontology. Think that the simplest atom can be in any of infinitely many energy states or in a linear combination thereof. *Remark 3* The next case in order of complexity is that of a thing with an infinite but denumerable state space, such as the collection of stationary states of an atom. In the simplest case of this kind a thing is characterized by a quadruple $\langle n, l, m, s\rangle$ of quantum numbers (radial, orbital, azimuthal, and spin). Beyond a certain energy threshold, though, the integer n must be replaced by a continuous parameter, so that the state space acquires a continuous component. *Remark 4* If, during a part of the existence of a thing (e.g. when it is being observed), only some components of its state function change their values, one says that the remaining components are *ignorable* and the study of the thing can be restricted to the state subspace spanned by the active state variables. This subspace may be called the *reduced state space*. Experimental science presupposes that such a reduction can always be effected. *Remark 5* While in most of general systems theory the state space is assumed to be finite dimensional, the state (or Hilbert) spaces occurring in the quantum theories are infinite dimensional. Every point in such spaces can be analyzed into infinitely many components, namely those along the axes constituted by the orthonormal eigenfunctions of an arbitrary hermitian operator in the Hilbert space.

2.6. *Equivalent Representations of States*

Recall that every law statement for a given thing may be regarded as the value of a certain function L which we called the corresponding *law function* (Sec. 2.3). For a fixed thing the arguments of this function are components of the state function or functions thereof. Therefore a law function may also be construed as a function transforming the state space into itself. That is, for any given thing, $L: S \to S$ is a function that assigns each state $s \in S_{\mathbb{L}} \subseteq S$ another state $L(s) \in S_{\mathbb{L}}$, not of course an arbitrary one but a state lawfully related to the former. Consequently the totality of laws of a given thing is contained in the set of all the transformations of its state space. This point will be amply exploited in Ch. 5 on change.

Surely not every transformation of a state space represents a law. For one thing the identity map on $S_{\mathbb{L}}$ does not represent any law. Besides, certain transformations of $S_{\mathbb{L}}$, though lawful or law abiding, represent not laws but just different choices of state functions or representations. Consider a certain state function transformation concerning a fixed thing: $\mathbb{F}^* = f(\mathbb{F})$, where f is a function subject to certain restrictions. (A typical case is that f be one-to-one bicontinuous, i.e. that its Jacobian be non-null.) Every such transformation ensues in a different representation of the states of the thing, i.e. in a different state space $S_{\mathbb{L}}^*$. In other words, if $f: S_{\mathbb{L}} \to S_{\mathbb{L}}^*$ is such a bijection then $f(S_{\mathbb{L}})$ is an alternative state space for the same thing: see Figure 3.2. Again: $X_m = \langle M, \mathbb{F} \rangle$ and $X'_m =$

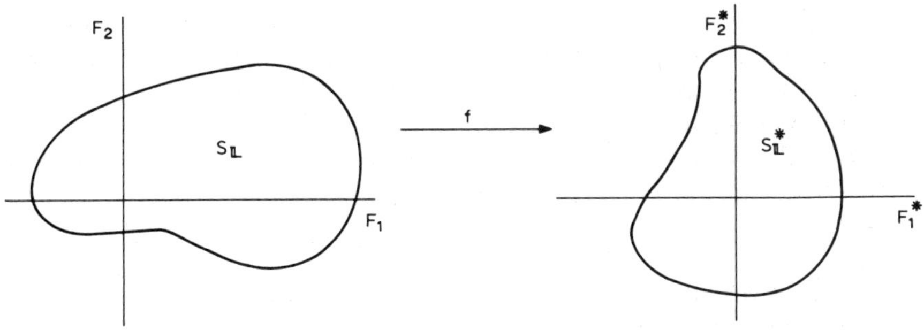

Fig. 3.2. Alternative state spaces for one and the same thing. Two such alternatives are equivalent just in case the law statements in $\mathbb{L}$ are invariant under the bijection f relating the two state spaces.

$=\langle M, f(\mathbb{F})\rangle$ are *alternative representations* of the same thing X iff f is a bijection on $S_{\mathbb{L}}$, where $S_{\mathbb{L}}$ is the space spanned by $\mathbb{F}$ with the restrictions $\mathbb{L}$.

Not every representation change is admissible: only those changes are admissible that keep the laws unchanged. We consecrate this important dichotomy by introducing

DEFINITION 3.12 Let $\mathbb{F}: M \rightarrow V$ be a state function for a given thing X in a given functional schema $X_m = \langle M, \mathbb{F}\rangle$. Then

(i) the *lawful transformations* of $\mathbb{F}$ are those transformations $f: V \rightarrow V^*$ such that $\mathbb{F}^* = f \circ \mathbb{F}$ and that leave the laws $\mathbb{L}$ of X form-invariant;

(ii) every transformed lawful state space $S^*_{\mathbb{L}} = f(S_{\mathbb{L}})$ is an *equivalent representation* of the states of X iff it is the image of $S_{\mathbb{L}}$ under a lawful transformation f;

(iii) the collection of all the equivalent representations of the states of X constitutes *the representation* of the states of X.

Example 1 In special relativistic theories the position and time coordinates transform according to the Lorentz formulas. These are not only laws but induce transformations of any state function such that the basic laws – e.g. the axioms of mechanics or of electrodynamics – retain their form, or are *Lorentz covariant.* And the components of the state function that do not change under a Lorentz transformation are called *invariant.* Examples of invariants: electric charge and entropy. *Example 2* In hamiltonian theories – whether they occur in physics or in biology or elsewhere – the state variables are the generalized coordinates q and momenta p (one set of each for every component of the system of interest). These variables span the state (or phase) space of the thing. New state functions $\mathbb{F}^* = \langle f(q, p), g(q, p)\rangle$ can always be introduced. But only those will be admissible that leave the canonical (or Hamilton's) equations invariant, i.e. those for which $q^* = f(q, p)$ and $p^* = g(q, p)$ are such that

$$\dot{q}^* = \partial H^*/\partial p^* \quad \text{and} \quad \dot{p}^* = -\partial H^*/\partial q^*.$$

Such transformations are called *canonical*. Every one of them induces an equivalent representation of the states of the thing concerned.

2.7. *State and State Preparation*

Whether a thing is in a given state (relative to some frame) as a result of its inner dynamics and its interactions with other things, or as a result of a sequence of human operations, is immaterial for the elucidation of the concept of an objective (though relative) state. After all, the manipulations leading to the preparation of a thing in a given state are strictly physical (or chemical or biological or social) processes even when they are controlled by a supposedly intelligent organism, and they are no less so when the thing concerned happens to be another organism – who is, say, being trained or brainwashed to behave in a certain way.

The previous platitude had to be stated because certain contemporary varieties of subjectivism claim that, unlike macrothings, microthings cannot be regarded as leading an autonomous existence because they are always prepared by some observer to be in (or jump into) pre-assigned states. More precisely, some quantum mechanics experts have argued as follows. (i) A microthing *is* in no definite state as long as it has not been forced into some state by some observer. (ii) Now, a state preparation is a human operation. (iii) Hence a microthing adopts a definite state only as a result of its being subject to certain human actions. The key formulas that illustrate this argument are the following. Before measuring the property represented by the operator Q, the system is in a superposition of states

$$\psi = \sum c_n \varphi_n, \quad \text{with} \quad Q\varphi_n = q_n \varphi_n,$$

where the φ_n are the eigenfunctions and the q_n the corresponding eigenvalues of the operator Q. Upon measuring whatever Q represents, ψ collapses into one of its components, say φ_m, with probability $|c_m|^2$.

The thing is only now in a definite Q-state φ_m, whereas before the measurement operation (which involves the design and reading of measuring instruments) it was in no definite (eigen)state.

The previous argument is a version of the old subjectivist thesis that properties, hence states, are dependent upon some observer rather than being objective. Moreover, it is not an argument proper, as the conclusion is just a reformulation of the premises (cf. Bunge, 1973b). Besides, the first premise happens to be false: from the fact that I happen not to know the state a thing is in, it does not follow that the thing is in no state at all – unless of course Berkeley's principle *esse est percipii* is adopted. The subjectivist thesis is moreover mathematically untenable. In fact properties are represented by functions (or by operators) and these concepts are not defined unless their entire ranges (of possible values) are given. Whether or not such a range is being sampled by empirical means is irrelevant to the objective (though relative) state of affairs, and so is the circumstance that a microthing may not be in a state that happens to be an eigenstate of the operator we may pick: a state is a state even when it is not an eigenstate of some operator. In other words, for the sake of mathematics one has to assume that the function has all of the values in its range even though only a few of them may be accessible to measurement. Consequently every thing is supposed to be in some state or other even if we have not prepared it to be in any one particular state. When an experimenter prepares a thing he picks a certain subset (occasionally even a point) of the state space of the thing. He thereby induces a real change in a real state of a real thing: he does not create either out of nothing. In sum, despite the subjectivist interpretation of quantum mechanics, states do not depend wholly on the way one looks at them but are objective (though relative) properties of real things.

2.8. *Concluding Remarks*

Every state is a state of some concrete object or other: there are no states in themselves. And conceptual objects are in no states whatsoever. Therefore a thing could be defined as *whatever is in some state or other.* Things differ by the states they are in, and their changes are changes of state. But all the things of a given (natural) kind share the same (lawful) state space – which is a way of saying that they share the same general properties. In sum, states serve to characterize not only individual things but also natural classes of things – hence the centrality of the state concept. But the matter of kinds deserves a special section.

3. From Class to Natural Kind

3.1. *Classes of Things*

In this section we shall elucidate the concepts of class, kind, and species (or natural kind) of things. To this end we shall make use of some of the ideas introduced in the theory of properties expounded in Ch. 2, Sec. 3 and shall borrow freely from Bunge and Sangalli (1977). To begin with the scope function $\mathscr{S}$ introduced by Definition 2.5 of Ch. 2, Sec. 3.2 can easily be redefined for things:

DEFINITION 3.13 The *scope* of a property is the set of things possessing it. I.e. the function $\mathscr{S}: \mathbb{P} \to 2^{\Theta}$ such that '$x \in \mathscr{S}(P)$', for $x \in \Theta$ and $P \in \mathbb{P}$, is interpretable as "*x possesses P*", is the *scope.*

Correspondingly the Definition 2.6 of a class of substantial individuals now reads

DEFINITION 3.14 A subset X of the set Θ of things is called a *class* (of things) iff there is a property $P \in \mathbb{P}$ such that $X = \mathscr{S}(P) \in 2^{\Theta}$.

And Postulate 2.6 now reads

POSTULATE 3.5 The intersection of any two classes of things, if nonempty, is a class: For any two compatible properties $P, Q \in \mathbb{P}$ there is at least a third property $R \in \mathbb{P}$ such that $\mathscr{S}(R) = \mathscr{S}(P) \cap \mathscr{S}(Q)$.

We proceed to investigate the algebra of classes of things, which differs from the algebra of sets if only because unions and complements are not defined in our theory. A useful tool for this investigation is the modified scope function introduced by Definition 2.12(iii) of Ch. 2, Sec. 3.4. In terms of the thing concept, it is the function $[\mathscr{S}]: [\mathbb{P}] \to 2^{\Theta}$ that assigns each set of concomitant properties the collection of things possessing them.

3.2. *Ideals and Filters of Classes of Things*

The modified scope function $[\mathscr{S}]: [\mathbb{P}] \to 2^{\Theta}$, which is injective, induces whatever structure its codomain has on its domain. Now, the power set of any nonempty set, such as 2^{Θ}, has the lattice structure under the inclusion relation. Indeed, 2^{Θ} is closed under the union and intersection operations, and intersections precede (are included in) unions. So, $\langle 2^{\Theta}, \cup, \cap, \subseteq \rangle$ is a lattice. And, being a lattice, we can define ideals and filters on it. Since these structures will turn out to have interesting ontological interpretations, we shall briefly recall the notions of an ideal and a filter.

An *ideal* of a lattice $\mathscr{L} = \langle L, \vee, \wedge, \leqslant \rangle$ is a structure $\mathscr{I} = \langle I, \vee, \leqslant \rangle$ such that

(i) $I \subseteq L$;

(ii) every ancestor is in: if $x \in L$ and $y \in I$ and $x \leqslant y$, then $x \in I$;

(iii) all suprema are in: if $x, y \in I$, then $x \vee y \in I$.

We designate by $\mathscr{I}\mathscr{L}$ the collection of all the ideals of $\mathscr{L}$. This set has a lattice structure. If $I, I' \in \mathscr{I}\mathscr{L}$, their infimum is $I \cap I'$, while their supremum, designated $I \sqcup I'$, is the intersection of all the ideals which contain both I and I'.

Dually, a *filter* of a lattice $\mathscr{L} = \langle L, \vee, \wedge, \leqslant \rangle$ is a structure $\mathscr{F} = \langle F, \wedge, \leqslant \rangle$ such that

(i) $F \subseteq L$;

(ii) every successor is in: if $x \in F$ and $y \in L$ and $x \leqslant y$, then $y \in F$;

(iii) all infima are in: if $x, y \in F$, then $x \wedge y \in F$.

We call $\mathscr{F}\mathscr{L}$ the collection of all filters of $\mathscr{L}$. $\mathscr{F}\mathscr{L}$ is a lattice.

It will soon become apparent that neither entities nor their properties are stray: substantial individuals come in filters while substantial properties come in ideals.

We introduce next a sort of inverse of the scope function, namely a function that assigns each family of sets of entities a collection of properties. More precisely, look at

DEFINITION 3.15 The *starred scope* is the function from the power set of the power set of the collection of things into the power set of the collection of properties,

$$\mathscr{S}^{*}: 2^{2^{\Theta}} \rightarrow 2^{\mathbb{P}},$$

such that $\mathscr{S}^{*}(\mathscr{X}) = \{P \in \mathbb{P} \mid \mathscr{S}(P) \in \mathscr{X}\}$, where $\mathscr{X} \subseteq 2^{\Theta}$ is a family of subsets of Θ.

It will be recalled that, since $\mathscr{S}(P \wedge Q) = \mathscr{S}(P) \cap \mathscr{S}(Q)$, the scope function takes suprema into infima. Well, the starred scope function does just the opposite: if $\mathscr{X}$ is a filter of subsets of Θ then $\mathscr{S}^{*}(\mathscr{X})$ is an ideal of properties. Hence $\mathscr{S}^*$ can be construed as a function

$$\mathscr{S}^*: \mathscr{F}(2^{\Theta}) \to \mathscr{I}\mathbb{P}$$

that takes filters of sets of entities into ideals of substantial properties.

Our next step requires

DEFINITION 3.16 If $T \subseteq \Theta$ is a set of things, then the *filter of* 2^{Θ} *generated by* T is the intersection of all the filters of 2^{Θ} that count T as a member, i.e. $[T) = \bigcap\{X \in \mathscr{F}(2^{\Theta}) | T \in X\}$, or also $[T) = \{X \in 2^{\Theta} | T \subseteq X\}$.

If we now apply the definition of $\mathscr{S}^*$ to the computation of $\mathscr{S}^*([T))$, we find

$$\begin{aligned}\mathscr{S}^*([T)) &= \{P \in \mathbb{P} | \mathscr{S}(P) \in [T)\} = \{P \in \mathbb{P} | T \subseteq \mathscr{S}(P)\} \\ &= \{P \in \mathbb{P} | \text{ For all } x \in \Theta\text{: } x \in T \Rightarrow x \text{ possesses } P\}.\end{aligned}$$

Thus $\mathscr{S}^*([T))$ turns out to be the set of all properties possessed by all members of T. But, by the remarks following Definition 3.15, $\mathscr{S}^*$ takes its values in the ideals of P. Hence $\mathscr{S}^*([T))$ is an ideal of P. Hence by the Definition 2.3 of p, we obtain

COROLLARY 3.3 Let $x \in T \subseteq \Theta$. Then

(i) $p(T) = \mathscr{S}^*([T)) =$ The ideal of all the properties of all T's;

(ii) $p(x) = \mathscr{S}^*([x)) =$ The ideal of all the properties of thing x.

To put it another way: Both the set of properties of a thing and the set of all the properties shared by all the members of an arbitrary set of things, far from being unstructured sets, are ideals. In particular if the set of things happens to be a class then the set of all the properties shared by all the members of the class consists of all the properties preceding the property defining the class. In fact we have

THEOREM 3.1 Let $T = \mathscr{S}(P)$ be the class defined by property $P \in \mathbb{P}$. Then the collection of properties of the members of T equals the set of all the properties preceding P:

$$p(T) = \{Q \in \mathbb{P} | Q \preccurlyeq P\}.$$

Proof For all $Q \in \mathbb{P}$, $T \subseteq \mathscr{S}(Q)$ iff $\mathscr{S}(P) \subseteq \mathscr{S}(Q)$ iff $Q \preccurlyeq P$.

Note that $p(T)$ is the *principal ideal* of P in $\mathbb{P}$, i.e. $(P)_{\mathbb{P}}$. (See Figure 3.3.) Note also that $p(T)$ is not an arbitrary notation but the value of the

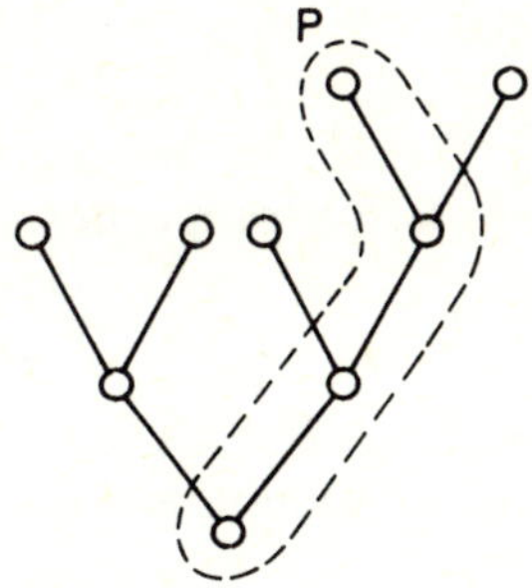

Fig. 3.3. The principal ideal of a property P is a fragment of the tree of properties. The dotted lines indicate $(P)_{\mathbb{P}}$, the bottom element represents a universal property.

function $p: 2^{\Theta} \to \mathscr{I}\mathbb{P}$ at $T \in 2^{\Theta}$. This function p reverses order, in the sense that the greater the set of things the fewer properties they share. In fact we have

COROLLARY 3.4 Let $T, T' \subseteq \Theta$. Then

$$\text{If } T \subseteq T' \text{ then } p(T') \subseteq p(T).$$

3.3. *Kinds and Species*

Whereas a single property determines a class (Definition 3.14), a set of properties will be said to determine a *kind*, and a set of lawfully related properties a *natural kind*. That is, the members of a kind are all those and only those things that share all the properties in the given set. (See Figure 3.4.)

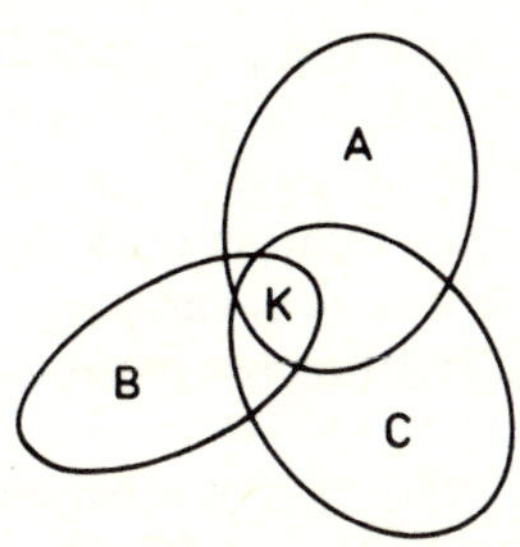

Fig. 3.4. Kind K is the intersection of classes A, B, and C.

More precisely, we make

DEFINITION 3.17 Let $k: 2^{\mathbb{P}} \to 2^{\Theta}$, be the function assigning to each nonempty set $\mathbb{R} \in 2^{\mathbb{P}}$ of substantial properties the set

$$k(\mathbb{R}) = \bigcap_{P \in \mathbb{R}} \mathscr{S}(P)$$

of things sharing the properties in $\mathbb{R}$. This value $k(\mathbb{R})$ is called the $\mathbb{R}$-*kind* of things. [Throughout this subsection and the next, $\mathbb{R}$ does not designate the real line.]

If $\mathbb{R}$ is finite, the corresponding $\mathbb{R}$-kind is a class. In fact calling $\mathbb{R} = \{P_1, P_2, \ldots, P_n\}$,

$$k(\mathbb{R}) = \bigcap_{i=1}^{n} \mathscr{S}(P_i) = \mathscr{S}(P_1 \wedge P_2 \wedge \ldots \wedge P_n),$$

so that $k(\mathbb{R})$ is the class of entities possessing the single (though compound) property $P_1 \wedge P_2 \wedge \ldots \wedge P_n$.

The notion of an $\mathbb{R}$-kind underlies all classing. In fact it allows one to introduce the notion of equivalence, or equality in certain respects, which is the very basis for the grouping of items:

DEFINITION 3.18 Let $\mathbb{R}$ be a set of substantial properties. Then two things $x, y \in \Theta$ are said to be $\mathbb{R}$-*equivalent*, or *equal in every respect* $P \in \mathbb{R}$, iff they possess precisely the same properties in $\mathbb{R}$:

$$x \sim_{\mathbb{R}} y =_{df} (P)(P \in \mathbb{R} \Rightarrow (x \text{ possesses } P \Leftrightarrow y \text{ possesses } P))$$

or, equivalently,

$$x \sim_{\mathbb{R}} y =_{df} p(x) \cap \mathbb{R} = p(y) \cap \mathbb{R}.$$

For example, since all the cells in an organism originate in a single cell, they are genetically identical. Even being different in other respects they are equal in the genetic respect, or G-equivalent. The totality of such cells, *qua* set not *qua* entity (organism), is called a *clone.*

Since $\sim_{\mathbb{R}}$ is an equivalence relation on S, it splits any set T of things into disjoint equivalence classes, variously called species, genera, orders, and so on. Such a splitting (or grouping) deserves a name:

DEFINITION 3.19 Let $\mathbb{R} \subset \mathbb{P}$ be a set of substantial properties and $T \subset \Theta$ a set of things. Then the $\mathbb{R}$-*classing* of T is the partition $T/\sim_{\mathbb{R}}$, and every member of it is called a *cell.*

Different equivalence relations induce different partitions. Of particular interest are those pairs of classings that, far from being on the same footing, are of different degrees of fineness (or coarseness). This notion is made precise by

DEFINITION 3.20 Let $\mathbb{R}$ and $\mathbb{R}'$ be sets of substantial properties, $\sim_{\mathbb{R}}$ and $\sim_{\mathbb{R}'}$ their respective equivalence relations, and $T/\sim_{\mathbb{R}}$ and $T/\sim_{\mathbb{R}'}$ their respective classings. Then $\sim_{\mathbb{R}}$ and $T/\sim_{\mathbb{R}}$ are *finer* than $\sim_{\mathbb{R}'}$ and $T/\sim_{\mathbb{R}'}$ respectively iff $\mathbb{R}' \subseteq \mathbb{R}$, in which case, for any $x, y \in S$,

(i) If $x \sim_{\mathbb{R}} y$ then $x \sim_{\mathbb{R}'} y$;

(ii) each cell of $T/\sim_{\mathbb{R}'}$ is the union of cells of $T/\sim_{\mathbb{R}}$.

This is precisely how the classings of items into species, genera, etc., are performed, namely with the help of equivalence relations of different power. (The higher powered of two equivalence relations is the finer. And the relation of fineness, whereby fine precedes coarse, is a partial order.)

The number of cells contained in each partition $T/\sim_{\mathbb{R}}$ depends upon $\mathbb{R}$, i.e. upon the number of respects that are being considered. Since there are at most as many $\sim_{\mathbb{R}}$-cells as there are subsets of $\mathbb{R}$, the following important result is obtained:

COROLLARY 3.5 If $\mathbb{R}$ is finite, so is the corresponding partition $T/\sim_{\mathbb{R}}$ of an arbitrary set T of entities.

Since in practice only a finite number of properties is handled, every classing of any collection of entities results in a finite number of classes. One of these classes is the scope of the conjunction of all the properties in $\mathbb{R}$.

Now, there are two ways of ascertaining whether two things are in the same $\sim_{\mathbb{R}}$-cell or are $\sim_{\mathbb{R}}$-equivalent. One is to note and match all their observable properties regardless of their weight: this is the pretheoretical method characteristic of classical taxonomy and also the one adopted by contemporary numerical taxonomy. It can be misleading, for a uniform appearance may hide important differences – and small differences may conceal basic kinship.

An alternative method, and the deepest, consists in grouping things by the laws they possess. When laws are made the *fundamentum divisionis* of a set of things, the resulting kinds are maximally natural – or, in Aristotelian jargon, accident is then unlikely to prevail over essence. The outcome is a set of natural kinds or species. Such sets – such as the chemical species and the biological species characterized in evolutionary terms – are typical of advanced modern science in contrast to purely empirical knowledge and to descriptive science. The definition of the concept of natural kind is obtained by just specializing that of an $\mathbb{R}$-kind (Definition 3.17). More precisely, we can get it by taking the restriction of the kind function k to the power set of the collection of all laws:

DEFINITION 3.21 Let $\mathbb{L}$ be the set of laws and let $k_{\mathbb{L}}: k|2^{\mathbb{L}}: 2^{\mathbb{L}} \to 2^{\Theta}$ be the function assigning to each $\mathbb{L}_i \subset \mathbb{L}$ of laws the set

$$k(\mathbb{L}_i) = \bigcap_{L \in \mathbb{L}_i} \mathscr{S}(L)$$

of entities sharing the laws in $\mathbb{L}_i$. This value $k(\mathbb{L}_i)$ is called the $\mathbb{L}_i$*-species* or *natural kind.*

Things have then properties of two kinds: the laws characteristic of their natural kind, and idiosyncratic properties: see Figure 3.5. That is, we can make

DEFINITION 3.22 Let $\mathbb{L}_i \subset \mathbb{L}$ be a set of laws and $k(\mathbb{L}_i)$ the corresponding natural kind. Further, call $p(x)$ the set of properties of a member x of $k(\mathbb{L}_i)$. Then

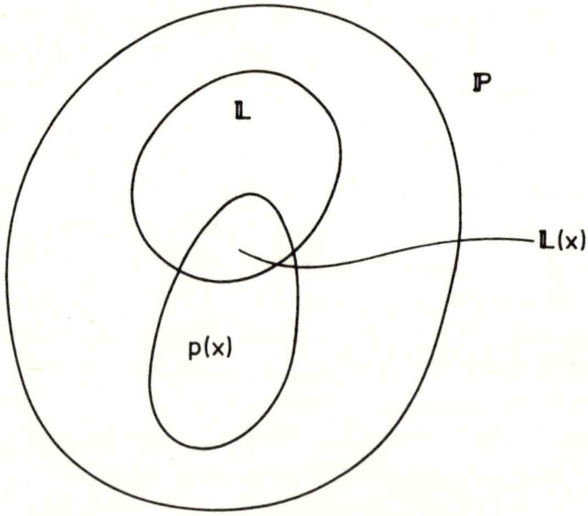

Fig. 3.5. Some of the properties of a thing x are its laws $\mathbb{L}(x)$; the rest, $p(x) - \mathbb{L}(x)$, are its idiosyncratic properties.

(i) the *species-specific properties* of x are all those in $\mathbb{L}_i$;

(ii) the *idiosyncratic properties* of x constitute the complement of its species-specific properties, i.e. $p(x) - \mathbb{L}_i$.

Two things sharing the properties characteristic of their common kind, though possibly different in every other respect, are equivalent with respect to the laws defining their kind. For example two chunks of material are golden iff, whatever their shape, weight, location, etc., they satisfy the laws defining the natural kind (chemical species) Au. In general:

DEFINITION 3.23 Let x and y be two things and $\mathbb{L}_i \subset \mathbb{L}$ a set of laws. Then x and y are *nomologically equivalent relative to* $\mathbb{L}_i$ iff x and y share all their species-specific properties, i.e. all the laws in $\mathbb{L}_i$:

$$x \sim_{\mathbb{L}_i} y =_{df} p(x) \cap \mathbb{L}_i = p(y) \cap \mathbb{L}_i.$$

And of course two things are *nomologically inequivalent* if they do not possess exactly the same laws – or, what amounts to the same, if they do not belong to the same natural kind. Hence just as individuation consists in the emergence (or the recognition) of a thing with idiosyncratic properties, so speciation consists in the emergence (or the recognition) of a collection of things with peculiar laws – i.e. a natural kind or species.

Finally we combine the notion of natural kind with that of composition (Definition 1.6). An entity may or may not be composed of parts of the same natural kind. If it is not so composed we say it is an atom of its kind. More precisely we propose

DEFINITION 3.24 Any thing belonging to a natural kind $k(\mathbb{L}_i)$ is an *atom* of that kind, or $k(\mathbb{L}_i)$*-elementary*, iff it has no proper parts of the same kind. Otherwise it is called *molecular*.

3.4. *The Algebra of Kinds*

Since properties do not satisfy the algebra of predicates – i.e. they do not form Boolean algebras – it may be surmised that kinds do not obey the algebra of sets. The following investigation will show that this is indeed the case.

Recall how the kind function k was introduced by Definition 3.17. Like p, the kind function k reverses order:

$$\text{If } \mathbb{R} \subseteq \mathbb{R}' \text{ then } k(\mathbb{R}') \subseteq k(\mathbb{R}).$$

I.e., as more conditions are added fewer individuals satisfy them. We have then two order reversing functions relating the family of sets of properties and the family of sets of things:

$$2^{\mathbb{P}} \underset{p}{\overset{k}{\rightleftarrows}} 2^{\Theta}.$$

While k assigns to every $\mathbb{R} \subseteq \mathbb{P}$ the set of all things possessing every one of the properties in $\mathbb{R}$, p assigns to a subset T of Θ the set of all the properties shared by all members of T. More precisely we find the following bridge between sets of properties and sets of things:

THEOREM 3.2 For all $\mathbb{R} \subseteq \mathbb{P}$ and all $T \subseteq \Theta$,

$$\mathbb{R} \subseteq p(T) \text{ iff } k(\mathbb{R}) \supseteq T.$$

Proof If $\mathbb{R} \subseteq p(T)$ then, for all $Q \in \mathbb{R}, T \subseteq \mathscr{S}(Q)$. So, $T \subseteq \bigcap\{\mathscr{S}(Q) | Q \in \mathbb{R}\} = k(\mathbb{R})$. Conversely, if $T \subseteq k(\mathbb{R}) = \bigcap\{\mathscr{S}(Q) | Q \in \mathbb{R}\}$, then $T \subseteq \mathscr{S}(Q)$ for all $Q \in \mathbb{R}$, so $\mathbb{R} \subseteq p(T)$.

This theorem has a number of interesting consequences – first of all

THEOREM 3.3 The composition pk (i.e. function k followed by function p) is a closure operator on $2^{\mathbb{P}}$. I.e., for all $\mathbb{R}, \mathbb{R}' \subseteq \mathbb{P}$

(i) $\mathbb{R} \subseteq pk(\mathbb{R})$;

(ii) $\mathbb{R} \subseteq \mathbb{R}' \Rightarrow pk(\mathbb{R}) \subseteq pk(\mathbb{R}')$;

(iii) $pk\,pk(\mathbb{R}) = pk(\mathbb{R})$.

Proof (i) follows from Theorem 3.2 by taking $T = k(\mathbb{R})$. (ii) follows from the fact that both k and p reverse order. Proof of (iii): $pk(\mathbb{R}) \subseteq$ $\subseteq pk\,pk(\mathbb{R})$ follows from (i). To check the inclusion the other way use first Theorem 3.2 with $\mathbb{R} = p(T)$ to get: For all $T \subseteq \Theta$, $kp(T) \supseteq T$. Now apply twice this last result, with $T = k(\mathbb{R})$ and $T = kp(k(\mathbb{R}))$, obtaining $k(\mathbb{R}) \subseteq kp(k(\mathbb{R})) \subseteq kp\,kp(k(\mathbb{R}))$. Hence by Theorem 3.2 $pk\,pk(\mathbb{R}) \subseteq pk(\mathbb{R})$.

Because of the symmetry described by Theorem 3.2 there is the further consequence that kp too is a closure operator on 2^{Θ}.

The subsets $\mathbb{R}$ of $\mathbb{P}$ such that $pk(\mathbb{R}) = \mathbb{R}$ will be said to be *closed* with respect to pk. By Theorem 3.1 only subsets of $\mathbb{P}$ which are ideals can be closed. We shall see in the sequel that these closed ideals play a special role. They are precisely those sets $\mathbb{I}$ of properties such that the things that share all the properties in $\mathbb{I}$ do not share any other properties. I.e. in general we have clause (i) of Theorem 3.3, which just states that, for a given $\mathbb{I}$, the things sharing all the properties in $\mathbb{I}$ (viz., $k(\mathbb{I})$) may also share other properties: $\mathbb{I} \subseteq pk(I)$. But if $\mathbb{I} = pk(\mathbb{I})$, then $\mathbb{I}$ is a set of properties characterized by the set of individuals which possess them.

We arrive thus at the next consequence of Theorem 3.2, which supplies alternative characterizations of the concept of a kind:

THEOREM 3.4 For any subset $T \subseteq \Theta$ of things the following statements are equivalent:

(a) T is a kind (i.e. there exists a subset $\mathbb{R}$ of $\mathbb{P}$ such that $T = k(\mathbb{R})$).

(b) T is determined by the properties shared by all of its members (i.e. $T = k(p(T))$).

(c) T is an $\mathbb{I}$-kind for a (unique) closed ideal $\mathbb{I}$ (i.e. there exists a necessarily unique closed ideal $\mathbb{I}$ of $\mathbb{P}$ such that $T = k(\mathbb{I})$).

Proof (a)$\Rightarrow$(b): apply kp to both members of $T = k(\mathbb{R})$ to get $kp(T) = kpk(\mathbb{R})$. Using Theorem 3.2 check that $kpk(\mathbb{R}) = k(\mathbb{R})$.

(b)$\Rightarrow$(c): it suffices to show that $p(T)$ is a closed ideal (the uniqueness assertion is easy to check). From the fact that kp is a closure operator we get $T \subseteq k(pkp(T))$. By Theorem 3.2 it now follows that $pkp(T) \subseteq p(T)$.

(c)$\Rightarrow$(a): trivial.

From the last clause of Theorem 3.4 it follows that, in order to generate all kinds, it is enough to consider the closed ideals of properties. Indeed, the correspondence between kinds and closed ideals is one to one. Furthermore we have

THEOREM 3.5 The collection of all kinds forms a complete lattice under inclusion, and this lattice is isomorphic to the dual of the lattice of all closed ideals.

Proof If χ is a family of kinds, the intersection $\bigcap_{X \in \chi} X$ is again a kind and it is the infimum of χ. The supremum is the kind $k\left(p\left(\bigcup_{X \in \chi} X\right)\right)$. The order reversing isomorphism is the function k that sends the closed ideal $\mathbb{I}$ into the $\mathbb{I}$-kind $k(\mathbb{I})$. Indeed, for any two closed ideals $\mathbb{I}$, $\mathbb{I}'$, applying Theorem 3.2 with $\mathbb{R} = \mathbb{I}$ and $T = k(\mathbb{I}')$ we get $\mathbb{I} \subseteq \mathbb{I}'$ iff $k(\mathbb{I}) \supseteq k(\mathbb{I}')$. (The inverse isomorphism is the function p.)

Note the difference between the structures of the family of kinds and the family of subsets of Θ (i.e. the classes). Whereas the latter forms a Boolean algebra, the family of kinds does not: for one thing the union of two kinds is not necessarily a third kind. This is of course a feature of the world and one that a purely extensional or nominalistic approach cannot possibly reveal.

Finally, Theorem 3.5 may be given the following twist. Designate with $\mathbb{P}^*$ the collection of all closed ideals of properties. This set contains not only the substantial (or really possessed) properties but others as well. In this larger set $\mathbb{P}^*$ the sup of arbitrary (possibly infinite) families of elements exist; they can be interpreted as the conjunction of (possibly infinite) families of properties. Theorem 3.5 tells us then that there is a 1:1 correspondence between the kinds of things and the sum total of (real and imaginary) properties. This correspondence is the function k, which may be regarded as a generalization of the scope function $\mathcal{S}$, for it agrees with $\mathcal{S}$ on the set $\mathbb{P}$ of substantial properties. (I.e. $k((P]) = \mathcal{S}(P)$, where $(P]$ is the closed ideal $\{Q \in \mathbb{P} | Q \leqslant P\}$.) Besides, k also assigns scopes $k(\mathbb{I})$ to the property ideals. Moreover the correspondence k is such that to the (possibly infinite) conjunction (i.e. the supremum) of a set of properties corresponds the intersection of their scopes.

So much for the algebra of kinds – which applies of course to natural or nomological kinds as a particular case.

We now have a theory of properties, distinct from the theory of predicates, and a theory of kinds, different from the algebra of sets. We can therefore use without qualms the concepts of a property and a kind. The differences between predicates and properties, and between sets and kinds, suffice to ruin the ontological interpretations of logic and of set theory. There is no reason to expect that pure mathematics is capable of disclosing, without further ado, the structure of reality.

3.5. *Variety*

We close this section by casting a glance at the concepts of variety. There are at least two of them. One consists in the diversity of kinds of thing "represented" (to speak platonically) in a restricted set of things, such as a random sample of the collection of organisms present in one cubic centimeter of top soil. There are several measures of this kind of variety, or *local variety*: the sheer number of different species, various indices of ecological variety, or variety conducive to ecological equilibrium (see Pielou, 1969), the information theoretic measure (see Ashby, 1956), and so on. These are all specific measures of the variety found, or to be expected, in some restricted set of things. Hence their study belongs in science.

Here we are interested in the concept of *universal variety*, or the diversity in the sum total of things. We have assumed that there are infinitely many things (Postulate 3.1). But of course we have empirical access to smallish samples of that totality. However we can transcend this limitation by dint of imagination – mainly by classing and by conjecturing patterns. Classing enables us to play down idiosyncrasies and it prepares the ground for finding regularities. And hypothesizing general patterns, or laws, allows us to play down particular circumstances as well as to refine and deepen the previous classings.

Now, we grasp laws only through our conceptual reconstruction of them, i.e. law statements. Among such scientific formulas there are some which discard both particular circumstances and idiosyncrasies: they are the *generic* or *basic* laws, in contrast to the specific or derivative ones. This difference is logically, methodologically and ontologically crucial and it boils down to this. Whereas every scientific theory contains at best a finite number of generic or basic laws, it also contains all the logical consequences obtained by conjoining such axioms with other premises of the theory – and these consequences are just as entitled to be called 'laws'. The basic/derived, or generic/specific dichotomy is so fundamental that only the basic (or generic) laws are required to possess covariance properties (Bunge, 1967b).

The consequence for the matter of universal variety is rather obvious, namely this. Whereas classing things according to any laws guarantees infinite variety (because there are infinitely many law statements), classing according to basic laws alone warrants the partition of the set of things into a finite number of natural kinds. These two modes of classing are so important that they warrant one more convention:

DEFINITION 3.25 The set of all things that share a basic law is called a *natural genus*, and that which shares a particular law, a *natural species*.

This law-natural kind correspondence, together with our metatheoretical knowledge of the structure of scientific theories, entails

THEOREM 3.6 There is a finite number of natural genera and there are infinitely many natural species.

We realize now that the question 'How varied is reality?' is ambiguous. The answer is: Whereas the specific variety of reality is infinite, its generic variety is finite. And, whereas the latter trait of reality renders scientific research possible, the former makes it open ended.

4. The World

4.1. *What Does the World Consist in and of?*

Most metaphysicians have asked, and some have answered, the following questions: (i) What does the world consist in – i.e. what basic kinds of thing are its constituents?, and (ii) What does the world consist of – i.e. what is there? We have an answer to the first question not to the second.

Our answer to the question about the kinds of thing the world is "made" of follows from Postulate 3.2 and Definition 3.5: *The world is the aggregation of its constituents, which are things.* Or, if preferred, the world is that thing which is the physical sum of all concrete or material existents. This is of course a materialist answer. However, most materialists seem to prefer the formula "The world is material". We do not adopt the latter formula because it suggests the physicalist thesis that the only realities are physical systems, while in point of fact living systems and supersystems (e.g. ecosystems and societies) are just as real – and concrete.

Our answer, being neither new nor striking, is not popular among the more sophisticated contemporary philosophers. Indeed consider but a few typical contemporary answers to question (i): The world consists in sensations (Mach, 1906) or in *Erlebnisse* or elementary experiences (Carnap, 1928); the world is the totality of facts (Wittgenstein, 1922); the world is the sum total of processes (Whitehead, 1929); the world is the sum of individuals of all kinds (Goodman, 1956); the world is made of bodies and persons (Strawson, 1959). Having to sketch our own view on the matter we shall not attempt to discuss any of these alternatives. Nevertheless we cannot help but submit that none of those views is continuous with science, hence none of them worthy of belonging in scientific metaphysics. In particular the subjectivist doctrines are inconsistent with natural science; the view that the world consists in facts, or events, or processes, is incompatible with physical cosmology and fails to clarify the very notions of fact, event, and process; the body/person dualism conflicts with physiological psychology and the evolutionary study of behavior; and the thesis that the world consists of individuals is unsatisfactory unless one adds the clause that such individuals are concrete. On the other hand, the thesis that the world is the aggregate of all things, is continuous with science, for things are what every factual science investigates.

As for the second question – namely What is there? – we shall abstain from answering it. That is, we shall not list the kinds of constituent of the world but shall leave the task to the special sciences. For, no sooner does the metaphysician pronounce the world to be "made of" such and such kinds, than the scientist discovers either that some of the alleged species are empty or that others are missing in the metaphysician's list.

4.2. *Individuals, Populations, Communities, and Species*

Scientists do not stop at individuals but group and interrelate them. And they gather things not into arbitrary sets but into kinds and, more particularly, into natural kinds. They recognize as many natural kinds as laws, yet they manage to keep down the count of natural species. This they do, not only by focusing on basic or generic laws (cf. Sec. 3.5) but also, when the latter are not available, by disregarding slight individual differences such as local (ecological) variations which, though real, are rated as secondary – rightly or wrongly so. In other words, scientists distinguish species from varieties and, though keeping an eye on the latter (for some of them may eventually evolve into separate species), they focus on the former. (For the methodological injunction *Species non sunt multiplicanda praeter necessitatem* see Hooker, 1853, Introductory Essay pp. xii–xvii, and Allen, 1870–71.)

Scientists, in particular biologists, recognize three levels of concrete entity: the individual thing, the aggregate of individuals of a given kind, or population, and the aggregate of individuals of different kinds, or mixture of populations (e.g. the ecosystem or biological community). Unlike both an individual thing and a kind, a population is a concrete aggregate of entities of one or more species. In particular a biological population is a group of organisms of a given species inhabiting a limited territory, interacting and interbreeding, and relatively isolated from other populations. According to modern evolution theory, the evolutionary unit – that involved in inheritance and selection – is neither the individual organism nor the species but the population (or perhaps even the ecosystem). This is so because the organisms in a population share a gene pool and an environment. (Different populations of the same species are detached: they share neither a gene pool nor exactly the same environment, and are thus likely to evolve exclusive characters whose accumulation and selection may result in speciation, i.e. the formation of a population belonging to a new species.) A species, on the other hand, has neither a uniform environment nor a common set of genes, hence no selection and survival mechanism: unlike a population, which is a thing, a species is a concept – albeit an indispensable one.

The logical relation between the concepts of individual, population, and species are elucidated by

DEFINITION 3.26 A thing X is

(i) a *monospecific population* iff X is the aggregation of some set A of individual things belonging to a single natural kind $K_{\mathbb{L}}$:

$$X=[A] \quad \text{with} \quad A \subseteq K_{\mathbb{L}};$$

(ii) a *polyspecific population* (or *community* or *mixture* as the case may be) iff X is the aggregation of two or more sets A_i of individuals belonging to different natural kinds $K_{\mathbb{L}_i}$:

$$X=\left[\bigcup_i A_i\right] \quad \text{with} \quad A_i \subseteq K_{\mathbb{L}_i}$$

$$\text{and} \quad K_{\mathbb{L}_i} \neq K_{\mathbb{L}_j} \quad \text{for} \quad i \neq j, 1 \leqslant i, j \leqslant n.$$

(Conversely, any given species or natural kind may be regarded as the union of the membership of all the possible populations of things of the kind. But this construal cannot be formalized in our system since we have defined a population with the help of the species concept.)

The preceding considerations are relevant to overcoming the nominalism/realism deadlock in biological taxonomy. The question is whether biology concerns individual organisms or species of such. Our answer is of course: Both plus populations, for taxonomic systems are relations among natural kinds, and a kind is a collection of individuals which, in the case of organisms, live in populations. Thus we escape being impaled in either horn of the dilemma: we do not have to choose between individuals belonging to no species (nominalism) and species transcending their membership (Platonism). Both individual organisms and populations of such are real, but both share certain clusters of properties (in particular laws) and thus belong to definite natural kinds. The Platonist is right in holding that every individual belongs to ("participates in") some species (or in the union of several species), wrong in assigning priority to the latter: a set consists of its membership. And the nominalist is right in claiming that kinds are things of the mind but wrong in holding that all kinds are arbitrary inventions: natural kinds are inventions adequate to reality insofar as they are sets of nomologically equivalent things.

In sum each of the philosophical poles, nominalism and realism, contains a grain of truth, none the whole truth, and both ignore the *tertium quid* that population genetics and ecology have interpolated between the individual organism and its species, namely the population. This conclusion reinforces our previous critique of Platonism in Ch. 2, Sec. 5.2: there are no universals in and by themselves. There are only things with definite properties. A general property (in particular a law) may be said to be universal in a given kind if every member of the latter possesses that property. (Recall Definition 2.18.)

4.3. *Existence Concepts*

The ordinary language expression 'there are' is ambiguous, as it designates two different concepts: the logical concept *something* and the ontological concept of *existence*. Logic takes care of the former, analyzing it as the existential quantifier, which it might be better to rechristen *particularizer*, or *indefinite quantifier*, to distinguish it from both the *universalizer* (or universal quantifier) and the *individualizer* (or descriptor).

Surely most contemporary philosophers hold that ∃ formalizes both the logical concept "some" and the ontological concept of existence. I shall argue that this is a mistake. Consider the statement "Some sirens are beautiful", which can be symbolized "$(\exists x)(Sx \ \& \ Bx)$". So far so good. The trouble starts when the formula is read "There are beautiful sirens". The existential interpretation is misleading because it suggests belief in the real existence of sirens, while all we intended to say was "Some of the sirens existing in Greek mythology are beautiful". The particularizer formalizes the prefix 'some' but not the phrase 'existing in Greek mythology'. (The juxtaposition of two particularizers bearing on the same variable makes no sense.) We need then an exact concept of existence different from ∃. Much to the dismay of most logicians we shall introduce one in the sequel. In fact we shall introduce an *existence predicate*, thus vindicating the age-old intuition that existence is the most important property anything can possess.

We shall introduce the concept of relative or contextual existence, exemplified by "Disjunction exists in logic but not in nature", and "Birds exist in nature but not in logic". We do that by making

DEFINITION 3.27 Let A be a well formed set included in some set X, and χ_A the characteristic function of A, i.e. the function $\chi_A : X \to \{0, 1\}$ such that $\chi_A(x) = 1$ iff x is in A, and $\chi_A(x) = 0$ otherwise. Then

(i) *x exists in A* $=_{df} (\chi_A(x) = 1)$;

(ii) *x does not exist in A* $=_{df} (\chi_A(x) = 0)$.

Remark 1 Note the condition that A be a well formed set, i.e. a set proper and not a meaningless array of symbols such as 'The set of all objects differing from themselves'. *Remark 2* It might be rejoined that Definition 3.27 could be replaced by a simpler one, namely

$$x \text{ exists in } A =_{df} x \in A.$$

True, but the membership relation is not a function, hence it does not allow one to take the next step, which is the introduction of an existence predicate:

DEFINITION 3.28 The *relative* (or *contextual*) *existence predicate* is the statement-valued function

$$E_A : A \to \text{Set of statements containing } E_A$$

such that $\ulcorner E_A(x) \urcorner$ is true if and only if $\chi_A(x) = 1$.

Therefore the old boring question whether existence is a predicate is seen to be ambiguous: the answer depends on whether $\exists$ or E_A is concerned. Whereas the particularizer or quantifier is not a predicate (statement valued function), the concept of relative existence we have just introduced, i.e. E_A, *is* a predicate.

The following examples show how to handle the concepts χ_A and E_A. In them 'M' stands for the set of characters in Greek mythology, 'c' for Chiron, 'C' for "is a Centaur", and 'W' for "is wise".

The centaur Chiron exists in Greek mythology.

$$Cc \;\&\; \chi_M(c) = 1, \quad \text{or} \quad Cc \;\&\; E_M c.$$

Some of the centaurs (existing) in Greek mythology are wise.

$$(\exists x)(Cx \;\&\; Wx \;\&\; \chi_M(x) = 1), \quad \text{or}$$
$$(\exists x)(Cx \;\&\; Wx \;\&\; E_M x).$$

All of the centaurs exist in Greek mythology and none of them is real.

$$(x)(Cx \Rightarrow (\chi_M(x) = 1 \;\&\; \chi_\Theta(x) = 0)), \quad \text{or}$$
$$(x)(Cx \Rightarrow (E_M(x) \;\&\; \neg E_\Theta(x))).$$

χ_A and E_A are concepts of unqualified relative existence as long as the reference set A remains unspecified. In tune with Postulate 3.4 in Sec. 1.3 we must distinguish two specific concepts of existence, namely those of conceptual existence (or existence in a conceptual context) and real existence (or existence in the world). The corresponding definitions are obvious:

DEFINITION 3.29
(i) *x exists conceptually* $=_{df}$ For some set C of constructs, $E_C x$;
(ii) *x exists really* $=_{df}$ For some set Θ of things, $E_\Theta x$.

For example the Pythagorean theorem exists in the sense that it belongs in Euclidean geometry. Surely it did not come into existence before someone in the Pythagorean school invented it. But it has been in conceptual existence, i.e. in geometry, ever since. Not that geometry has an autonomous existence, i.e. that it subsists independently of being thought about. It is just that we make the indispensable pretence that constructs exist provided they belong in some body of ideas – which is a roundabout fashion of saying that constructs exist as long as there are rational beings capable of thinking them up. Surely this mode of existence is neither ideal existence (or existence in the Realm of Ideas) nor real or physical existence. To invert Plato's cave metaphor we may say that ideas are but the shadows of things – and shadows, as is well known, have no autonomous existence. Nevertheless conceptual existence must be reckoned with not only to account for logic, mathematics and semantics, but also to better understand real existence.

Definition 3.29(ii) asserts that *all* things, and *only* things, possess the property of existing really – a property represented by E_Θ. This vindicates Aristotle's principle that *real existence is singular.* There are no general things: every real existent is an individual. (Therefore 'general systems theory' is a misnomer for 'general theory of systems'.) Whatever is general is either a property (e.g. a law) or an attribute (in which case it may be called a universal) or a proposition or a set of propositions (e.g. a theory).

A first consequence of Definition 3.29 is the non-Aristotelian principle that only fully qualitied objects are real. In particular neither the bare substance without properties nor the property without a substratum are real: they are so many fictions. Also: *pace* Quine, being the value of a bound variable does not guarantee existence – not even shadowy conceptual existence.

A second consequence of our identifying being with being a thing (or being a member of the set of things) is

THEOREM 3.7 (i) The world exists really. (ii) Every part of the world exists really.

Proof Part (i) follows from Postulate 3.3, Definition 3.4, and Definition 3.29. Part (ii) is entailed by Corollary 3.2 jointly with Definition 3.29.

Remark 1 Our postulated identity of existents and things amounts to stating that the world is *the world of things* and that no other world exists really. The extra worlds invented by the idealist philosophers as well as by Chwistek (1949) and Popper (1968) are so many fictions. *Remark 2* The identity of existents and things does not debase thoughts: it just denies the latter an independent existence. In Vol. 4, Ch. 10, we shall construe "person x thinks idea y" as "x brains y" much in the way as "x does y". *Remark 3* We are not identifying reality with actuality: for all we know reality might be the union of actuality and real possibility. More on this in Ch. 4, Sec. 2.4. *Remark 4* The existence function χ_Θ and the existence predicate E_Θ allow us to characterize very succinctly the main contenders over concrete or material existents:

Materialism $(\exists x)(\chi_\Theta(x)=1$, or $(\exists x)E_\Theta x$,
Immaterialism $(x)(\chi_\Theta(x)=0)$, or $(x)\neg E_\Theta x$.

4.4. *Nothingness and Virtual Existence*

To be, to exist really, is to be a thing. Consequently nonbeing, or nothingness, is failing to be a thing. There are several ways of handling the ontological concept of nothingness, depending on the context. These are some:

(i) b does not exist really: $\neg(b \in \Theta)$, or $\chi_\Theta(b)=0$, or $\neg E_\Theta b$;
(ii) there is nothing of kind K: $K=\varnothing$, or $(x)(\chi_K(x)=0)$, or $(x)\neg E_K x$;
(iii) there exists nothing with property P: $\mathscr{S}(P)=\varnothing$, or

$$(x)(\chi_{\mathscr{S}(P)}(x)=0), \text{ or } (x)\neg E_{\mathscr{S}(P)}x.$$

In each case nothingness or nonbeing is identified with nonexistence, not with some positive entity or even with some positive property. All three statements are negative, as they amount to either "There are no elements in the given set", or "The given individual does not belong to the given set". Therefore it is nonsense to regard nothingness as an entity, the way Heidegger and Sartre do, and even more to claim that existence is the synthesis of being and nothingness, as Hegel did.

The concepts of being (a thing) and nonbeing (or nothingness) are dichotomic. In other words, χ_A is a two valued function, and E_A is a unary predicate. That is, there are no intermediate degrees of existence, whether real or conceptual: the expression 'degree of existence' (or 'degree of reality') is a *flatus vocis*. Aquinas and some of his followers (e.g. Maritain) notwithstanding, there is nothing between being and nothingness: there are no degrees of being. For the same reason it is nonsense to hold that, while perceptible things may well exist, their atomic constituents have no autonomous existence but are either mere mathematical formulas or else an outcome of repeated human observations – as the Copenhagen interpretation of the quantum theory used to maintain.

However, there are two puzzling concepts in contemporary theoretical physics that, *prima facie*, denote neither existence nor nothingness. One is the concept of a vacuum occurring in quantum field theories, the other is the concept of a virtual quantum (e.g. particle, or photon). We shall presently show that their status is not the same: that the physical vacuum is a thing all right while a virtual particle is a fiction.

The vacuum that quantum field theories refer to is the field in its lowest stationary energy state. When the field is in this state it does not cease being a field: only, it has no quanta (photons in the electrodynamic case). Thus it is a hypothesis of quantum electrodynamics that, although there may be no electrically charged matter in a given region of space, there is always an electromagnetic field – the zero field or ever present fluctuating background. So much so that the vacuum is assigned definite physical properties, which become manifest when an intruder (matter or radiation) disturbs it. In sum, *vacuum field* $\neq$ *nothingness*. All that happens is that the word 'vacuum' has become inadequate, for it has acquired a new signification – it designates a new concept. The new picture of the physical world synthesizes atomism and plenism, by assuming that every cranny of the world is filled with things, although matter and radiation exist only in discrete units or quanta and not everywhere.

As for virtual particles and virtual photons, they are alleged to be extremely short lived and in principle unobservable, and to take part in (virtual) processes that do not conserve energy. For example, the interaction between a proton p and a neutron n is usually pictured as being embodied in, or transmitted by, a virtual pion π^+ emitted by p (i.e. $p \rightarrow n + \pi^+$) and then reabsorbed by n (i.e. $n + \pi^+ \rightarrow p$). The energy gained in the first virtual process is lost in the second, so that there is no net energy change. Besides, the period is so short (of the order of 10^{-24} sec) that the process is unobservable, whence the hypothesis is declared to be out of reach of experimental criticism. Strangely enough, the same people who accept these assumptions (*a*) recognize that the basic formulas of the theory entail the theorem of the conservation of energy, and (*b*) usually pay lip service to the operationalist philosophy.

Fortunately there is no need to take virtual particles seriously, i.e. to assign them a degree of reality midway between full reality and nothingness. The whole idea of a virtual particle stems from two mistaken philosophical assumptions. One is the naive realist belief that every term in an infinite series (or every branch in a Feynman diagram) must have a real counterpart. The other is the corpuscularian hypothesis that particles are the only realities, whence any interaction among them must be mediated by some of them – even if they have to be virtual. If both opinions are shed then the virtual particles and the shady transactions in which they are allegedly involved are seen to be idle fictions (Bunge, 1970a).

4.5. *Existence Criteria*

Our theory of things supplies no criterion for either establishing or refuting any hypothesis to the effect that such and such an object really exists. It is not the business of metaphysics to offer existence criteria, just as it is not within the province of the philosophy of mathematics to propose criteria for the existence of definite mathematical objects such as, e.g., the solutions to a given equation. However, a general indication can be given in either case.

In mathematics an object is said to exist in a given context if it satisfies certain conditions – e.g. an equation – i.e., if it has certain properties and holds certain relations to the other mathematical objects. Likewise in factual science something can be inferred to exist if it holds certain connections (not just relations) to something else whose existence has already been established or at least assumed. (Cf. Peirce, 1909, 6.318.) We sum up this idea in the following methodological

CRITERION 3.1 An object other than the entire world exists really if it is shown to be connected to some real object other than it.

This is a criterion of relative existence. Absolute existence cannot be established even though it need not be excluded. In order to show that *x* exists we must exhibit its connections with some thing *y* whose existence is not questioned in the given investigation. In particular, but not necessarily, this second object can be a human observer. In this case, if *x* is real then *x* will be a link in a chain ending up in an observer *y*, and the existence of *x* will make itself felt in some change in *y* – such as e.g. *y*'s sudden perception of *x*.

Our reality criterion must not be taken for a *condition*, let alone a *definition*, of reality. Our definition of "reality" cannot be other than this:

DEFINITION 3.30 Let Θ be the set of all things and $[\Theta]$ its aggregation. Then

$$\textit{Reality} =_{df} [\Theta] = \Box = \textit{the world}.$$

The reality of an object consists in its being a part of the world. And the conjecture of the reality of an object must be *tested* through its immediate or remote perceptible effects but it does not *consist* in the latter. To say that it does condemns the greater part of reality to nonexistence and amounts to mistaking being for a being criterion – as operationism has been doing since the days of Peirce (cf. 5.406). If the world were to consist of a single indivisible physical thing, it would still exist or be real but there would be nobody to experience it and use or even reject Criterion 3.1.

Real existence statements should be taken in metaphysics just as seriously and as critically as they are taken in science. They must be regarded as pointing (rightly or wrongly) to actual or possible external referents, and also as being susceptible to correction in the light of fresh knowledge (i.e. new data or new conjectures or alternative theories). While methodical scepticism is indicated in this regard as in everything else, systematic scepticism won't do. That is, we may well doubt the existence of some particular thing but, if we want to check – i.e. to test the existence hypothesis – then we must, according to Criterion 3.1, assume without blinking, at least for the time being, the existence of something else. In practice we assume the existence of the entire world and we do not identify the latter with its explored part.

Assuming my own existence is necessary but insufficient for testing any existence hypothesis concerning some other thing. For one thing solipsism does not even account for my own birth, let alone for that of my enemy. Nor does solipsism explain the many disturbances – some agreeable, others disagreeable – impinging upon the subject. Not even phenomenalism suffices if we want to attain objectivity and efficiency. Consider the sentences 'There are cars' and 'Cars have at least three wheels'. To a realist, whether naive or critical, these sentences signify exactly and literally what they say. Not so to a phenomenalist: he will regard them as abbreviations of 'A subject has a car-like perception', and 'Whenever somebody has a car-like perception he can count at least three wheel-like perceptions' – or something of this sort. This subjectivism of his alienates him from science and can get him into trouble in everyday life. For, if the phenomenalist's car breaks down in the midst of a desert, when asking for a tow truck he might not say that *there is* a car stuck in the desert, let alone that there is nothing wrong with its wheels – that *they are* still on the car though unperceived. Unless he asks the Deity to do the perceiving for him in the meantime, the phenomenalist will have to contrive some complicated counterfactual sentences that no car mechanic would be interested in unravelling (e.g., 'If I were there now, I would perceive...'). For all practical purposes the phenomenalist will have to adopt a realist philosophy – unless he is willing to let his car go up in philosophical smoke. In short, the phenomenalist is bound to do double talk.

Finally, as for the concept of existence in a possible world and, even worse, that of existence in several different possible worlds, we shall leave them to the possible worlds metaphysician. The real world ontologist, busy as he is trying to figure out what the real world is like, has not time to waste with those figments of the idle imagination.

5. Concluding Remarks

We have built a concept of thing out of the notions of a bare individual (Ch. 1) and of a property (Ch. 2). We have defined a thing as a substantial individual possessing substantial properties. And we have identified being, or real existence, with being a thing. Moreover we have assumed that the world contains only things. In fact we have defined reality as the aggregation of all things and indeed nondenumerably many of them.

There are conceptual objects to be sure but not as constituents of the world, much less constituting a realm of their own. Constructs, whether busy or idle, are fictions. Hence although in our ontology every class of objects is split into a class of things and one of constructs, the latter are not assigned an independent existence. It is not just that only individuals are real: only individual things – whether simple or aggregate – exist really.

Things can be represented schematically by what we have called functional schemata. A functional schema is a certain set equipped with a state function having a certain number of components. This state function is subject to restrictions – the law statements supposed to represent objective patterns. These laws limit the conceivable states a thing can be in: they restrict the domain of the state function for the thing. The set of all nomologically possible states of a thing constitutes its state space – or rather the state space for the thing in the given representation or functional schema.

Things come in natural kinds or species, i.e. classes of things possessing ("obeying") the same laws. The family of natural kinds or species does not have a Boolean structure but one of its own: it is a semilattice. This is due to the lack of isomorphism between the set of substantial properties and the set of predicates. A natural kind constitutes a natural grouping because it rests on a bunch of laws, but it is not a real thing: it is a construct – only, not an idle one.

The world consists in and of things but not every member of the class of things is actual. There are possible things, such as my great-great-grand children. Hence there are possible properties and possible facts. Besides, every actual thing has some actual properties and others which it may – or may not – acquire. In other words we are tacitly acknowledging real possibility. And we need this concept before we tackle the problem of change, as nothing will happen unless it is possible to begin with – i.e. unless there are things with possible states. Let us then turn to the problem of possibility.

CHAPTER 4

POSSIBILITY

We have assumed that the world is constituted by things only (Ch. 3). But things change and, as Aristotle saw, if a change occurs then it was possible to begin with (*Metaphysics* IX, 3). The seed germinates because it has the potentiality to do so. *A* becomes *B* only if it is in the nature of *A* that *A can* turn into *B*: this possibility of *A*'s is one of its properties. (The converse is false: *A* may have the capacity to become *B* but this possibility may not be actualized due to adverse circumstances.)

Possibility, then, is inherent in reality because reality is changing. In other words, there are real possibles not just conceptual ones. An immutable world would be deprived of real possibility – and so would a changing world subject to inscrutable and unworldly Fate. But reality is neither unchanging nor chained to Fate, so there are real possibles. That is, we may partition reality into actuality and real possibility. (If on the other hand reality were identical with actuality, there could be no real possibility.) Therefore our metaphysics is possibilist not actualist.

Still, recognizing possibility, even objective possibility, does not entail regarding it as irreducible. Indeed an actualist may acknowledge possibility but will attempt to construe it in terms of actuality. He is bound to mistake real possibility for its test or for its measure. Thus he will tend to define the possible as that which either is or will be (Diodorus Chronos). Or he will assert that 'possible' means "sometimes" while 'necessary' means "always" (Russell, 1919). Or at least he will slight the importance of possibility by rendering it dependent upon actuality or somehow grounded in it – as is the case with Sellars (1963, Ch. 3) and Smart (1963, p. 23).

We shall see in this chapter that there is indeed a sense of 'possible' which is partially reducible to actuality, as when a child given the opportunity to eat a candy won't fail to actualize his natural bent or disposition to do so. This kind of possible property, that becomes actualized whenever the necessary conditions occur, will be called *causal disposition.* However there is another, deeper concept of possibility that is not so reducible and seems to have escaped the attention of most philosophers. An atom in an excited state can decay – with or without an external disturbance – into any of a number of lower energy levels none of them being predetermined. And the shuffling of genes that occurs in the fertilization of an egg cell is likewise random: any of a huge number of possible genomes can be realized. This kind of real possibility, to be called *chance propensity*, is not accounted for by actualism.

Real possibility will then be admitted as an ontological category not to be confused with either conceptual possibility or uncertainty. And we shall begin by distinguishing various kinds of possibility. First of all we shall study the concepts of conceptual possibility – alas, only to show that they are philosophically secondary because thoroughly reducible. Thereupon we shall analyze the two concepts of real or physical possibility. In the course of our study we shall have no occasion to employ modal logic, which will turn out to be too coarse a tool.

1. Conceptual Possibility

1.1. *Possibility Concepts*

There are several possibility species. They fall into either of two genera: *conceptual possibility* and *real possibility*. The former concerns formulas (in particular propositions) while the latter concerns factual items. Having such utterly different referents one may well wonder whether there is such a thing as a neutral concept of possibility subsuming those two, as modal logic presupposes. In any case, before analyzing the various concepts of possibility let us exhibit them globally: see Table 4.1.

1.2. *Four Concepts of Conceptual Possibility*

In the following 'K' will designate a body of knowledge – a set of data, conjectures, theories, etc. We propose first of all

DEFINITION 4.1 Let p be an arbitrary formula and A a subset of a body of knowledge K. Then

(i) p is *logically possible relative to* $A =_{df} A$ does not entail $\neg p$;

(ii) p is *mathematically possible relative to* $K =_{df}$ There is a model M in K such that p is satisfiable (formally true) in M;

(iii) p is *epistemically possible relative to* $K =_{df} p$ and K are mutually compatible (i.e. p contradicts no member of K);

TABLE 4.1
Concepts of possibility

Genus	Species	Referents	Meaning sketch
Conceptual or *de dicto*	Logical	Statements	Non contradiction
	Model theoretic	Partially interpreted formulas	Satisfiability in some model
	Epistemic	Factual statements	Consistency with what is known
	Methodological	Formulas	Provability or confirmability
Real or *de re*	φ-ontic	Physical facts	Lawfulness
	ψ-ontic	Psychical facts	Conceivability (compatibility with psychological laws and with background of subject)
	Deontic	Human actions	Not prohibited (by moral or positive law)
	Pragmatic	Human actions	Feasibility

(iv) p is *methodologically possible relative to* $K =_{df}$ There is no method m in K such that tests run with m disconfirm p relative to K.

(v) p is *conceptually possible relative to* $K =_{df}$ p is either logically or mathematically or epistemically or methodologically possible relative to K.

Example 1 Any noncontradictory proposition is logically possible, for all systems of logic admit the principle of noncontradiction. *Example 2* The formula "$x^2 = -1$" is mathematically possible since it is satisfied e.g. in the field of complex numbers (namely by $x := i$). *Example 3* For all we know Oparin's theory of the origin of life might be true. *Example 4* The predictions calculated with the help of specific scientific theories are methodologically possible since their negates are refutable in principle.

Remark 1 Our definition reduces conceptual possibility to conceptual actuality. Consequently it renders the concept of conceptual possibility redundant – in principle though not in practice. *Remark 2* The various concepts refined by Definition 4.1 are concepts of *conditional or relative possibility*. Absolute possibility can be construed as hidden conditional possibility, by just adding suitable existential quantifiers. For example, on the basis of Definition 4.1(i) we could make the following definition:

p is logically possible $=_{df}$ There is no set of premises entailing $\neg p$. But one wonders what the use of this concept could be, as every responsible judgement is made by taking some definite premises for granted. *Remark 3* Our concepts of conceptual possibility satisfy no system of modal logic. In particular logical possibility, as defined above, satisfies only one of the two axioms common to all modal logics, namely "If p then it is possible that p". (See Hughes and Cresswell, 1968.) Besides, the standard systems of modal logic deal with absolute not conditional (relative) possibility.

The only ingredient of modal logic we do need is Aristotle's definition of necessity ($\Box$) in terms of possibility ($\Diamond$), namely:

p is necessary $=_{df}$ It is not possible that $\neg p$, or $\Box p =_{df} \neg\Diamond\neg p$.
This definition, together with Definition 4.1, entails

DEFINITION 4.2 Let p be an arbitrary formula and A a subset of K. Then

(i) p is *logically necessary relative to* $A =_{df} A$ entails p;

(ii) p is *mathematically necessary relative* to $K =_{df} p$ is satisfied in every model contained in K;

(iii) p is *epistemically necessary relative to* $K =_{df} K$ implies p.

(iv) p is *methodologically necessary relative to* $K =_{df}$ For all methods m in K, tests on p run with m confirm p relative to K.

(v) p is *conceptually necessary* relative to $K =_{df} p$ is either logically or mathematically or epistemologically or methodologically necessary relative to K.

Again, none of these concepts of conceptual necessity coincides with those handled by modal logic. To begin with, they are concepts of relative (contextual) not absolute (context free) necessity. Besides, our notion of logical necessity, which reduces to the entailment relation, violates the modal axiom: $\Box(p \Rightarrow q) \Rightarrow (\Box p \Rightarrow \Box q)$. Instead it satisfies $\Box q \Rightarrow \Box(p \Rightarrow q)$, as can be seen with the help of the principle of conditional proof. (Anyway, what would be gained by constructing a deviant modal calculus?) In sum we have no use for modal logic because it is incapable of elucidating any useful notion of conceptual possibility and – to anticipate a result to be obtained in the next section – because it is equally impotent in tackling the notion of real or physical possibility.

Though for different reasons, we concur with the verdict that "Modal logic is of no philosophical significance whatsoever" (Bergmann, 1960).

1.3. *Conceptual Possibility*: *Relative*

We close by emphasizing the relativity of conceptual possibility as well as its difference from factual or real possibility. Formulas, just like actual facts, are neither possible nor impossible in themselves: they just are. What is possible or impossible with regard to a formula is some relational property of it such as exemplifying it, or proving it, or confirming it, or showing that it is compatible with a certain body of knowledge. In other words possibility is not inherent in formulas but is a relation between them and certain other conceptual items. That is, a statement of the form "It is possible that *p*" is not on the same level as *p* itself but is a metastatement. Moreover it belongs in pragmatics not in semantics, let alone in ontology.

Not so states of affairs: these are either actual or possible absolutely, in the sense that their possibility does not depend upon any body of knowledge. Propositions such as "That atom may disintegrate within the next minute", and "This organism is viable", are object statements not metastatements; they involve the notion of real possibility not that of conceptual possibility. Moreover these two concepts are unrelated except linguistically. To begin with they have nonoverlapping reference classes: while the referents of the concept of conceptual possibility are propositions, those of the concept of real possibility are facts. Hence their meanings are radically different. Therefore the attempt to have a single theory (namely modal logic) cover both concepts is bound to fail just as miserably as the attempt to describe wedding rings and algebraic rings with the same theory.

So much for conceptual possibility.

2. Real possibility

2.1. *Fact*

Unlike the concept of conceptual possibility, that of real (or physical or ontic) possibility refers to factual items. That is, the statement that *p* is really possible assigns the factual referent(s) of the proposition *p*, not *p* itself, the property of being really possible. So much so that a proposition of the form "p is really possible", where p describes some fact, may in turn be conceptually impossible relative to some body of knowledge.

Our referents are then factual items: things, properties of things, kinds of thing, states of things, and changes of state of things. Factual items are represented by definite descriptions and propositions such as the following:

Thing b	Thing b exists.
Property P	Thing b possesses property P.
Kind K	Thing b is of kind K.
State s	Thing b is in state s.
Event e	Thing b undergoes change e.

The concepts occurring in the above, with the exception of that of event, have been elucidated in the previous chapter; the notion of change will be clarified in the next chapter. We proceed to single out two kinds of factual items, namely being in a given state and going from one state to another:

DEFINITION 4.3 Let X be a thing. Then f is a *fact* involving X iff either

(i) f is a *state* of X, i.e. there is a state space $S_{\mathbb{L}}(X)$ for X such that $f = s \in S_{\mathbb{L}}(X)$, or

(ii) f is a *change of state* of (or *event* in) X, i.e. there is an $S_{\mathbb{L}}(X)$ such that $f = e = \langle s, s' \rangle \in S_{\mathbb{L}}(X) \times S_{\mathbb{L}}(X)$.

Note that, as here conceived, a state is always a state of some thing, and likewise an event is a change in the state of some thing. Note also that the states in question are only those which are lawful, as indicated by the subscript $\mathbb{L}$. (However, there is an ambiguity with regard to 'event', since although a pair $\langle s, s' \rangle$ may be lawful it may well happen that there is more than one way to go from s to s'. In other words, $\langle s, s' \rangle$ does not represent a single event but rather a whole collection of events. This ambiguity, to be removed in Ch. 5, won't do any harm at present.)

Let us now turn to a structural characterization of the notion of a fact. Call F the set of facts as defined by Definition 4.3. This set is of course the domain on which the predicates "is really possible" and "is actual" (or "is the case") are defined. (For example, we have "State s of thing b is really possible" and "Thing b is actually in state s".) These two concepts can be characterized in simple terms, namely thus. Assume, as is usual in the applications of probability theory, that every single factual item is construed as (represented by) a set – even if the latter is just a singleton or even void. That is, the set F of facts will be regarded as a family of subsets of some set – never mind which one. Stipulate that, if x is in F and is nonvoid, then x is an unspecified possible fact, whereas if $x = \varnothing$ then x is impossible. Assume further that possibles can conjoin, disjoin and invert. That is, if x and y are really possible factual items (i.e. nonempty members of F) so are $x \cap y$, $x \cup y$, and $x^{-1} = F - x$. The intersection of x and y represents the compossibility of x and y, the union represents the alternative possibility, and the inverse or complement x^{-1} of x represents whatever is possible when x is not. (Note that x^{-1} is not a fact but a set of facts.)

On the other hand actuals either occur or they don't, and some of them occur jointly but never disjointly. Indeed there is no such thing as the actual fact that fact *a or* fact *b* happens, even though the corresponding proposition may be true. Nor does x^{-1} happen. There is no such thing as the unique complement or inverse of an actual fact. Complementation and disjunction are marks of possibility (as well as of constructs) not of actuality. Further, if x and y are really possible, we call $A(x)$ the actualization or occurrence of x, $A(x^{-1})$ the nonoccurrence of x, $A(x \cap y)$ the occurrence of both x and y, and $A(x \cup y)$ that of at least one of the facts. (See Table 4.2.) And we lay down

POSTULATE 4.1 Let F be a collection of sets, $\cap$, $\cup$, and $^{-1}$ the boolean operations on F, and A a function from F into a set of propositions. Then the structure $\mathscr{F} = \langle F, \cap, \cup, ^{-1}, A\rangle$ is a *space of facts* iff

(i) $\langle F, \cap, \cup, ^{-1}\rangle$ is an algebra of sets;

(ii) F represents the set of facts (actual or possible) involving some thing or things X;

(iii) A is a function from F into the set of propositions concerning the thing or things X, such that

TABLE 4.2

Real possibility and actuality. Proviso: $x, y \neq \varnothing$

Formula	Interpretation	Formula	Interpretation
$x \in F$	x is a possible fact	$A(x)$	x occurs
$x \cap y \in F$	x and y are compossible	$A(x \cap y)$	x and y occur
$x \cup y \in F$	x or y is a possible fact	$A(x \cup y)$	x occurs or y does
$x^{-1} \subset F$	that which is possible when x is not, is a possible fact	$A(x^{-1})$	x does not occur

(a) for any $x \in F$, $A(x^{-1}) = \neg A(x)$;

(b) for any $x, y \in F$, $A(x \cap y) = A(x)$ & $A(y)$ – i.e. A is a morphism of opposition and multiplication;

(iv) for any $x, y \in F$ other than the empty set, $A(x)$ represents the occurrence (actualization, realization) of (the possibility) x, $\neg A(x)$ the nonoccurrence of x, and $A(x)$ & $A(y)$ the occurrence of both x and y.

It is easily shown that the actualization function A has the following further properties:

THEOREM 4.1 Let A be the propositional function on F defined by Postulate 4.1. Then for all $x, y \in F$

(i) $A(x \cup y) = A(x) \vee A(y)$;

(ii) $A(x^{-1} \cup y) = A(x) \Rightarrow A(y)$;

(iii) $A(x \cap x^{-1}) = A(x) \mathbin{\&} \neg A(x)$.

The last result is of course the reason for interpreting $\varnothing$ as impossibility.

Our Postulate 4.1 is at variance with an alternative set of axioms of occurrence (Suppes, 1970, p. 38), supposed to characterize the notion of occurrence or actualization. These axioms do not discharge this function if only because the first of them states that, if x and y are events, and x occurs, then their union $x \cup y$ occurs as well. In our ontology there are no disjunctive facts: there are only disjunctive possibilities and disjunctive propositions about facts whether possible or actual.

Postulate 4.1 constitutes only a partial characterization of the possibility/actuality contrast. It tells us just that (*a*) the set of occurrences is a proper subset of the set of possibilities, and (*b*) the algebraic structure of the latter is richer than that of the former. The actualization process is somehow representable as a collapse of the richer structure into the poorer one: in this process union (or disjunction) and complementation (or negation) are forgotten. However, this characterization in terms of a forgetful functor is insufficient because it fails to specify (*a*) which subset of the conceptually possible facts (for this is what F is) constitutes the *really* possible facts and (*b*) which subset of the collection of really possible facts gets actualized. These defects cannot be repaired by proposing a more complex theory in the same vein: no a priori theory, independent of considerations concerning laws and circumstances, can single out the set of really possible facts, let alone that of actuals. Let us then get in touch with reality.

2.2. *Chrysippian Possibility*

The profound Stoic philosopher Chrysippus defined the possible as "that which is not prevented by anything from happening even if it does not happen". Contrast this view to that of Diodorus Chronos, according to whom "the possible is that which either is or will be true". (Cf. Kneale and Kneale, 1962; Rescher and Urquhart, 1971.) According to this other view the concepts of actuality and possibility are coextensive: that which never happens is impossible. (See also Hartmann, 1938.) Consequently the notions of unrealized potentiality and of missed opportunity find no room in this conception. Yet, as Aristotle and Chrysippus saw, not all possibles are realized: "thus it is possible that this gem will break even if it never does" (Cicero, *De fato* IX).

Chrysippus' definition of real possibility hinges on the notion of lack of constraint or absence of inhibition. The latter notion can be construed as follows:

DEFINITION 4.4 If x and y are possible facts (i.e. $x, y \in F$) then x ***inhibits*** (or ***prevents***) the occurrence of $y =_{df} A(x) \Rightarrow \neg A(y)$.

Chrysippus' idea is then perhaps rendered by

DEFINITION 4.5 A fact is ***really possible*** iff there is no actual fact the occurrence of which prevents it: If x is a fact then

$$\Diamond_r x =_{df} \neg(\exists y)(y \text{ is a fact } \& \ y \text{ occurs } \& \ y \text{ prevents } x)$$
$$\Leftrightarrow \neg(\exists y)(y \in F \ \& \ A(y) \ \& \ (A(y) \Rightarrow \neg A(x))).$$

In other words, anything is (really) possible that is unhampered or free to happen. Freedom, whether physical or moral, is thus equated with real possibility. (Physics and ethics acquire therefore a common metaphysical basis.) The concept of real impossibility is obtained by negating the two sides of the previous definition: a fact is ***really impossible*** iff it is inhibited by some other fact. We infer then that if anything is impossible then it is not the case. This corollary is just the contraposition of "Whatever is the case is possible".

We shall not elaborate on the Chrysippian notion of possibility because we shall adopt an alternative conception of possibility, namely the nomological one to be expounded presently.

2.3. *Real Possibility as Lawfulness*

Our point of departure will be Bolzano's conception of possibility as lawfulness: "physically possible is that which does not contradict any of the so called laws of nature" (Bolzano, 1821, art. 'Möglich', p. 65). However, we cannot adopt this definition literally because we regard laws as objective patterns not as propositions and consequently cannot make use of the concept of consistency occurring in Bolzano's statement.

To begin with recall the Definition 4.3 of a fact as either a lawful state or a lawful change of state of a thing. We shall assume that these are the only really possible facts. That is, we identify real possibility with nomic possibility. More explicitly, we assume

POSTULATE 4.2 A fact f is *really possible* iff f is a lawful fact, i.e. iff either $f \in S_{\mathbb{L}}(X)$ or $f \in S_{\mathbb{L}}(X) \times S_{\mathbb{L}}(X)$ for some thing X and some state space $S_{\mathbb{L}}(X)$. In symbols,

$$\Diamond_r f \text{ if and only if } f \text{ is a lawful fact.}$$

Note that this is not a definition but a hypothesis and moreover one used surreptitiously when framing Definition 4.3. In fact Postulate 4.2 can be ignored or even negated. An empiricist ontology – one recognizing no objective laws, as was the case with Hume's and James' – finds no use for that assumption. And a religious metaphysics, unless it be of the Cartesian or Leibnizian kind where the Deity is always strictly law-abiding, will negate our axiom.

The following consequences follow immediately from our postulate:

(i) *Lawless individual facts are impossible.*

(ii) If, with Hume, we define the miraculous as the lawless, then we infer that *miracles are impossible.* If this conclusion is rejected then the concept of real possibility must be redefined – or just dropped, perhaps together with that of an objective law or pattern. Either alternative would be consistent with an empiricist ontology.

(iii) Randomness too has been defined as lawlessness, but this is not satisfying. First, whereas the predicate "lawful" applies to individual facts, "random" applies only to sets of facts – such as e.g. a sequence of hits and misses of shots aimed at an object with unknown location. Second, randomness is a kind of lawfulness not its opposite. What is the opposite of lawfulness is chaos or the absence of all laws, even stochastic (probabilistic) laws. But chaos too is a property of sets of facts not of individual facts. For example, by collecting haphazardly a number of facts (possible or actual) concerning unrelated things we can form a lawless collection of individually lawful facts. (Think of a set made up of an arbitrary collection of astrophysical, genetic, and political events.) In sum randomness – a basic feature of reality if we are to believe contemporary science – is, unlike chaos, compatible with lawfulness or nomic necessity. More on these matters in Sec. 6.4.

(iv) We conclude the falsity of the identification of lawfulness and necessity, popular since the 17th century, made obsolete by the birth of statistical theories from the 1870's on, assailed by Peirce, and recently revived by Montague (1960, repr. in 1974). That identity is in fact a cornerstone of Montague's possible-world semantics and metaphysics – of which more in Sec. 6. In short lawfulness is not identical with necessity but with (real) possibility. More on necessity anon.

2.4. *Factual Necessity*

Conceptual necessity, it will be recalled, can be introduced in terms of conceptual possibility by way of Aristotle's definition, namely: $\Box p = \neg \Diamond \neg p$. We cannot use this definition to obtain the notion of nomic necessity out of that of nomic possibility as coextensive (though not cointensive) with that of lawfulness. For one thing negation applies to propositions, not to facts: the expression '$\neg x$' makes no sense when x denotes a fact. We might of course reinterpret '$\neg x$' as "It is not the case that x", i.e. as $\neg A(x)$. But even so Aristotle's definition would not help us, because real or factual necessity has a component absent from lawfulness, namely *circumstance*.

Indeed for anything to actually happen and thus be (really) necessary it must not just "obey" certain laws but also "count on favorable circumstances". Not even a deterministic (nonstochastic) law statement describes only what is actually the case: *every law statement describes possibles* – without of course the help of modal operators. (Think of the bundle of trajectories in the state space representing the set of solutions of an equation of evolution.) A fortiori, stochastic laws describe even weaker possibles, namely chance facts, i.e. facts occurring only with a fixed frequency. In sum we may keep Aristotle's definition of possibility for the conceptual realm – as we did in Sec. 1.2 – but it is inapplicable in science and in scientific ontology. Here we need rather

DEFINITION 4.6 Let $x \in F$ be a really possible (i.e. lawful) fact. Then
(i) x is *necessary* iff there is another fact $y \in F$, called the *circumstance attending* x, such that $A(y) \Rightarrow A(x)$;
(ii) x is *contingent* iff x is not necessary.

For this reason actual states of affairs, or situations, are best accounted for (described, explained or predicted) with the help of both law statements and propositions representing the circumstances (idiosyncrasies, initial conditions, boundary conditions, etc.) jointly necessary for the situations of interest to actually occur either always or with a fixed frequency. In other words, the preceding considerations constitute the ontological ground for the familiar schema of nomological explanation:

$$\text{Laws, Circumstances} \vdash \text{Facts.}$$

At first sight the necessity or contingency of a fact or situation depends exclusively on the kind of law involved: deterministic laws would cover necessary facts, nondeterministic laws contingent facts. For better or worse this simple schema is false. In fact deterministic laws can cover contingent facts and stochastic laws can cover necessary facts. A familiar example of the former kind is the impact of a classical particle on a wedge: see Figure 4.1. In this case there are two equally probable trajectories, i.e. the problem has two different solutions on an equal footing. And a familiar example of the second type is the repeated

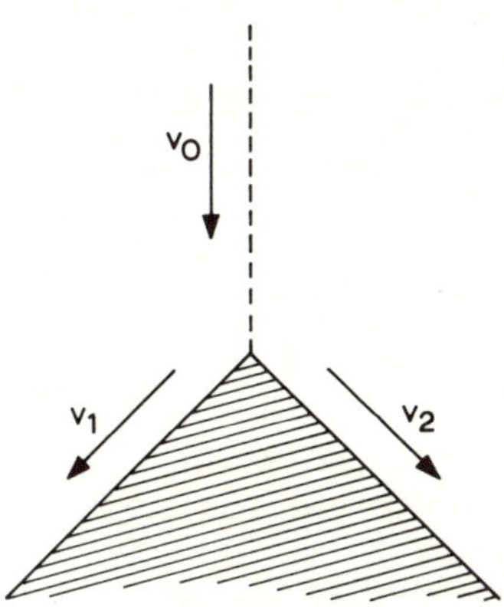

Fig. 4.1. A particle (or a fluid jet) collides against a rigid wedge and is deflected either left or right with the same speed. For a classical discussion see Truesdell (1974).

tossing of a penny: in the long run certain regular sequences, such as five heads in a row, are bound to occur and in fact they occur with a fixed probability.

In sum, the explanatory schema $\ulcorner L, C \vdash Fact \urcorner$ lumps together four accounts:

Deterministic law, C_1
$\vdash$ Necessary fact (occurs whenever C_1 or C_2)

Stochastic law, C_2

Stochastic law, C_3
Contingent fact (occurs with fixed probability whenever C_3 or C_4)

Deterministic law, C_4

To conclude this subsection. Real possibility is identical with lawfulness. But real necessity is not definable in terms of possibility and negation. Never mind, for the concept of necessity plays hardly any role in factual science except as a synonym for actuality: whatever happens to be the case must be, and conversely. Little is gained by replacing "It is snowing" with "Necessarily it is snowing", or "The snowfall probability is $\frac{3}{4}$" with "Necessarily the snowfall probability is $\frac{3}{4}$", let alone with "The probability that it snows is necessarily $\frac{3}{4}$". *The prefix 'necessarily' is unnecessary.* An actual particular fact is a fact is a fact. Calling it 'necessary' is just a way of saying that, far from being merely possible, it is the case.

In sum, we have the following partition of the factual realm:

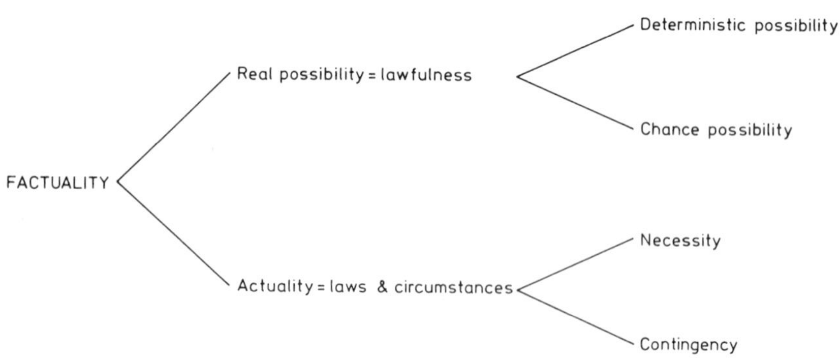

2.5. *Possibility Criteria*

If scientific research is conceived of as being ultimately the search for objective laws (Bunge, 1967a, Vol. I, Ch. 6), and if lawfulness is equated with nomic possibility (Postulate 4.2), then science must be regarded as the study of the really possible. In fact every scientific activity of the theoretical kind deals not only with actual facts but also with real possibilities. Thus theoretical mechanics – unlike experimental mechanics and applied mechanics – studies all possible motions of all possible bodies. Theoretical chemistry – unlike experimental chemistry and chemical engineering – studies in principle all possible chemical compounds. And theoretical genetics studies all possible gene recombinations and mutations.

In other words all theoretical science utilizes the concept of a possible thing and even that of a possible state of a possible thing, much to the dismay of actualist philosophers. But of course the notion of a possible individual of this kind is not characterized with the help of either modal logic or model theory – let alone with that of any system of possible worlds metaphysics. In theoretical factual science a possible thing, such as a possible compound or a possible organism or a possible community, is an arbitrary member of the class of hypothetical referents of the theory of interest – i.e. an individual possessing ("obeying", "satisfying") the law statements of the theory. On the other hand the notion of a possible world (nowadays often identified with that of a model of an abstract theory) is alien to factual science: we have access to no substantial world other than the real world. True, possibility is inherent in the world, but there is no special carrier of possibility – no possible substantial world in addition to the actual world. Every concrete thing possesses certain possibilities to the exclusion of others.

What holds for theoretical possibility holds also, *mutatis mutandis*, for theoretical impossibility. Every theory rules out a number of possibilities just as it countenances others. Of course any theory can go wrong. But as the number and variety of theories agreeing on the impossibility of something increases, the likelihood of a "prohibition" increases. Even so one should keep an open mind. For example relativistic mechanics rules out the possibility that there be superluminal particles (tachyons). But the theory has nothing to say about noncorpuscular entities travelling faster than light. Moreover a good theory about such entities would help in searching for them.

Any theory hypothesizing the existence of a thing or a fact not envisaged by antecedent theories should respect the latter in their own "domains of validity" (extensions) and should enjoy the support of independent evidence or else stimulate the search for it. And any theory hypothesizing the existence of things "forbidden" by well corroborated theories should say whatever those theories tell us about the things those theories allow. Needless to say, even logic must be taken into account when listing impossibles. Metaphysics, on the other hand, is hardly in a position to admit or rule out any fact. What metaphysics can do is to clarify some of the concepts involved in scientific judgments of possibility or impossibility.

When shifting our attention from reality to science, talk of facts becomes talk of the knowledge of facts and, in particular, a definition of possibility may become a possibility criterion. If we were to adopt Chrysippus' concept of real possibility (Sec. 2.2) we would have to adopt the following criterion: *If x is a fact, then x is known to be really possible iff nothing is known to prevent x.* But since we have proposed a modification of Bolzano's notion of possibility (Sec. 2.3) we cannot accept the Chrysippian possibility criterion. We shall instead suggest a criterion inspired in the following example.

A theoretical chemist will declare an unknown compound to be really (chemically) possible iff (*a*) its components exist, (*b*) they can assume a stable configuration, and (*c*) a reaction mechanism resulting in the hypothesized molecule is theoretically possible, i.e. consistent with what we know. In short, the chemist's possibility statement is the conclusion of a (rather complex) argument involving items both theoretical and empirical. We generalize this procedure by proposing

CRITERION 4.1 Let T be a theory, and E a body of empirical evidence couched in the language of T and relevant to T, and suppose that both T and E refer to a fact x described by a proposition $p[x]$. Then x is *really possible according to* T *and* E iff $T \cup E$ fails to entail $\neg p[x]$.

This is a theoretical possibility criterion and it refers us back to what is already known (or presumed to be known), not to new observations or future theories. If a theoretically possible fact occurs it may count as a confirmer of the theory. If it fails to occur the theory remains unaffected and we have no case against the possibility of the fact. And if a theoretically impossible fact does occur then we shall have to revise either the theory or the evidence employed in inferring the impossibility.

Consider now the possibility criterion suggested by the Diodorean concept of possibility, namely: *Fact x is known to be possible iff x is observed to happen at some time.* This criterion is quite useless, as it works only *ex post facto*: it does not allow me to regard my own demise as possible until it is too late for doing any thinking of this kind. And once the fact is observed to happen it may as well be pronounced necessary. But if it is not observed to happen then the reasonable attitude is to wait and see rather than to declare it impossible. These are not defects that can be remedied while retaining the spirit of Diodorus' conception of possibility: they are inherent in any attempt to frame an *empirical* possibility criterion. Indeed such a criterion must refer to actuals and is therefore impotent in distinguishing possibility from actuality. In other words, the observation that something is the case only shows that it was indeed possible to begin with but does not suggest whether it was just possible or necessary. In conclusion, empirical criteria of possibility are impossible. We must remain content with a theoretical criterion like the above. Which is fitting, since the concept of possibility is theoretical not empirical.

3. Disposition

3.1. *Intuitive Idea*

Thus far we have been concerned with the general concept of real possibility. We shall study henceforth two special concepts of real possibility. One is *disposition*, or *causal propensity*, as exemplified by fragility and inherited susceptibility to TB. The other concept is that of *chance propensity*, as exemplified by a photon that may go through either slit of a two slit screen. Let us start with the former, which is the more familiar of the two – therefore probably not the more basic.

A force may cause a change in relative position, a lump of sugar may dissolve, a muscle cell may divide, a literate person may read, a viable society can survive. These potentialities will be realized provided the proper environment or means is supplied: otherwise they will not. (To repeat: Such potentialities are realized *whenever* suitable conditions are present.)

Moreover, as the Philosopher saw, potency precedes act. (And conversely, as is equally obvious. Thus, being born with a normal brain and being adequately fed are necessary conditions for a person to be capable of learning and doing certain things, even if he never actually learns or does them.) The disposition to do x is then prior to doing x. Thus a piece of matter will actually refract light provided it was refractive to begin with; and if it was refractive then it was so whether or not it was exposed to light. (Do not mistake a property for the tests for it.) Recall Newton's *Opticks* (1782 ed., Vol. IV, p. 6): "Refrangibility of the Rays of Light, is their disposition to be refracted or turned out of their way, in passing out of one transparent Body or Medium, into another". A light ray has all the time the refrangibility property, even while it propagates in a vacuum. Likewise a crystal retains its refrangibility while locked up in a chest.

A peculiarity of dispositions (or causal propensities) is that they come in pairs. Thus (all) light rays are refrangible iff (some) bodies are refrangible. And every lock can be opened by some key(s). Let us keep this complementarity in mind, for it will occur in the very definition of a dispositional property.

The peculiarity of a *causal* propensity, in contrast to a chance propensity, is that the former never fails to be actualized when conjoined with suitable circumstances. Solubility becomes actual dissolution, divisibility division, viability continued existence, and so on. In clumsy but few words: *Disposition & Circumstance = Actuality*. Equivalently: Disposition equals actuality minus certain circumstances. More precisely: A disposition is a condition that, though necessary, is not sufficient. Accordingly, if x is a fact, then:

$$\text{If } (\exists y)(y \in F \;\&\; y \neq x \;\&\; (A(y) \Rightarrow A(x)) \quad \text{then} \quad \Diamond_r x.$$

(The converse is false, as real possibility is not exhausted by causal propensity.) Real possibility of this kind, i.e. disposition, is then an insufficient condition: actuality ensues as soon as this shortcoming is remedied, i.e. when the missing conditions appear. (For a similar independent construal see Raggio, 1969.)

The circumstances favorable or unfavorable to the actualization of a potentiality of a thing x involve some thing y other than x and forming part of the environment of x. This other thing, to be called the *complement* of the first, must have a disposition matching that of the thing of interest, for the first disposition to actualize. Thus a certain key may open provided it is joined to a proper lock with the unlocking disposition. What exhibits an actual or manifest property is the whole formed

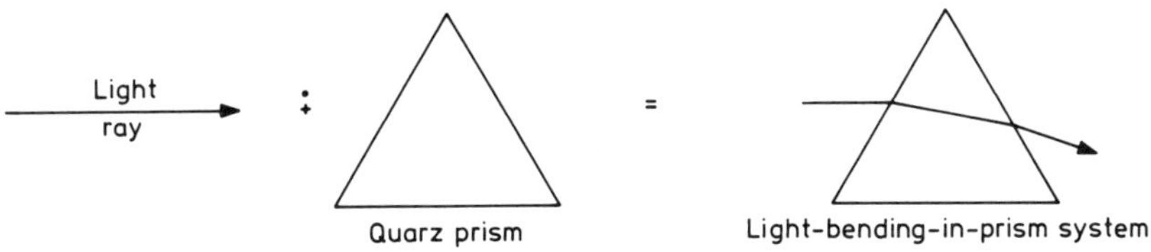

Fig. 4.2. Two entities, each with a certain disposition, aggregate into a third entity with a certain manifest property.

by the thing of interest and its complement. (See Figure 4.2.) The general schema is as follows:

> *Thing x with disposition P joins thing y with disposition Q to form thing* $z = x \dotplus y$ *with manifest property R.*

A few more examples should do no harm. *Example 1* If a piece of magnetizable material is placed in a magnetic field, it becomes magnetic. *Example 2* A fertilized bird's egg lying long enough in a nest at a suitable temperature becomes a chick. *Example 3* The inherited susceptibilities to TB, schizophrenia and other illnesses are parallel: the environment brings about, or else inhibits, such an inborn disposition.

We close this subsection by noting that the degree of a predicate is no indication of whether the property it represents is manifest or dispositional and, if the latter, whether it is causal or stochastic. Thus human abilities are often intrinsic properties (though of course displayed only under certain circumstances). On the other hand the velocity of a macrosystem is a mutual property (of thing and reference system) and a manifest one. Likewise with chance propensities. Thus the localizability of a quantum mechanical system is intrinsic in the case of a single "particle" and mutual for a pair of correlated "particles".

3.2. *Elucidation*

The preceding informal considerations are summed up and exactified in the following convention, which employs only concepts that have been clarified in the previous chapters:

DEFINITION 4.7 Let $z = x \dotplus y$ be a thing composed of different things x and y. Further, let P be a property of x and Q a property of y, with Q possibly equal to P. Then P is said to be a *disposition* (or *causal propensity*) of x, and Q the *complement* of P with respect to y, iff there is a third property R such that

(i) $z = x \dot{+} y$ possesses R;

(ii) $z = x \dot{+} y$ possesses neither P nor Q;

(iii) neither x nor y possesses R.

Note that what is called a *disposition* and what its complement is a matter of viewpoint or convention. Thus one says that sugar is soluble (in water), but this is because water has the disposition to dissolve sugar.

Example An aggregate of reactants won't start reacting unless the required pressure and temperature conditions become available – which conditions are properties of the immediate environment of the aggregate. When they do, the disposition to react in a certain way is actualized, and eventually new things (the reaction products) emerge that do not have all the properties of the reactants but are characterized instead by new properties determined by the former.

We are now in a position to elucidate a notion that has become incorporated into common knowledge:

DEFINITION 4.8 The totality of dispositions (causal propensities) of a thing is called the *causal potentiality* of the thing. I.e., if x is a thing then

$$\Pi_c(x) = \{P \in p(x) | P \text{ is a disposition of } x\}.$$

DEFINITION 4.9 One thing has a *greater causal potentiality than* another iff the potentiality of the former includes that of the latter. I.e., if x and y are things, then

$$\Pi_c(x) > \Pi_c(y) =_{df} \Pi_c(x) \supseteq \Pi_c(y).$$

Example 1 The potentiality of an individual organism consists of its genetic make-up or program. A poor genetic endowment yields few possibilities of a rich experience, just as a poor environment cuts out many possibilities of using the genetic make-up. *Example 2* Shortly before the birth of molecular biology it was considered preposterous to regard the human zygote as a potential man, or the acorn as a potential oak tree. Both assertions would have been laughed away as remnants of preformationism. But contemporary biology is preformationist insofar as it holds the whole mature organism to be programmed in the zygote. *Example 3* The psychological faculties or abilities, rejected by behaviorism, are back. Thus N. Chomsky, G. A. Miller and other psycholinguists stress the difference between ability and performance – an instance of the potentiality-actuality difference. Thus verbal behavior does not coincide with linguistic ability although it is the only overt hence observable manifestation of it. The ability is in this case the capacity to utter and understand sentences new to the speaker. But this potentiality can be either actualized or frustrated by the environment.

3.3. *Potency and Act*

To Aristotle potency was prior to act: actuality was the unfolding of potentiality. The latter was in turn left unexplained. The distinction between potency or disposition, on the one hand, and act or manifest property, on the other, has been kept by modern science – a fact suppressed by the positivist philosophy of science. Thus Newton: "Colours in the Object are nothing but a Disposition to reflect this or that sort of Rays more copiously than the rest" (Newton, 1782, Vol. IV). Other physical properties can be described with paraphrases of the latter statement: just think of conductivity, magnetic permeability, refrangibility, or viscosity. They are all powers or dispositions of the causal kind. Such examples have suggested the conjecture that "all physical (and psychological) properties are dispositional" (Popper, 1957, p. 70). But this would be impossible if only because the very concept of potency makes sense only relative to that of act – as shown e.g. by Definition 4.7.

Actualists have tried to ignore the long list of dispositions exhibited by modern science. They often argue against objective dispositions on the strength of an empiricist epistemology. They claim, and rightly so, that only actual things and manifest properties are publicly observable. (But they do not seem to realize that we cannot possibly do without dispositionals such as "observable".) They will try hard to define, and thus eliminate, dispositions in terms of observable properties. A favorite gambit of theirs is Carnap's method of introducing dispositionals by way of bilateral reduction sentences (Carnap, 1936–37), namely thus:

> Thing x subjected to test condition C is assigned disposition D just in case x exhibits behavior B: $Cx \Rightarrow (Dx \Leftrightarrow Bx)$.

But of course this is a *criterion* not a *definition*. Moreover in scientific theories, while some dispositionals (like solubility) are defined others (like electric conductivity) are not: they are taken as undefined. Likewise with psychological properties: even though actual aggression is an (ambiguous) indicator of rage, we should be ill advised to define anger as a disposition to attack, as has been suggested (Ryle, 1949) – if only because this fails to account for attacks in cold blood. Nor can we define thought as a disposition to talk, itching as a disposition to scratch, and so forth: this is just mistaking evidence for reference (Feigl, 1967). Thinking, feeling an itch and the like are just as actual as talking and scratching: that they are not public or intersubjective is another matter – one for methodology not ontology.

Dispositions are just as important as manifest properties even though the latter can explain the former, as shown by Definition 4.7. The possibility of dissolving is not the same as actual dissolution: while the latter is a property of a complex system (solute *cum* solvent), the former is a property of one of its components. For example, the solubility of salt in water consists in a certain feature of the crystal structure of dry salt which, when joined with water, gives rise to actual dissolution. We may then assert that dispositions of the causal kind are rooted in manifest properties. But this is not the same as eliminating the former. Besides, as will be seen in Sec. 5, noncausal dispositions are not so explainable: on the contrary, they help explain manifest properties. In conclusion, actualism cannot hold the fort.

3.4. *Unrealized Possibilities and Counterfactuals*

It will be noted that we have not employed counterfactuals in our elucidation of disposition. However, it is widely believed that dispositional expressions must be defined in terms of subjunctive conditionals. Thus it is commonly held that

> x is fragile $=_{df}$ If x *were* thrown against a rigid body, then x *would* break.

and

> x can do $y =_{df}$ If x *were* placed in suitable circumstances, x *would* do y.

This view seems to have originated in the operationist confusion between meaning and test: since actual breaking is evidence for fragility, and performance for ability, it is inferred that that is what they are. Science has paid no attention to this doctrine: if definable at all, i.e. if not primitive, dispositionals are defined with the help of declarative sentences only. Witness any scientific treatise or research paper.

There are good reasons for avoiding the subjunctive mood. Firstly to avoid ambiguity. Indeed one and the same subjunctive sentence may be construed in at least two different ways (Bunge, 1968d). In fact 'If A were the case then B would obtain' can be interpreted as representing either the statement

If A then B. But not A. (No inference at all.)

or the inference

If A then B. But not B. (Tacit inference: not-A.)

Such an ambiguity renders subjunctive conditionals unsuitable for scientific communication, except of course at the level of heuristic language. A second reason for avoiding subjunctive conditionals is that they are not propositions, hence they do not fall under the rules of logic. Therefore the "definitions" of fragility and ability at the beginning of this subsection are phoney.

It should not be concluded that subjunctives are junk: indeed they have considerable suggestive or heuristic power. But (a) they should be used with caution and (b) they should be avoided altogether in reconstructing scientific theories as well as in reporting experimental results. In any case, far from being capable of shedding any light, subjunctive conditionals are in need of clarification.

So much for dispositions of the causal kind. Let us now turn to a kind of disposition that is not reducible the way causal disposition is, namely chance propensity. But before doing so it will be convenient to play the probability interlude.

4. Probability

4.1. *Abstract Concept*

Let us recall the gist of the theory underlying any technical (as opposed to ordinary language) statements of the form "The probability of a equals b", or "$Pr(a)=b$" for short. Pr is a real valued function on a certain set F characterized only by its structure. To begin with F is a family of subsets of a certain abstract set S, i.e. $F \subseteq 2^S$. Further, F is closed under finite unions and intersections, and also under complementation. Therefore F is closed under finite unions and symmetric differences. So, F has a ring structure. Moreover, F is a *σ algebra* because it is a ring and $S \in F$. In other words, F is a σ algebra in the sense that its members obey the algebra of sets extended to countably finite unions. We shall call F the *probability space* or *support* of the *Pr* measure. (F is often called the *space of events* or the *sample space*, names as suggestive as they are out of place in pure mathematics.) We are now ready for

DEFINITION 4.10 Let F be a σ algebra on a nonempty set S, and *Pr* a real valued function on F. Then *Pr* is a *probability measure* on F iff

(i) *Pr* is a non-negative real valued function on F. [I.e., for every member A of the collection F of subsets of the basic set S, $Pr(A) \geqslant 0$.];

(ii) *Pr* is completely additive in F. [I.e., for any countably infinite collection of pairwise disjoint sets in F, the probability of their union equals the sum of their individual probabilities. In particular, if A, $B \in F$, and $A \cap B = \varnothing$, then $Pr(A \cup B) = Pr(A) + Pr(B)$.];

(iii) *Pr* is normed. [I.e., $Pr(S) = 1$.]

Note the following points. Firstly, the theory based on these sole assumptions is *semiabstract* insofar as it does not specify the nature of the elements of the probability space F. On the other hand the range of *Pr* is fully interpreted, hence the *semi*. This is why probability theory finds applications everywhere, from physics to metaphysics. Secondly, an obvious way in which one can satisfy the condition that F be a σ algebra, is by taking $F = 2^S$. Indeed, any power set is a σ algebra. Thirdly, the first two axioms define a *measure*, and the third makes it a probability measure. This shows that the foundations of the theory of probability constitute but a chapter of measure theory – not one of philosophy. However, such foundations, like any others, are insufficient for any applications. Indeed, as long as the probability space F is not specified, i.e. as long as no model is constructed, probability has nothing to do with possibility, propensity, or randomness. Let us dwell for a while on this important methodological point.

An *application* of any abstract or semiabstract theory to some domain of facts consists in joining to the theory two different items: (*a*) a model or sketch of the object or domain of facts to which the theory is to be applied, and (*b*) an interpretation of the basic concepts of the theory in terms of the object to which it is applied. In particular, an application of probability theory consists in joining Definition 4.10 (or some of its consequences) with (*a*) a stochastic model – e.g. a coin flipping model or an urn model-, and (*b*) a set of interpretation (or semantic) assumptions sketching the specific meanings to be attached to a point x in the probability space F, as well as to its measure $Pr(x)$. As long as these additional assumptions are not made, the theory is indistinguishable from measure theory: only those specifics turn the semiabstract theory into an application of probability theory or part of it. (For an elucidation of the notion of degree of interpretation or its dual, degree of abstraction, see Vol. 2, Ch. 7, Sec. 3.4.)

Now, before we proceed to build a model of a class of facts we must identify the latter, i.e. choose a definite ("concrete") probability space F. The latter may but need not represent a collection of states or of changes of state of a thing. (Whether or not a point x in F represents a state or an event, it is usually called either of the latter by most probability theorists.) Next we must make sure that F is a σ algebra, or else we must manufacture one out of the given basic set S. If S is the state space of some thing, and if we wish to speak of the probability of a state, we must make a détour: since the elements of S are points, not sets, we construct the power set 2^S. In other words, we construe the idiom 'the probability of state s' as "the probability of the singleton $\{s\}$". Once this algebraic matter has been attended to we can proceed to interpret the primitives F and Pr, for instance in terms of states and their propensities. Let us turn to this latter problem.

4.2. *Probability State Space*

In some cases the states of a thing can be assigned definite probabilities and some pairs of states of the same can be allotted definite conditional probabilities. As we saw in the last subsection, in order to assign probabilities to states it is expedient to construe the latter as sets. The cheapest way of doing so is to take the set of all the subsets of the set $S(X)$ of states of thing X, i.e. the power set $2^{S(X)}$, as the probability space. In this fashion we can manufacture

DEFINITION 4.11 Let X be a thing and $S(X)$ a state space for X. Then the structure $\langle 2^{S(X)}, Pr\rangle$ is a *probability state space* for thing X iff

(i) $S(X)$ is denumerable and

(ii) $Pr: 2^{S(X)} \rightarrow [0, 1]$ is a probability measure defined for every set of states and satisfying in particular the condition: $Pr(S(X)) = 1$.

Caution: the states in a probability state space are not ordered naturally according to their increasing probability, for it may well happen that different (e.g. succeeding) states have the same probability. Thus according to statistical mechanics the evolution of a closed system follows on the whole (i.e. with but a countable number of exceptions) the line of increasing vulgarity: see Figure 4.3. But the fact that there can be occasional decreases in probability values prevents us from identifying prior states with improbable states. In other words, the direction of a process is not given unambiguously by the direction of increasing probabilities.

If all the states of a thing are equally probable, the thing may be said to be *state homogeneous*. And, since no states are preferred, once the thing has attained a given state it remains there, i.e. it does not change any further. But this is of course a fiction: there are no unchanging things, hence no state homogeneous things. Real things are *state heterogeneous*, in the sense that some of their states are definitely more probable than others. In other words, a state heterogeneous thing is a thing with states that deviate from the homogeneity or equilibrium value p_0. (For a thing with a finite number N of equiprobable states, $p_0 = 1/N$.) This suggests adopting the number $h(s) = Pr(s) - p_0$ or its square as a measure of the deviance or distinction of state s. For the entire collection $S(X)$ of states of a thing X we adopt

DEFINITION 4.12 The *state heterogeneity* of a thing with state space S equals

$$H(S) = \sum_{s \in S} h^2(s), \quad \text{with} \quad h(s) = Pr(s) - p_0,$$

where p_0 corresponds to maximal (and ideal) state homogeneity.

Figure 4.4 shows three typical heterogeneity values. It will be noted that any given state heterogeneity value can be realized in diverse manners.

The following consequences follow easily because of the normalization of the total probability to unity:

(i) The state heterogeneity of a thing with N states is

$$H(S) = \sum_i p_i^2 - 1/N, \quad \text{with} \quad p_i = Pr(s_i) \quad \text{and} \quad s_i \in S.$$

(ii) As the total number of states of a thing approaches infinity, the state heterogeneity approaches the average probability:

$$\lim_{N \to \infty} H(S) = \sum_i p_i^2 = \bar{p}.$$

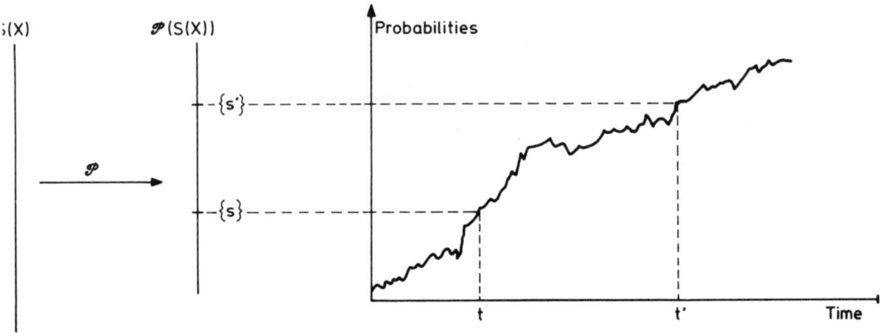

Fig. 4.3. The thermodynamic probabilities of the states of a statistical mechanical system: on the whole later stages have higher probabilities – but only on the whole, i.e. as a tendency.

The last quantity is sometimes regarded as a suitable measure of the degree of organization of a thing. This interpretation does not seem to be adequate in view of Definition 4.12. If anything, $H(S)$ measures the degree of mutability or mobility of a thing – provided S is a probability state space for the thing.

The last notion to be elucidated in this subsection is that of thing integration. Let X and X' be two different things that join to form a third thing $X \dotplus X'$. If the combination is of the conglomeration kind, the individual state spaces do not alter and the total state space is the union of the partial state spaces. And the probability that the components of the thing be in states s and s' respectively is the product of the probabilities of s and s', i.e. $Pr(s, s') = Pr(s) \cdot Pr(s')$. These probabilities change if the two components interact in the process of joining. And the stronger the interaction or correlation the greater will be the departure of $Pr(s, s')$ from the independence or noninteraction value $Pr(s) \cdot Pr(s')$.

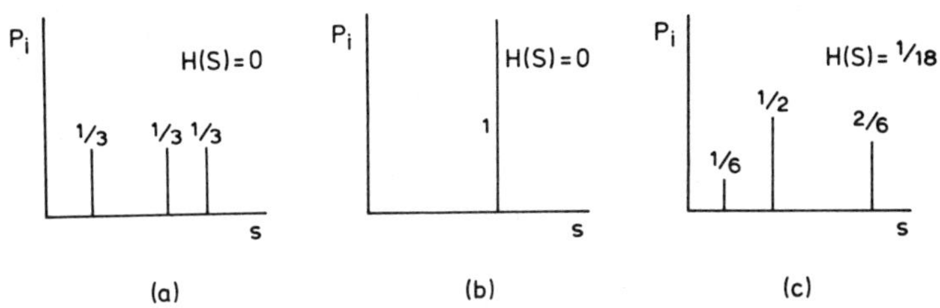

Fig. 4.4. State homogeneity (cases *a* and *b*) and heterogeneity (*c*).

We can then choose such a deviation, or $Pr(s, s') - Pr(s) \cdot Pr(s')$, as a measure of the interaction among the things in states s and s' respectively. The sum of the absolute values, or of the squares of these differences, for all states, seems to constitute a suitable measure of the degree of integration of the system:

DEFINITION 4.13 Let X and X' be two things with probability state spaces $\langle 2^{S(X)}, Pr \rangle$ and $\langle 2^{S(X')}, Pr \rangle$ respectively, and let $X \dotplus X'$ be the thing composed of X and X'. Then the *degree of integration* of $X \dotplus X'$ (or *strength of the interaction* between X and X') is

$$K(X, X') = \sum_{s \in S(X)} \sum_{s' \in S(X')} \left[\frac{Pr(s, s') - Pr(s) \cdot Pr(s')}{Pr(s) \cdot Pr(s')} \right]^2$$

provided this double sum converges.

Clearly, the above definition cannot be generalized to a system the components of which fail to have probability state spaces.

4.3. *Propensity Interpretation*

There are several interpretations of probability. The main contenders are the subjectivist (or personalist or Bayesian) interpretation and the objectivist interpretation, of which there are at least two variants: the empirical and the factual. The following tree summarizes these views.

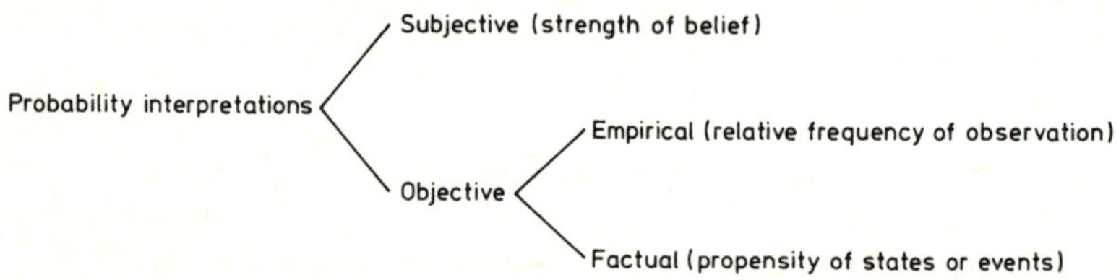

According to the subjectivist school probability values are not objective individual properties of states of things or events in things but rather degrees of certainty about our information. Clearly, this interpretation is not the one occurring in scientific theories such as quantum mechanics or genetics, which do not concern our beliefs, let alone our beliefs about the strength of our beliefs. Science, and therefore also scientific ontology, employ an objective interpretation of probability. There are two main variants of this interpretation: the frequency and the propensity interpretations. According to the former, probability values are limits of observed frequencies: they are properties of sets of data rather than properties of the referents of such data. It has been shown repeatedly that this view is mathematically untenable, if only because frequencies fluctuate erratically around stable probability values. Strictly speaking then there is no such thing as a consistent frequency interpretation of probability: there are only frequency estimates of probability values (Bunge, 1973b, 1976a). We are then left with the strictly objective, or factual interpretation.

Recall from Sec. 4.1 that the probability calculus has two specific undefined notions: those of probability space F and probability measure Pr. And remember from Vol. 2, Ch. 6 that a full interpretation of a mathematical formalism involves interpreting all of its primitives. Now, a *factual interpretation of probability theory* consists in assigning F and every value $Pr(x)$, for $x \in F$, factual meanings. One such possible interpretation consists in regarding the basic set S, out of which F is manufactured, as the state space of a thing, hence every element of the probability space F as a bunch of states, and $Pr(x)$ as the strength of the thing's propensity to be in the state (or states) x. Similarly, if x and y are states (or sets of states) of a thing, the conditional probability of y given x, i.e. $Pr(y|x)$, is interpreted as the strength of the propensity or tendency for the thing to go from the state(s) x to the state(s) y.

Given the structure of the probability function and the interpretation of its domain F as a set of facts, the propensity interpretation is the only possible one. Indeed, if $x \in F$, then $Pr(x)$ cannot be but a property of the individual fact x. That is, contrary to the frequency view, probability is not a collective or ensemble property, i.e. a property of the entire F, but a property of every individual fact – its propensity to happen. (Hence the expression 'single case propensity interpretation', used by Giere, 1974, is redundant.) What *are* ensemble properties are derived functions such as the moments of a probability distribution (in particular its average), its standard deviation if it has one, and so on. This consideration suffices to ruin the frequency school, according to which probability is a collective or ensemble property.

As for the subjectivist (or personalist or Bayesian) interpretation, it is untenable even assuming that the domain F of Pr is a collection of mental facts. For, in the expression '$Pr(x) = y$', there is no room for a variable denoting the person who feels that the number y measures the strength of his belief in x. The personalist must interpret not only F in subjective terms: he should also have a variable representing his attitude towards his own mental facts *F*. If this variable is the generic probability value *y*, then it is the same for all persons, so there is no occasion for a subjectivist interpretation. And if it is a different variable then it does not occur in probability theory but in some psychological theory concerning uncertainty as a mental state. That different persons may come up with different estimates or guesstimates of the probability of one and the same fact is obvious but does not countenance the subjectivist interpretation any more than the uncertainties concerning electric charge values would justify a subjectivist electrostatics.

To return to the propensity interpretation. This must be distinguished from the probability elucidation or exactification of the presystematic notion of propensity, tendency, or ability. In the former case one attaches factual items to a concept, whereas in the latter one endows a factual concept with a precise mathematical structure. (Cf. Vol. 2, Ch. 6, Sec. 3.8.) In science and in ontology we need both factual interpretation and mathematical elucidation. Note also that an interpretation of probability is incomplete unless it bears on the two primitives of the theory, i.e. *F* and *Pr*. Thus the factual interpretations of Poincaré, Smoluchowski, Fréchet and others are incomplete for being restricted to assuming that a probability value is a "physical constant attached to an event *E* and to a category *C* of trials" (Fréchet, 1939). This is like saying that *e* is a physical constant without adding that *e* happens to be the electric charge of the electron. The propensity interpretation is not open to such a charge of incompleteness, for its asserts explicitly that $Pr(x)$, for $x \in F$, is the propensity or tendency for *x* to happen.

It is instructive to contrast the propensity to the actualist (or frequency) interpretations of probability values, assuming that both agree on the nature of the support set *F*. (This assumption is a pretence: not only a frequentist like von Mises but also Popper, the champion of the propensity interpretation, have emphasized that facts have no probabilities unless they occur in experimentally controlled situations.) This contrast is displayed in Table 4.3.

Note the following points. Firstly, although a probability value is meaningful – i.e. it makes sense to speak of the single fact propensity – it is so only in relation to a definite probability space *F*. Likewise a frequency value makes sense only in relation to a definite sample-population pair. For example, the formula "*x* is rare" presupposes a certain set of occurrences, to which *x* belongs, and among which *x* happens to be unfrequent.

TABLE 4.3

Potentialist vs. actualist interpretation of probability

$p=Pr(x)$	Propensity	Frequency
0	x has (almost) nil propensity	x is (almost) never the case
$0<p\ll 1$	x has a weak propensity	x is rare
$0\ll p<1$	x has a fair propensity	x is fairly common
$p\doteq 1$	x has a strong propensity	x is very common
$p=1$	x has an overpowering propensity	x is (almost) always the case

Secondly, in the case of continuous distributions, zero probability is consistent with very rare happenings: i.e. even if $Pr(x)=0$, x may happen, though but rarely as compared with other events represented in the probability space. (All breakthroughs and all emergents – hence the most important events – have low probabilities, perhaps vanishing ones.) Consequently a fact with probability 1 can fail to happen. (Recall that the rationals have zero Lebesgue measure. Entire sets of states and events are assigned zero probability in statistical mechanics for this very reason even though the system of interest is bound to pass through them. This is what 'almost never' is taken to mean in that context, namely that the states or the events in question are attained only denumerably many times.)

Thirdly, the frequency column should be kept alongside the propensity interpretation though in a capacity other than interpretation or definition. Indeed although it fails to tell us what "$Pr(x)=y$" *means* it does tell us under what conditions such a formula is *true*. Long run frequency is in short a *truth condition* for probability statements. (For the problem whether truth conditions determine meanings, see Vol. 2, Ch. 8.)

Fourthly note again that our propensity interpretation differs from Popper's in that the latter requires the system of interest to be coupled to an experimental arrangement. No such hang-up from the frequency or empiricist interpretation remains in our own version of the propensity interpretation. Nor do we require that only events (i.e. changes of state) be assigned probabilities, as an empiricist must (since states may be unobservable): states too may be assigned probabilities, and in fact they are assigned in many a stochastic theory, e.g. statistical mechanics and quantum mechanics. (The statistical mechanical measure of entropy is a function of the probability of a state or, as Planck put it, it measures the preference [*Vorliebe*] for certain states over others.) In other words not only transition probabilities but also absolute probabilities can be factually meaningful. The requirement that only transition probabilities be considered in physics (Strauss, 1970) is untenable for the following reasons. First, since a transition probability is a conditional probability, and the latter is defined in terms of absolute probabilities – not the other way around – the former cannot have a factual meaning unless the latter has one. Second, an electron cloud (or position distribution for an electron) has a definite physical status, so much so that it can often be objectified by means of X rays.

So much for the propensity interpretation of probability. We shall see in the next section that the propensities of this kind, unlike the causal propensities, must be regarded as primary (irreducible) properties of things of certain kinds.

5. Chance Propensity

5.1. *Irreducible Potentialities*

Classical physics assigns every point particle a definite position at each instant. On the other hand a quantum mechanical "particle" possesses at each instant a definite position distribution, i.e. a whole range of possible positions, each with a given weight or probability: see Figure 4.5. Likewise with other quantum mechanical properties, such as linear and angular momentum, spin, and energy: save in exceptional cases

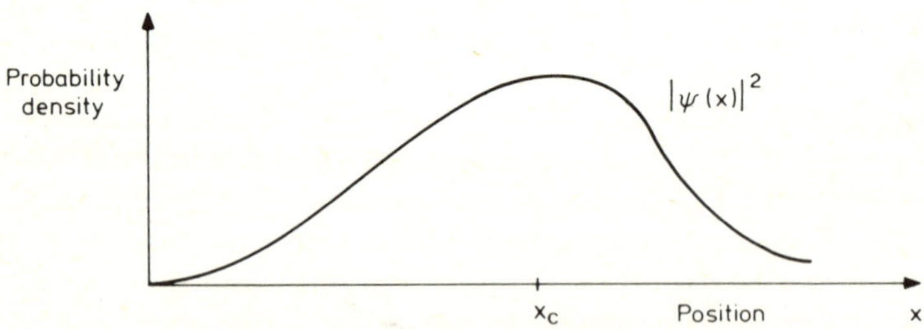

Fig. 4.5. The classical position value x_c and the quantal position distribution $\rho = |\psi(x)|^2$ of a "particle" at a given instant. The weighted average of x, i.e. $\langle x \rangle_{AV}$, coincides with x_c in the nonrelativistic limit.

each "particle" has a whole bunch (interval) of values of each of these properties. (The exceptions are constituted by the states of the "particle" that happen to be eigenstates of the corresponding property. For example, a thing in an energy eigenstate has a sharp energy value not a whole distribution of energy values. But such states, i.e. stationary states, are privileged.)

An alternative way of describing this contrast between classical and quantal properties is as follows. Where classical physics offers a nonstochastic dynamical variable Q_c, quantum physics introduces a distribution $\psi^* Q\psi$ (a "local observable"), where Q is an operator acting on the state function ψ, a place and time dependent function that determines the position probability density. (It may be assumed that these densities or bilinear forms represent the basic dynamical variables of a thing, whereas the operators Q are syncategorematic symbols serving to construct the distributions $\psi^* Q\psi$.) Every one of these random variables ("local observables") is a property of an individual thing, not a collective property of a whole bunch of similar things.

From a classical point of view this situation is intolerable. Hence the classicist will often interpret ψ as representing the state of a statistical ensemble rather than as a property of an individual thing. (This is in fact the interpretation favored by Einstein and Blokhinzev.) Yet the theory does not tolerate this interpretation of an expression such as '$\psi^* x\psi$': since it contains a single position variable x, it cannot be forced into representing a whole aggregate of particles. As long as we accept quantum mechanics we must become reconciled with its nonclassical features, such as that position is a random variable with a number of possible values, each with its own probability, for each microthing. Only the average of such a distribution is a sharp point and it coincides with the classical value. In other words, classical physics gives only a global or superficial account of microentities by ignoring the fact that the dynamical properties of the latter are random variables smeared over the entire space accessible to those entities. (See Bunge, 1977c.)

It must be stressed that the $\psi^* Q\psi$ do not just point to possible states and events: they represent properties possessed all the time by a microthing. That is, the $\psi^* Q\psi$ represent *manifest* (though not directly observable) properties rather than latent ones. On the other hand a sharp value of Q, such as any of the eigenvalues of Q, is a dispositional property of the thing concerned. In other words a thing with a Q-distribution $\psi^* Q\psi$ has the *propensity* to acquire this or that sharp Q

(eigen)value. For example, a photon travelling in free space is not concentrated at a point but, when bumping into an atomic electron, it may suddenly contract – only to be absorbed the next moment. And an electron may, under the influence of an external field, acquire a definite momentum value, though at the expense of its localization. Here the distribution is primary while the sharp values are exceptional. On the other hand in classical physics sharp values are basic and distributions derivative: they result from the interplay of numerous entities. (See Figure 4.6.) Only a classical foundation of quantum mechanics could reinstate the primacy of the sharp actual value, but then at the price of keeping the primacy of probability. (See de la Peña (1969) and de la Peña and Cetto (1975).)

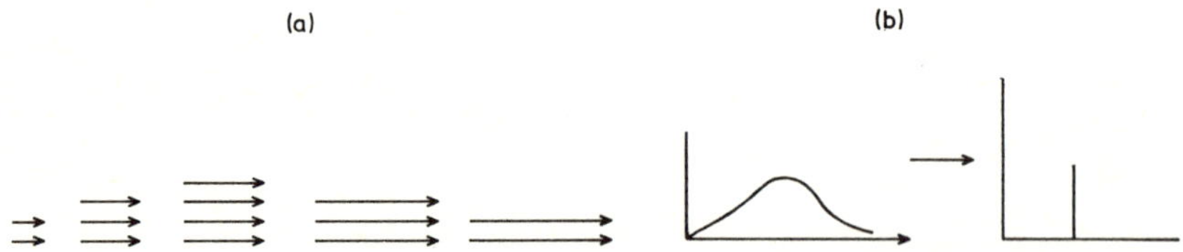

Fig. 4.6. (*a*) Classical physics: from sharp individual velocity values to distributions. (*b*) Quantum physics: from distributions to sharp individual values.

It should be stressed that the stochastic properties represented by the quantum mechanical operators Q (or their corresponding densities $\psi^* Q \psi$) are usually intrinsic properties, not properties of the thing interacting with an experimental set up. Indeed the Q's are defined as dynamical variables of the microthing of interest regardless of its circumstances: the apparatus, if present at all, has its own dynamical variables distinguished from the former by some index and acting only on the part of the state function that refers to the apparatus. The alleged ubiquity of the latter is not warranted by the mathematical formalism (Bunge, 1967b and 1973b). Besides, as emphasized by Mellor (1971), there is no point in talking of dispositions if one ascribes them to the whole "chance setup" or "experimental situation" rather than to the thing of interest, for the occurrence of a given situation is an opportunity for the disposition to display itself, not to be confused with the disposition itself. (For an excellent and detailed discussion of this matter see Settle (1974).)

5.2. *Analysis*

It will be remembered from Sec. 3 that a causal disposition or propensity is a potency that becomes actualized under suitable circumstances. The environment offers opportunities for actualization or curtails them and it is necessary and sufficient for actualization. In the case of chance propensities the possibilities lie wholly in the thing itself and the environment takes its pick. Moreover the environment may be nonexisting, in that the occurrence of an actual property may be spontaneous or uncaused, as is the case with natural radioactivity. In other words, whereas in the case of causal dispositions actualization requires that the thing of interest join another real entity, in the case of chance propensities actualization may occur if the thing joins the null thing – which is of course an artifice introduced to gain uniformity. More precisely, we make

DEFINITION 4.14 Let $z = x \dotplus y$ be a thing composed of different things x and y, where y is possibly the null thing. Further, let Q be a property of x. Then Q is said to be a *chance propensity* of x iff

(i) Q is representable nontrivially by a random variable or a probability distribution, so that each value (or interval) q of Q has a definite probability p;

(ii) y has a property R such that, when x joins y to form $z = x \dotplus y$, the probability that x acquires the individual property q (a value of Q) equals p. [I.e., $Pr(Q = q|R) = p$.]

Example 1 x = electron, y = diffraction system, Q = position distribution of x, R = letting the electron pass through a slit. *Example 2* x = person, y = learning environment, Q = learning ability, R = = allowing the person to learn.

Finally we make

DEFINITION 4.15 The *chance potentiality* of a thing x is the collection of its chance propensities:

$$\Pi_X(x) = \{P \in p(x) | P \text{ is a chance propensity}\}.$$

A thing has a greater chance potentiality than another just in case the chance potentiality of the former includes that of the latter. For example, an atom has a greater chance potentiality than any of its components; likewise a community has a greater chance potentiality than any of its members. Finally, the *potentiality* of a thing equals the union of its causal potentiality and its chance potentiality:

$$\Pi(x) = \Pi_c(x) \cup \Pi_\chi(x).$$

5.3. *Upshot*

In the present chapter we have introduced a trichotomy in the set $p(x)$ of properties of any thing x. Indeed we have assumed that $p(x)$ splits into three disjoint subsets: those of

(i) *manifest properties*, i.e. those possessed by the thing under all circumstances as long as the thing exists and remains in the same natural kind – e.g. the electric charge of an electron and the protein synthesizing activity of an organism;

(ii) *causal dispositions*, or propensities to acquire certain manifest properties under certain circumstances – e.g. the chemical reactivity of an atom and the impregnability of a mature female mammal;

(iii) *chance propensities*, or dispositions to acquire with probability certain manifest properties, depending (or not) on the circumstances.

Moreover we have drawn a radical distinction between causal and chance propensities, two categories that are usually conflated. To begin with, chance propensities are not ordinary or causal dispositions such as fragility, as there is nothing necessary about the former. In fact a distribution may collapse into any of a number of values, each possible transition having a definite probability. (Conversely, a sharp property may end up by being blurred.) Secondly, chance propensities are often basic, as shown by quantum physics and genetics. But not always: see Table 4.4. It is not true that every manifest property is rooted in a chance propensity, as extreme possibilism would have it. Nor is it true that every disposition and every chance propensity is rooted in a manifest property, as actualism claims. Both kinds of reductionism are at variance with science.

Real possibility is objective or absolute in an epistemological sense, i.e. for being independent of the knowing subject. But it is not absolute otherwise. Potencies, whether causal (like the viability of a seed) or stochastic (like the ability to learn an item on first presentation), are actualized under certain circumstances, frustrated under altered circumstances. Moreover the possibilities themselves depend upon the circumstances just as much as upon the thing concerned. Thus a peasant who migrates to a town acquires certain job and educational possibilities

TABLE 4.4

Examples of actual or manifest property, causal disposition, and chance propensity. Arrows indicate property precedence

Manifest property	Causal disposition	Chance propensity
Electric charge ——————————————————→		Ionizing capacity
Sharp position value ←——————————————	——————————————	Position distribution
Body mass ——————→	Body weight in a gravitational field	
Display or performance of an inheritable ability ←	Possession of an inborn ability ←	Acquisition of an inheritable ability by random gene shuffling.

that were formerly denied him, and on the other hand loses a number of traits, such as integration in a community. Hence, rather than as a monadic operator possibility should be construed as a dyadic one: fact x is possible (or impossible) under circumstance (state of the environment) y. If only for this reason modal logic is impotent in handling real possibility, whereas the relativity of possibility is incorporated into the notion of conditional probability. But modal logic deserves a separate subsection.

6. Marginalia

6.1. *Modal Logic and Real Possibility*

Modal logic was built in order to elucidate both the logical and the ontological concepts of possibility (see e.g. Lewis and Langford, 1932). For this reason it has often been regarded as a prerequisite for both logic and metaphysics (see e.g. Marcus, 1968). Having disposed of the former claim in Sec. 1 let us now examine the latter.

One of the most vigorous and well known defenses of the thesis that modal logic elucidates the concept of real possibility is Montague's (1960, 1974). He claimed that the principles of a certain modal calculus hold under the interpretation of '$\Box$' as "it is physically necessary that", which in turn is defined as follows: "ϕ is physically necessary iff ϕ is deducible from a certain class of physical laws specified in advance". In other words, Montague identified necessity with (theoretical) lawfulness, and disregarded circumstances. (Besides, he attributed formulas, not their referents, physical necessity, thus conflating the latter with logical necessity.) But, as we saw in Sec. 2.4, law statements, whether stochastic or not, specify only possibilities: they are so universal that they concern all possibles of a kind rather than just necessary facts. In other words, a really possible fact is one referred to by a law statement. Necessity cannot be specified without the help of the notion of circumstance, which is not incorporated into that of law. Just think of the simplest case of an equation of motion: the actual (necessary) trajectories cannot be determined unless the initial conditions are adjoined to the solutions.

Furthermore no system of modal logic is a calculus of real possibles for the simple reason that it concerns propositions not facts. For this very reason the modal logician is forced to twist "It may rain" into "It is possible that the proposition ⌜It will rain⌝ be true", or "⌜It will rain⌝ is true under some conceivable circumstances". That is, he shifts possibility from reality to statements about reality – so much more docile – thereby concealing real possibility under the rug of conceptual possibility and truth. He does the same, of course, with regard to necessity. (Worse: he has no notion of truth independent of that of satisfiability in a possible world, which he identifies with a model. And he does not know of the existence of factual and partial truth as distinct from formal and total truth.)

In science and in ontology we have little use for the notion of absolute (unconditional) possibility as well as for the modal principle ⌜$p \Rightarrow \Diamond p$⌝. What would be important is an altogether different thesis, namely: "Whatever is now the case was possible at some earlier time". But this statement includes a concept of time, which is not yet available to us and in any case is alien to modal logic. Finally ⌜$\Diamond p$⌝ is ontologically indistinguishable from ⌜$\Diamond \neg p$⌝. Thus, "That tile may fall down" has intuitively the same meaning as "That tile may not fall down" – not so in modal logic. The probabilistic versions of these statements too have the same meaning: "The probability of that tile's falling down is $\frac{1}{2}$" means the same as "The probability of that tile's not falling down is $\frac{1}{2}$" because these statements are interdeducible. Modal logic ignores all this. So we can ignore modal logic.

Most mathematical logicians ignore modal logic because they do not need it. And some, notably Quine, dislike modal logic for the wrong reason, namely because they distrust the very notion of possibility. My own evaluation of modal logic has a different rationale: I believe it is a futile game because it fails to deliver the goods it promised, namely a clarification of the notions of conceptual and of real possibility – which, to make things worse, it confuses. It is not that these notions are unimportant; on the contrary, they are much too important to be left in the care of so poor a theory as modal logic. In particular the notion of real possibility is rampant in science, which manages to handle it without the help (or rather hindrance) of modal logic. In fact every scientific theory concerns possible facts: its very referents are possible things in possible states, and every such state is representable as a point (or a region) in the state space of the thing. (To meet actuals, which theory treats as merely special cases of possibles, one has to visit observation, measurement or experiment.) But science, as we have seen, handles possibles in a way which is at variance with modal logic. For one thing in science a possible fact is a fact compatible with the laws, constraints, and circumstances (e.g. initial and boundary conditions). If the circumstances are given then the fact is either actualized or probabilified. In either case the possibility notion is construed in a fashion alien to modal logic. In other words the possibles of science are just members of some set or other – e.g. the set of nomologically possible states. Hence they are treated in a fully truth-functional way, without the assistance of modal logic. (See Figure 4.7.)

What holds for traditional modal logic holds also for the model

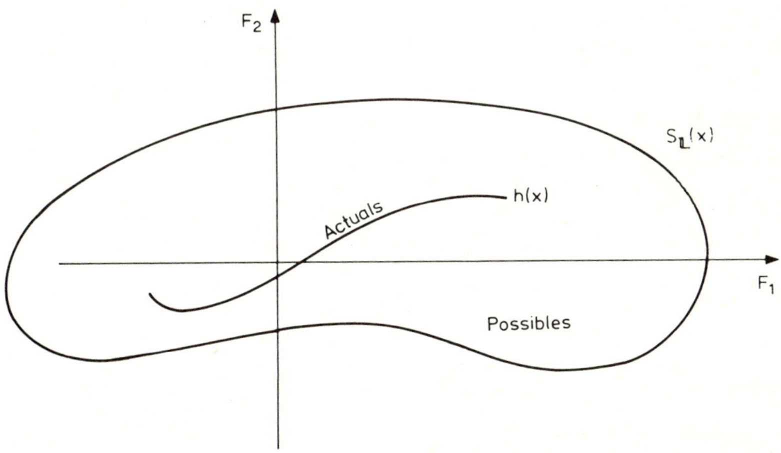

Fig. 4.7. Possibility and actuality in science and in ontology. An arbitrary point in the nomological state space $S_{\mathbb{L}}(x)$ of a thing x represents a possible state of x. The history $h(x)$ of x represents the sequence of actual states of x. No modal logic is involved.

theoretic construal of modalities (Kripke, 1959; Cresswell, 1973). According to this view possibility is satisfiability (in some model) whereas necessity is satisfiability in every model. We have admitted this idea as providing an elucidation of *one* of the four notions of *conceptual* possibility (Sec. 1.2). But the predicate "is satisfiable" refers to abstract or semiabstract (i.e. not fully interpreted) formulas – not to facts. Hence the contemporary view of modality is just as irrelevant to ontology as was Leibniz' construal of possibility as truth in some possible world.

The related notion of a model set does not fare any better with reference to ontology. In fact recall its definition (Hintikka, 1969, p. 72): "a model set [i.e. a set of formulas that are wholly true under some interpretation of the extralogical constants occurring in them] may be thought of as a partial description of a possible state of affairs or as a possible course of events ('possible world')". Here again we are concerned with abstract or semiabstract formulas not with factual law statements, let alone with factual circumstances.

In sum, neither modalities nor models for modalities elucidate the notion of real possibility. If they did, subjective idealism would be true and science redundant: we would ask the philosopher not the biologist whether sparrows could mutate into daffodils. In conclusion, modal logic fails to elucidate the concepts of real possibility – in particular those of causal disposition and chance propensity –, so it is just as useless in ontology as in semantics. Nevertheless, modal logic has promoted a new metaphysics, or rather metaphysics fiction – to which we turn presently.

6.2. *Possible Worlds Metaphysics*

If real possibility is construed as lawfulness (Sec. 2.3), then the question 'Which are the possible worlds?' becomes 'Which are the real laws?' Now, the answer to the latter is of course: 'The real laws are those inherent in the real world'. Hence the solution to the original problem is: "The set of really possible worlds is the singleton formed by the real world". Hence ours is neither the best of all possible worlds (as Leibniz and Wolff professed to believe) nor the worst of them (as the Hebrew and Christian traditions have it). Ours is the only (nomologically, really) possible world. Any alternative worlds are imaginary and belong in nonscience fiction, such as mythology. All of the (logically) possible worlds but one are (really) impossible.

Philosophers with little taste for the real world like to indulge in daydreaming about conceptually possible worlds. These provide a cozy shelter and do not have the rude manners of the real world – in particular the habit of refuting conjectures about real possibilities. A quick look at three typical problems handled by such speculations will suffice.

A first problem faced by any system of possible worlds metaphysics is of course that of defining the very concept of a possible world. This task has been performed in an elegant way by Cocchiarella (1974), who has refined, with the help of contemporary formal tools, some of the intuitions of C. I. Lewis (1923). A possible world is defined as a certain set of propositions – not of things or of facts. But this is of course a possible world of the conceptual kind (Sec. 1) – one abiding by the laws of logic not by those of factual science. Hence it is of no interest to ontology, which is concerned with the general concept of real possibility and with the question: What makes certain factual items really possible and others impossible?

As soon as one assumes the existence of a multiplicity of possible worlds one faces the question of what if anything they share. In particular, assuming that a certain individual belongs to two possible worlds, how are we to identify it? (This is the problem of transworld identification.) A special technical tool has been devised to attack this problem, or rather pseudoproblem, namely the concept of a rigid designator (Kripke, 1971). This technicality is of course unnecessary in our ontology for the simple reason that it deals only with the one real world.

Finally, some systems of possible worlds metaphysics have been applied to certain (mini)problems such as that of the counterfactuals (David Lewis, 1973). Lewis contends that "there are possible worlds other than the one we happen to inhabit" and elucidates with their help the ambiguous notion of a counterfactual. This is a typical case of what the schoolmen called an explanation of the obscure by the more obscure. Besides, who needs counterfactuals in science except as heuristic props – and even so restricted to the case when the antecedent, though contrary to actual fact, is really possible? (See Sec. 3.4.)

The following objections can be raised against possible worlds metaphysics in general:

(i) It employs modal logic, which is perfectly dispensable in all contexts except perhaps in ethics. (Whether modal logic would be needed to cope with a world different from ours we do not know. Perhaps it would be relevant in a chaotic, i.e. lawless, world – but such a world would be unlivable anyhow.)

(ii) Because it is concerned with a general concept of possibility, instead of with real possibility, possible worlds metaphysics is indifferent to worldly matters, which it is not prepared to handle. (Just think of the questions 'Is it possible to stop inflation?' and 'Is it possible for a butterfly to turn into a frog?') Hence it constitutes no ontology proper but is just an exercise in modal logic or in model theory.

(iii) Possible worlds metaphysics does not take real possibility seriously enough for, far from regarding it as inherent in the real world, it confines it to imaginary worlds. The scientific ontologist, on the other hand, has learned from quantum physics, genetics, learning theory, the theory of social mobility, and others, that possibility is inherent in reality. He has also learned that the philosopher should face the world and learn from it instead of retreating from the world by building escapist systems.

If possible worlds metaphysics is philosophically just as powerful as chess, so is the epistemology based on it. I mean the system of inductive logic where hypotheses are assigned probabilities on the basis of a count of possible worlds. Granted that counting such ghostly objects may appeal to the like of those who used to count the number of angels that could stand on the tip of a needle, one cannot help but ask two nasty questions. First: Even assuming there is a reasonable criterion for such a count, how does one go about verifying the outcome? Second: How does one allot probabilities to the various possible worlds?

6.3. *Modality and Probability*

Probability exactifies possibility, but not everything possible can be assigned a probability. There are possibilities with reference to which probability makes no sense, for the simple reason that they are not the object of a stochastic theory. For example forces may be characterized as "tendencies towards change of relative position" (Maxwell, 1892, II, pp. 211, 215). Likewise, solubility and magnetic permeability are (causal) dispositions that classical physics assigns no probabilities. Even in atomic physics, which is basically stochastic, we have similar cases. For example, a hydrogen atom at a given energy level occupies a $(2\ell + 1)$-fold degenerate state, in the sense that the given energy value is consistent with $2\ell+1$ different possibilities of spatial orientation. An external field acting on the atom may remove the degeneracy by throwing the atom into any one of $2\ell+1$ nondegenerate states. This number measures the potential variety of the original (degenerate) state, or its potentiality. But as long as such possible nondegenerate states (or the possible transitions to them) are not assigned any probabilities, they remain individually nonquantified. In short, possibility does not warrant probability, be it in classical or in contemporary science.

The converse does hold of course: *wherever there is probability there is possibility*. In such cases, and in these only, "Probability is the *quantitative measure* of possibility" (Terletskii, 1971, p. 15). In other words probability provides a numerical exactification of possibility if and only if (*a*) one is dealing with chance propensity and (*b*) a stochastic theory of the facts of interest is available (Bunge, 1976a). Needless to say, a probabilistic or stochastic theory is one containing at least one random variable – i.e. a function every value of which is assigned a definite probability. In such a theory a particular probability value will occur only by postulation or by recourse to experience. For example, given a law of evolution of a probability p in the course of time, such as e.g. $\ulcorner dp/dt = kp \urcorner$, we may compute the probability value $p(t)$ at an arbitrary time t provided we give (by assumption or by observation) the initial value $p(0)$. If this additional piece of information is added, then we are allowed to make the inference to any (past or future) value of p, namely thus: $p(t) = p(0) \,.\, exp\,(kt)$. (This result is mathematically necessary but it represents only real possibility.)

In other words, we adopt the principle that anything probable according to a scientific stochastic theory T is deemed to be possible on T and moreover that the strength or weight of a possibility is equal to the corresponding probability. (Note that facts, not propositions, are assigned probabilities by scientific theories.) This idea is perfectly consistent with our nomological concept of possibility (Sec. 2.3) because, by definition, a scientific theory contains law statements (Bunge, 1967a, Chs. 6 and 7). When dealing with a stochastic theory we may use a possibility criterion more specific than the general Criterion 4.1 of Sec. 2.5 namely

CRITERION 4.2 Let T be a stochastic theory referring to a domain F of facts, and E a body of empirical data couched in the language of T and moreover relevant to T. Then if x is in F, *x is really possible according to T and E* iff $T \cup E$ contains the formula $\ulcorner Pr(x) \geqslant 0 \urcorner$.

Note that we are not equating zero probability with impossibility. Impossibility (relative to a given T-E pair) may be construed instead as the absence of a probability value. That is, anything that fails to be assigned a probability with the help of T and E will be regarded as *impossible on T and E* – though not necessarily impossible according to different premises. And anything assigned the probability value 1 with the help of T and E may be called *necessary according to T and E*. Further, two facts will be held *compossible* on T and E if the probability of their joint occurrence is defined in $T \cup E$.

We can also introduce the notion of a possibility field, namely as follows.

DEFINITION 4.16 Let F be a set of facts and let T be a stochastic theory referring to F. Then $F_0 \subseteq F$ is a *possibility field according to T* iff T defines a probability measure Pr on F_0, i.e. if, for every $x \in F_0$, $Pr(x) \geqslant 0$.

Consider now some subset F_0 of F. If the possible facts in F_0 are mutually exclusive – i.e. if the actualization of one of them precludes that of all others – they are said to constitute a *possibility fan*. In general the rays in such a fan are not equally probable. The spread of the fan, i.e. the *indeterminacy* of the bundle of possibilities, can be measured in various ways. (The best known measures are the mean standard deviation and the information-theoretic "entropy" or average information content of the bundle. However, we shall make no use of these notions here.)

In sum, whenever a set of facts constitutes a possibility field that is the concern of a stochastic theory, such a theory exactifies the coarse modal notions – not however any of the modal calculi. Table 4.5 exhibits the correspondence between the handful of modal expressions not involving iterated modalizers, and the infinitely many probabilistic statements. Out of generosity we have included the notions of near impossibility and near necessity, which are not actually handled by modal logic. On the other hand the latter makes certain distinctions, e.g. between $\Diamond p$ and $\Diamond \neg p$, and between $\Box p$ and $\Box\Box p$, which are unintelligible elsewhere.

Whereas in modal language there are only possibility and contingency between the poles of impossibility and necessity, in any probabilistic context there is a rich spectrum between them – often a nondenumerable one. Moreover what is merely possible in the short run (for a small

TABLE 4.5

Exactification of modal notions by a stochastic theory involving discrete probabilities only (i.e. excluding possibles of zero probability)

Ordinary modal language	Exact probabilistic language
x is possible x is contingent	$0 < Pr(x) < 1$
x is just possible x is almost impossible	$0 < Pr(x) \ll 1$
x is impossible	$Pr(x) = 0$
x is necessary	$Pr(x) = 1$
x is almost necessary	$0 \ll Pr(x) < 1$

sample or a short run of trials) may become necessary or nearly so in the long run. Thus the probability that a single hydrogen atom will emit a certain 21 cm line is one in 10^7 per second, yet we detect as many such waves as we wish because of the huge numbers of hydrogen atoms. The most basic physical events are extremely improbable, but they have an impact because of their sheer numbers. Give chance a chance and it will become necessity. For this reason particle physicists, atomic physicists, molecular physicists and geneticists cannot accept Borel's injunction – in the name of mathematical empiricism – to discard improbable events for being "meaningless" (Borel, 1949). Quite on the contrary, microphysicists accept the so-called "principle of compulsory strong interaction", originally formulated with tongue in cheek as follows: "Anything that is not forbidden is compulsory" (Gell-Man, 1956; see also Melvin, 1960, p. 481). A somewhat more careful wording of the above principle reads thus: All possible repetitive chance events (in particular those consistent with the conservation laws) are likely to occur in the long run. Biologists have come to accept a similar principle, namely this: Any given ecological niche where life is possible, ends up by being inhabited.

To sum up, the classical modal principles are much too poor to account for the continuum lying between impossibility and necessity. Not even the ordinary concept of probability suffices, if only because of the ambiguity of the ordinary language phrase 'x is not probable', which may stand either for "$Pr(x)$ is small" or for "Pr is not defined at x". Chance propensity can be caught only with the fine mesh of stochastic theories, which presuppose the mathematical theory of probability.

There is no better trap for catching what Hermann Weyl (1940) called "the evasive ghost of modality".

6.4. *Randomness*

Calls arrive at a telephone exchange roughly at random rather than regularly because they are made (roughly) independently from one another. However, there is nothing haphazard or chaotic about such an irregular sequence of events. In fact there is a definite probability for an arrival to occur within any given time interval. That is, the set of arrivals is what we have called (Definition 4.16) a *possibility field.* (See Figure 4.8.) Consequently a number of (probabilistic) predictions can be computed – e.g. concerning the most probable number of calls over a certain long time interval.

Fig. 4.8. Random arrival times – of e.g. customers at a counter, or of cosmic rays at a Geiger counter. The probability of an arrival within any given time interval is unaffected by arrivals in other intervals. See e.g. D. R. Cox, *Queues* (Methuen, London, 1961), p. 6.

In other words, because the set of arrivals has a definite probabilistic structure, a theory about it can be built; this theory is called *queuing theory*. Likewise with all other random events. For example the zig-zag walk of a drunkard is random and so is the trajectory of a pollen corpuscle immersed in a fluid, and of a particle of smoke moving in mid air. Randomness consists in this case in that there is a definite probability for each segment and each angle. Surely the precise trajectory is unpredictable. However, a number of predictions can be made for the entire set of steps because of the stochastic lawfulness of the set. For example one can predict what will be the average total distance travelled by the particle (or the drunkard) over a given time interval.

The preceding considerations suggest the following elucidation of the concept of full randomness:

DEFINITION 4.17 A set F of facts is *fully random* iff $\langle F, Pr\rangle$ is a possibility field such that the probability of any member of F is independent of the occurrence or nonoccurrence of any other possibilities in F.

If on the other hand the probability of each fact is conditional on the occurrence of some other fact, then the set of facts is not fully random.

Thus a Markov chain, though stochastic, is not fully random but is somewhere between a fully random chain and a fully determinate one. This suggests that randomness comes in degrees. In fact we can construct a measure of the degree of randomness, namely as follows. (We shall do it for a finite possibility field but the generalization to the infinite case is straightforward.) Here it is:

DEFINITION 4.18 Let F be a set of n facts and suppose that the absolute probabilities $Pr(x)$ and the conditional probabilities $Pr(y|x)$ are defined for all $x, y \in F$. Then the *degree of randomness* of F is

$$r = 1 - (1/n) \sum_{x,y \in F} |Pr(y|x) - Pr(y)|.$$

The extreme values of r are

(a) $r = 1$, i.e. *full randomness*, if the occurrence of any one fact makes no difference to that of any other fact, i.e. if $Pr(y|x) = Pr(y)$ for all $x, y \in F$;

(b) $r \cong 0$, i.e. *near determination*, if, for every x and y in F, the differences between the conditional probabilities and the absolute ones are near unity.

One might think of identifying causation with $r = 0$, since in this case whatever happens is conditioned by some other fact. ("Every event has a cause.") But this would be mistaken for the following reasons: (*a*) whereas the degree of randomness (r) is defined for facts in general, the causal relation holds only between events; (*b*) in the case of causation we do not need, and usually do not have, any probabilities; (*c*) as defined above r never becomes exactly zero.

Randomness, being a special case of stochasticity, is a type of order and should therefore not be mistaken for chaos or the total absence of law. Chaos may be defined as follows:

DEFINITION 4.19 A set of facts is *chaotic* iff no probability measure can be defined on it.

For example, the set of facts

G = {The last shot in World War II, The reader's latest dream, My first car crash, The independence of Angola} is chaotic because it is impossible to allot a probability to each member of the set in such a way that $Pr(G) = 1$ – and this because it is not possible to build a theoretical model of G.

Of course the particular set G we have just introduced to illustrate the concept of chaos or haphazardness is an artifact: the members of G do not form a natural sequence. In other words such facts cannot compound to form a real fact. We surmise therefore that chaos, though conceivable – in a nonchaotic way – is unreal. That is, none of the chaotic sequences we can imagine will be found in reality. Reality is to a large extent stochastic and even random but not chaotic, i.e. lawless.

Chaos is difficult to diagnose. Take for instance a time series or any other sequence. If the sequence is finite then, no matter how long it is, there exists at least one random variable that maps the sequence. (Think of a sequence of the form $\langle x_i \mid f(x_i) = p_i \;\&\; i \in \mathbb{N} \rangle$.) So, the sequence cannot be said to be chaotic. If on the other hand the sequence is infinite, then it cannot be drawn from observation since all experience is finite. Such a sequence must be the outcome of a calculation with some formula or algorithm. Hence it is not chaotic either. In sum even if there were chaotic sets of facts, other than those we make up for the sake of an argument, we would not be able to recognize them by their lawlessness. (See however the Kolmogoroff–Chaitin definition of a "random" – in our view chaotic – sequence in information theoretic terms: Chaitin (1974). A number sequence would be said to be completely chaotic iff the smallest algorithm capable of specifying it to a computer has about the same quantity of information as the sequence itself – i.e. if there is no general rule compressing the given information.)

6.5. *Probability and Causality*

Some philosophers, among them Suppes (1970) and Popper (1974), have offered a probabilistic elucidation of causality. It boils down to the following definition: For any events a and b,

> a is a *cause* of $b =_{df}$ The probability of b given a is greater than the absolute probability of b, i.e. $Pr(b \mid a) > Pr(b)$.

In this way those philosophers hope to reduce causal determinacy to a particular case of what may be called stochastic determinacy.

However, the reduction of causation to probabilistic correlation won't do, as shown by the innumerable counterexamples of the following kind (cf. Bunge, 1973c). The conditional probability of the reader's reading these lines, given that he was born, is greater than the absolute probability of that event, yet his birth cannot be regarded as a cause of his reading these lines. A necessary condition of course but not a cause – unless one believes in predestination. Hence causation cannot be defined in terms of probability. Both concepts are used by the scientist who computes the *probability* that a given event will *cause* a certain effect.

Whatever causes probabilifies but not conversely. Anything that probabilifies (increases chance propensities) may be called an *influence*. More precisely, if the occurrence of a state (or of a change of state) is more probable given the occurrence of another, then the latter may be said to influence the former. The concept of influence can then be elucidated in probabilistic terms provided all the state spaces of the patient are probabilistic. But this is not so in general, except of course at the level of quantum physics. Therefore we shall not adopt this definition of influence.

6.6. *The Many Worlds Interpretation of Quantum Mechanics*

We have claimed throughout this chapter that contemporary science, in particular quantum mechanics, espouses possibilism and eschews actualism. However, there is a certain interpretation of the relative state formulation of quantum mechanics (Everett, 1957) that is definitely actualist, namely the many worlds interpretation. This view, taken by DeWitt (1970), Stapp (1971), Hart (1971) and a few others, is best discussed with reference to the expansion of the state function ψ in orthogonal functions φ_n, namely thus: $\psi = \sum_n c_n \varphi_n$, where (in the simplest cases) n runs over the natural numbers. The main conflicting interpretations of this series expansion are shown in Table 4.6.

In the two main possibilist interpretations the coefficients of the expansion are probability amplitudes. In the actualist interpretation there are no probabilities: the series expansion represents the actual branching of the individual thing into infinitely many actual and noninterfering copies of it. Every branch φ_n is deemed to be actual: there is no actualization process (such as, e.g., the collapse $\psi \to \varphi_n$) for the simple reason that there is no potency to begin with. And, since these branches do not interfere with one another, there is no way of knowing that they are all simultaneously real – except of course on authority. The thing ramifies infinitely and infinitely many times, and so do any of its observers – unknown to him, as he cannot communicate with any of his copies.

TABLE 4.6

Alternative interpretations of the expansion of a state function into an infinite set of orthogonal eigenfunctions of a dynamical variable

	Possibilism		Actualism (Many worlds)
	Operationism	Realism	
ψ	Actual state before measurement	Actual state	Actual state
φ_n	Possible state before measurement	Mathematical auxiliary	Actual state
$\sum c_n\varphi_n$	Superposition of possibilities	Mathematical trick	Branching of thing into infinitely many non interfering things
$\|c_n\|^2$	Probability of finding thing in state φ_n when measuring property represented by operator	Weight of nth eigenvalue in spectrum of dynamical variable	

The obvious objections to this piece of science fiction are these. First, such infinite ramifications would violate all of the conservation laws; in particular there would not be enough energy to distribute among the infinitely many copies of any given thing. Second, the ramification is in principle undetectable, hence empirically just as irrefutable as it is unconfirmable: it takes an act of faith to accept it. Third, in the case of a dynamical variable with a nondenumerable spectrum, the world would branch into a nondenumerable set of identical copies of itself. If we wish to avoid this absurdity we must not try to interpret quantum mechanics in actualist terms and must reckon with chance propensities. We must take real possibility seriously.

7. Concluding remarks

We have acknowledged real possibility and distinguished two kinds of it: causal disposition (e.g. fragility) and chance propensity (e.g. the propensity of an atom to occupy a definite energy level). Moreover we have regarded these two kinds of property as radically different:

whereas a causal disposition of a thing is unthinkable without a disposition in some other thing (think of the lock and key pair), a chance disposition is an irreducible property of the individual thing possessed by the latter even in the absence of an experimental set-up.

Our view on possibility is incompatible with actualism, according to which possibilities are only in the mind. We have met crypto-actualism in the guise of possible worlds metaphysics (Sec. 6.2) and open actualism in the many worlds interpretation of quantum mechanics (Sec. 6.6). A more common expression of actualism is the frequency interpretation of probability. According to the latter there is no such thing as a chance propensity for a single thing: there are only limiting frequencies defined for entire ensembles of things or for whole sets of states of a single thing, such as a sequence of throws of a coin. The phrase '$Pr(a) = b$' would then be short for something like "The relative frequency of a in a large ensemble (or a long sequence) of similar trials approaches b". This view is refuted by the existence of microphysical theories concerning a single thing, such as a single electron, to be sharply distinguished from a theory concerning an aggregate of coexisting electrons. Another example: genetics is in a position to calculate the probability of any gene combination, which, given the staggering number of possibilities, is likely to be a one time event. A relative frequency is a frequency of actuals, hence it cannot be identical with a possibility although it can measure the strength or weight of the latter. Unlike frequencies, probabilities do measure real possibilities.

If we take real possibility seriously then we must reject the narrow determinism espoused by Spinoza, Hobbes, the French *philosophes* and *géomètres*, Kant, Łukasiewicz, and many other necessitarians (as Peirce called them), to whom possibility was an epistemological category not an ontological one. (For example Kant held that all of the categories of modality express only relations to the body of knowledge: Kant (1781, p. A219). And Łukasiewicz stated that, if possibility were an ontological category, it would contradict causality and even the excluded middle principle, which he regarded as concerning actualities not possibilities: Łukasiewicz (1970, p. 35).) We have embraced instead, and tried to work out and update, the possibilism inherent in such diverse thinkers as Epicurus, Aristotle, Chrysippus, Cournot, and Peirce. However, we reject unconditional possibilism, according to which anything is possible (Naess, 1972). We have advanced rather *nomological possibilism*.

Possibilism is not only a metaphysical doctrine but, like many other ontological ideas, has implications for other disciplines, in particular action theory and ethics. Take for instance the much celebrated formula "Human freedom is the same as the knowledge of necessity", put forth by Spinoza and popularized by Engels. This formula needs correction on several counts. Firstly, by construing freedom as an epistemological category, it denies real or ontic freedom. Secondly it is obvious that knowledge, though certainly necessary to attain freedom in certain respects, is not sufficient for it. Thus the person who knows that he or she must die is not free to alter the course of events so as to avoid the final outcome. Thirdly a knowledge of real possibility is just as important as a knowledge of necessity – the more so since the latter is a particular case of the former. The person who wishes to get hold or rid of x in order to attain y will be well advised to use the bit of knowledge (or conjecture) that, in fact, the means x brings about (with a certain probability) the end y. Whether the law relating x to y is deterministic or stochastic makes little practical difference if multiple trials are allowed. What is important is that the relation exist and be known. In sum, freedom is not to be equated with the knowledge of necessity. Rather, factual knowledge of all kinds is just (and no less than) a means, among others, for attaining freedom in some respects.

So much for the concepts of possibility. We are now ready to face the general concept of change.

CHAPTER 5

CHANGE

So far we have not dealt with change. But, if we are to believe science, we must uphold the ontological postulate that all things are in flux. In fact the sciences describe, explain, predict, control or elicit changes of various kinds – such as motion, accretion, division, and evolution. Therefore ontology should analyze and systematize these various types of change.

While some metaphysicians have been busy praising change others have denied it and none have described it correctly. We shall endeavor to build a concept of change capacious enough – hence poor enough – to accommodate all of the concepts of change, particularly those occurring in the sciences, and to sketch general theories of change of some typical kinds. Such are the aims of the present chapter, devoted to change in general. The matter of qualitative change will be taken up in the companion volume, *A World of Systems*.

A change is an event or a process, whether quantitative or qualitative or both. Whatever its nature, a change is a modification in or of some thing or things: more precisely, it consists in a variation of the state of an entity. To put it negatively, there is no change separate from things – nor, indeed, are there changeless things even though some change slowly or only in certain limited respects. The world, then, consists of things that do not remain in the same state forever. This metaphysical hypothesis is an extrapolation from both ordinary experience and scientific knowledge. The contrary hypothesis, that nothing changes, is a philosophical extravagance hardly worthy of a sane person's consideration.

Since a change is a transition of a thing from one state to another, we must base the study of change on the concept of state analyzed in Ch. 3, Sec. 2. It will be recalled that a state of a thing is a list of individual properties of the thing, each of the latter being represented by a value of a state function or variable. And the collection of all possible states of a thing is called the (lawful) *state space* of the thing. Any change (of state) of a thing can be represented as a trajectory in its state space. This approach has the distinctive advantage that it does not require a detailed knowledge of the nature of the thing concerned: it is therefore ideally suited to ontology or the general theory of things. Let us then begin by recalling briefly the notions of state and state space, plunging immediately thereafter into the heart of the matter.

1. Changeability

1.1. *Preliminaries*

Every thing is in some state or other relative to a given reference frame. And every (relative) state of a thing is uniquely determined by the properties of the latter (relative to some frame). Thus a state of a thing with n properties, each represented by a (state) function, is an n-tuple of values of the functions representing those properties. Different states correspond to (are represented by) different n-tuples. A change in the representation of properties, or in the choice of reference frame, ensues in a different representation of states. This shows that our knowledge of a state depends partly upon ourselves and the state of the art – not however that the state itself is an artifact of the beholder. States, in sum, are relative.

Pretend, for the sake of definiteness, that a certain thing x, such as a camera shutter or an electrical switch, has a single dichotomic general property: that it is either open (1) or closed (0). The possible states of x are then 0 and 1. In other words, the state space of x is $S(x) = \{0, 1\}$. Actually this is an extreme oversimplification: 0 and 1 are just the states of interest for some definite purposes. Any real thing has a large number n of properties P_i, each representable by a function F_i, where $1 \leqslant i \leqslant n$. The possible degrees or intensities of property P_i are represented as so many values of the function F_i. Hence the conceivable state space of the thing will be the cross product of the images of these functions. And the nomological state space will be a subset of the former space, resulting from the laws involving the P_i: see Figure 5.1. All of this holds, of course, for a given representation of properties, hence states.

Every point s belonging to the lawful state space $S_{\mathbb{L}}(x)$ of a thing x represents a possible state of the thing. The actual state of the thing is represented by what is called its *representative point* in the state space. An actual change of thing x is represented by a *trajectory* of the representative point; this trajectory is of course the graph of a certain function on $S_{\mathbb{L}}(x)$. It is convenient, though not indispensable, to express

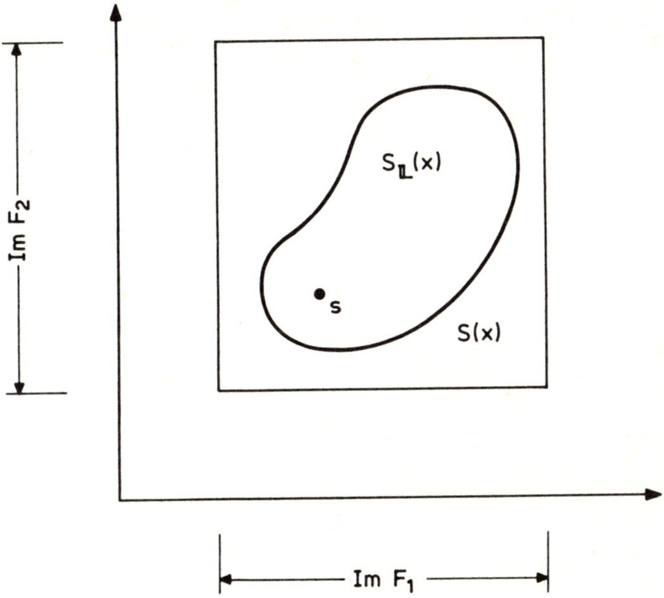

Fig. 5.1. The conceivable state space $S(x)$ and the lawful state space $S_{\mathbb{L}}(x)$ of a thing x with two properties represented by functions F_1 and F_2. '*Im F*' is read 'the image of *F*' and is the set of values, or range, of *F*.

the curve with the help of a parameter, the standard interpretation of which is time. But one of the advantages of the state space representation of change is that it requires no explicit use of the time concept.

To fix ideas consider the state space of the gene pool of a population of organisms. That set is included in the cartesian space whose axes are the gene frequencies of the pool. Every point in this space represents a possible state of the gene pool. The representative point – the one representing the actual state of the system – will remain fixed only if the organisms suffer no mutations – which is never the case, not even if the environment of the population is highly stable, such as a tropical lagoon. Otherwise the representative point moves (in state space not in ordinary space) describing a trajectory. This curve – which in the case of biological evolution has no loops – represents the genic *history* of the gene pool, or the evolution of the corresponding population at the genic level. Any segment of the total history may be called an *event* or a *process* involving the gene pool. (See Figure 5.2.)

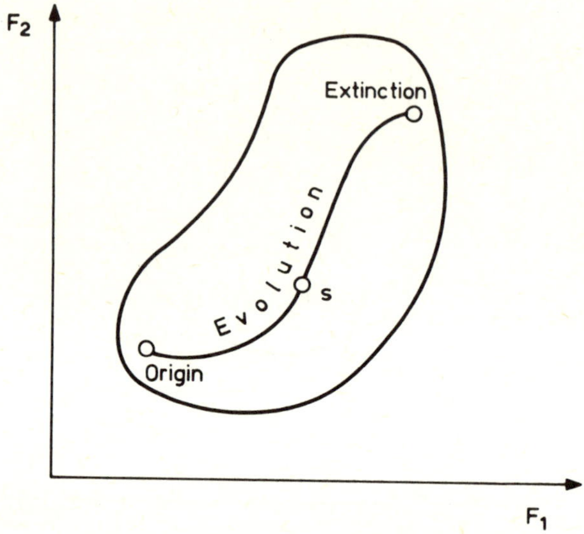

Fig. 5.2. The shifting of gene frequencies in a gene pool, in response to the joint action of mutation and selection. The end points of the curve represent the origin and the extinction of the given population.

So much for an intuitive background.

1.2. *Changeability*

Before tackling actual change we may as well elucidate the intuitive notion of mutability or changeability:

DEFINITION 5.1 Let x be a thing. Then

(i) x is *unchangeable* iff $S_{\mathbb{L}}(x)$ is a singleton for all choices of state function for x;

(ii) x is *changeable* (or *mutable*) iff $S_{\mathbb{L}}(x)$ has at least two distinct members for all choices of state function for x.

And now our first hypothesis:

POSTULATE 5.1 Every (concrete) thing has at least two distinct states, and the state space of any construct is empty. That is,

(i) if x is a thing then, for all choices of state function for x, $|S_{\mathbb{L}}(x)| \geqslant 2$;

(ii) if y is a construct then there are no state functions for y – or, equivalently, $S(y) = \varnothing$.

An immediate consequence of Postulate 5.1 and Definition 5.1 is

COROLLARY 5.1 All (concrete) things are changeable, and constructs are neither unchanging nor changeable.

Remark 1 We are not (yet) stating that every thing is actually undergoing some change or other, but only that it can change. *Remark 2* The second part of Corollary 5.1 states that the categories of change and of immutability do not apply to constructs. Constructs are neither eternal objects (Plato) nor changeable ones (Hegel). What do change from person to person are the brain processes occurring when constructs are thought.

Our next concept is that of overall change irrespective of the order in which it occurs:

DEFINITION 5.2 Let $S_{\mathbf{L}}(x)$ be a lawful state space for a thing x, and let $S_i(x), S_j(x) \subset S_{\mathbf{L}}(x)$ be two collections of states of x [e.g. two regions of $S_{\mathbf{L}}(x)$]. Then the ***overall change*** of x between $S_i(x)$ and $S_j(x)$ equals the symmetric difference between the given subsets:

$$\chi_{ij}(x) = S_i(x) \mathbin{\Delta} S_j(x).$$

An overall change may be slight or large, superficial or deep. If there are considerable changes in the individual properties of a thing, but the thing neither acquires nor loses any general property, we may say that it undergoes a *large* change. If on the other hand a thing gains or loses general properties then it can be said to undergo a *deep* change. Whereas in the first case the axes of the state space remain fixed, in the second some are added or removed. We obtain a smooth description of this sort of change if we build the state space with all the necessary axes and use at any given moment only those portions of the total space that concern the properties actually possessed by the thing at the moment. In Figure 5.3 we depict the evolution of an imaginary thing undergoing a deep change. The first part of the process is described by a curve on the horizontal plane $Im\, F_1 \times Im\, F_2$; the second part is described by a curve on the vertical plane $Im\, F_2 \times Im\, F_3$. The qualitative change occurs at the junction of the two curves. The entire curve is continuous even though its tangent has a discontinuity.

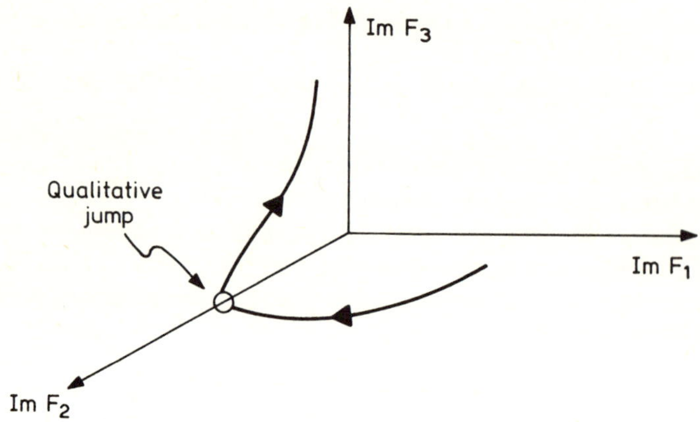

Fig. 5.3. Each arc of the curve describes a quantitative change. A qualitative jump happens at the junction of the two arcs.

In other words, we adopt

DEFINITION 5.3 Let $\mathbb{F}$ be a state function that spans the entire state space $S_{\mathbb{L}}(x)$ for a thing x. The thing undergoes a *qualitative change* iff $S_{\mathbb{L}}(x)$ equals the union of at least two subspaces, each of which is spanned by a different projection of $\mathbb{F}$. Otherwise [i.e. if none of the components can be ignored during any stretch of the process], the thing undergoes only a *quantitative change.*

An immediate consequence of this definition is

COROLLARY 5.2 Every qualitative change is accompanied by a quantitative change. [I.e., for any thing x, if x changes qualitatively then x changes quantitatively.]

The converse does not follow, which is just as well since it is false. Note that the above is not an independent ontological hypothesis but a mere corollary of a convention. The catch is that the latter has been framed so as to yield that consequence.

So much for the qualitative/quantitative distinction. The present chapter will be devoted to change in general; qualitative change will be treated in the following volume of this treatise.

2. Event

2.1. *The Ordered Pair Representation of Events*

Whatever changes may be thought of either as turning into a different thing or as going into a different state. In the former case the change of interest can be construed as the ordered couple $\langle x, x' \rangle$, where x and x' are the initial and the final things respectively. However, since names or singular terms such as 'x' and 'x'' are not descriptive, this representation of change is unilluminating. Besides, it forces an unnecessary multiplication of the number of things. For this reason it is not used in science and we shall not employ it in ontology.

Far more information is conveyed if the name 'x' is replaced by the sentence 'thing x is in state s', where s is a point (or a set of points) in a state space $S(x)$ for x. We can then construe a change of x as a transition from state s to some other state $s' \in S(x)$. In other words, we adopt what may be called the *principle of nominal invariance*, or permanence of names through change, and describe changes as changes of state. The principle, a designation rule, may be stated as follows:

PRINCIPLE 5.1 A thing, if named, shall keep its name throughout its history as long as the latter does not include changes in natural kind – changes which call for changes of name.

In order for this principle to work we must include in the state space of a thing all of the latter's possible states, from beginning to end. This allows us to designate a given person at age 50 by the same name as at age 5 even though in between the person may have renewed every single atom in his/her body. In short, there are no self-identical things but only constant names helping us keep track of the changes undergone by things.

As for the representation principle, a semantic assumption, it can be formulated thus:

PRINCIPLE 5.2 Let $S(x)$ be a state space for a thing x, and let $s, s' \in S(x)$ be two states of x. Then the *net change* in x from state s to state s' is representable by the ordered couple of these states, i.e. by $\langle s, s' \rangle \in S(x) \times S(x)$.

In the simplest nontrivial (yet imaginary) case, $S(x) = \{a, b\}$, and there is a single function $g: S(x) \rightarrow S(x)$ such that $g(a) = b$ and $g(b) = a$, to represent the two possible nontrivial changes of the thing, namely $\langle a, b\rangle$ and $\langle b, a\rangle$. Permanence is represented by the function $i_S: S(x) \to S(x)$ such that $i_S(a) = a$ and $i_S(b) = b$, i.e., the identity function on S.

The above principles are the clue to the theory of change to be built in the sequel. We first tackle the simpler case of the finite state space, thereafter the general case.

2.2. *The Event Space*

Let $S(x)$ be a state space for a thing x. Any pair of points in this set will represent unambiguously, though perhaps not exhaustively, a conceivable event, or change, in x. (Conceivable rather than really possible because (*a*) $S(x)$ itself is the collection of conceivable states of x, and (*b*) even if $S(x)$ is taken to be a lawful state space, not all of the transitions from or to a given state may be really possible. Just think of the state space $S(x) = \{\text{alive, dead}\}$.)

In this section we shall investigate the precise membership and structure of the space $E(x)$ of conceivable events. We start by examining the simple case of a three state thing, such as an electric switch with three states: open, closed, and transient. Calling a, b, and c the states, we have $S(x) = \{a, b, c\}$. We now form the ordered pairs

$\langle a, a\rangle$ thing x stays in state a (the identity event at a)
$\langle a, b\rangle$ thing x goes from state a to state b, etc.

Every one of these pairs represents a conceivable change in x, i.e. a conceivable event involving x. Assume further, for the sake of expediency, that in this particular case all such events can happen, i.e. are lawful. In other words, set $E_L(x) = E(x) = S(x) \times S(x)$. We have thus 9 elementary or noncomposite events in $E(x)$:

$$\begin{array}{lll} e_1 = \langle a, a\rangle \equiv u_a, & e_4 = \langle a, b\rangle, & e_7 = \langle b, a\rangle \\ e_2 = \langle b, b\rangle \equiv u_b, & e_5 = \langle b, c\rangle, & e_8 = \langle c, b\rangle \\ e_3 = \langle c, c\rangle \equiv u_c, & e_6 = \langle a, c\rangle, & e_9 = \langle c, a\rangle. \end{array}$$

That is, $E(x) = S^2(x) = \{u_a, u_b, u_c, e_4, e_5, e_6, e_7, e_8, e_9\}$. The first three members of this set are nonevents or null changes. Thus u_a is the nonevent consisting in that the thing stays in state a. All others are proper events. A standard pictorial representation of this space is shown in Figure 5.4.

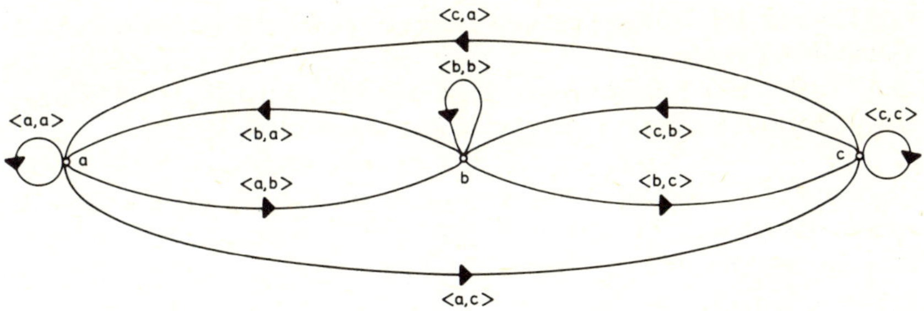

Fig. 5.4. The transition or Moore graph for a three state thing.

We assume next that the composition of certain events is possible while that of others is not. For example, event $e_4 = \langle a, b\rangle$ can be followed by event $e_5 = \langle b, c\rangle$, and the complex event may be regarded as a decomposition of event $e_6 = \langle a, c\rangle$. To symbolize the composition of events we use the asterisk $*$ and write: $e_4 * e_5 = e_6$ or, explicitly, $\langle a, b\rangle * \langle b, c\rangle = \langle a, c\rangle$. (That is, intermediate states do not show up in the net change: they are absorbed.) On the other hand the complex event $\langle b, c\rangle * \langle a, b\rangle$, which is the reverse of the former, does not occur, and so is left undefined. In other words, to represent the combination of events we introduce a (partial) binary operation $*$ in $E(x)$. If e and f are in $E(x)$, then $e * f = g$ is another event in $E(x)$ consisting in event e being *followed* by event f – with the proviso that in certain cases the composition is not defined.

The possible binary combinations of events occurring in a three state thing are shown in the following table. Every entry in this multiplication table shows the net change not the change process. But of course the

$*$	$\langle a, a\rangle$	$\langle a, b\rangle$	$\langle a, c\rangle$	$\langle b, a\rangle$	$\langle b, b\rangle$	$\langle b, c\rangle$	$\langle c, a\rangle$	$\langle c, b\rangle$	$\langle c, c\rangle$
$\langle a, a\rangle$	$\langle a, a\rangle$	$\langle a, b\rangle$	$\langle a, c\rangle$						
$\langle a, b\rangle$				$\langle a, a\rangle$	$\langle a, b\rangle$	$\langle a, c\rangle$			
$\langle a, c\rangle$							$\langle a, a\rangle$	$\langle a, b\rangle$	$\langle a, c\rangle$
$\langle b, a\rangle$	$\langle b, a\rangle$	$\langle b, b\rangle$	$\langle b, c\rangle$						
$\langle b; b\rangle$				$\langle b, a\rangle$	$\langle b, b\rangle$	$\langle b, c\rangle$			
$\langle b, c\rangle$							$\langle b, a\rangle$	$\langle b, b\rangle$	$\langle b, c\rangle$
$\langle c, a\rangle$	$\langle c, a\rangle$	$\langle c, b\rangle$	$\langle c, c\rangle$						
$\langle c, b\rangle$				$\langle c, a\rangle$	$\langle c, b\rangle$	$\langle c, c\rangle$			
$\langle c, c\rangle$							$\langle c, a\rangle$	$\langle c, b\rangle$	$\langle c, c\rangle$

table allows us to analyze certain events as processes, i.e. sequences of elementary events.

A number of regularities emerge clearly from this table. We assume that they hold for every event space and summarize them in

DEFINITION 5.4 Let $S(x) \neq \varnothing$ be a state space for a thing x and let $E(x) = S(x) \times S(x)$. The triple $\mathscr{E} = \langle S(x), E(x), *\rangle$, where $*$ is a partial (not everywhere defined) binary operation in $E(x)$ such that, for all a, b, c, d in $S(x)$,

$$\langle a, b\rangle * \langle c, d\rangle = \begin{cases} \langle a, d\rangle \text{ if } b = c \\ \text{not defined if } b \neq c, \end{cases}$$

is the *event space* of x associated with $S(x)$ iff

(i) every element of $E(x)$ represents a conceivable change of (or event in) thing x;

(ii) for any $e, f \in E(x)$, $e * f$ represents the event consisting in that event e *composes* with event f in the indicated order;

(iii) for any $s \in S(x)$, $\langle s, s\rangle \in E(x)$ represents the *identity event* (or nonevent) at s, i.e. the staying of x in state s.

An immediate consequence is that every conceivable event $\langle s, s'\rangle$, where $s, s' \in S(x)$, has a unique converse, viz., $\langle s', s\rangle$. (Which only goes to show that $E(x)$ is the totality of *conceivable* events not of really possible ones.) Another is

COROLLARY 5.3 For every state $s \in S(x)$ there is an identity event $i_s = \langle s, s\rangle$ such that, for each $e \in E(x)$ for which $*$ is defined, $e * i_s = = i_s * e = e$.

In other words, the structure $\mathscr{E} = \langle S(x), E(x), *\rangle$ is a *category* with set of objects $S(x)$, set of morphisms $E(x)$, and identity morphisms i_s, for all $s \in S(x)$; the composition is $*$. The transition graph in Figure 5.6 is a vivid pictorial representation of this category in the case where $S(x)$ has only three members.

Note that in the category $\mathscr{E}$ the following holds: For any two objects (states) $s, s' \in S(x)$ there is exactly one morphism $s \mapsto s'$, namely $\langle s, s'\rangle$. This in turn implies that every morphism in the category is an isomorphism, i.e. the inverse of $\langle s, s'\rangle$ is the unique morphism $\langle s', s\rangle$. This entails

COROLLARY 5.4 Let $s, s' \in S(x)$ be states of x. Then

(i) no event other than an identity event is immediately repeatable: $\langle s, s'\rangle * \langle s, s'\rangle$ is not defined in $E(x)$;

(ii) if an event is followed by its converse then no net change results: $\langle s, s'\rangle * \langle s', s\rangle = \langle s, s\rangle$.

Hence for an event to be repeated the thing must first go back to the initial state, either through the converse event (when possible) or through intermediate events, as in the case of $\langle s, s'\rangle * \langle s', s''\rangle * \langle s'', s\rangle = \langle s, s\rangle$.

From Definition 5.4 it follows that intermediate states do not show up in the net result. Thus the two processes

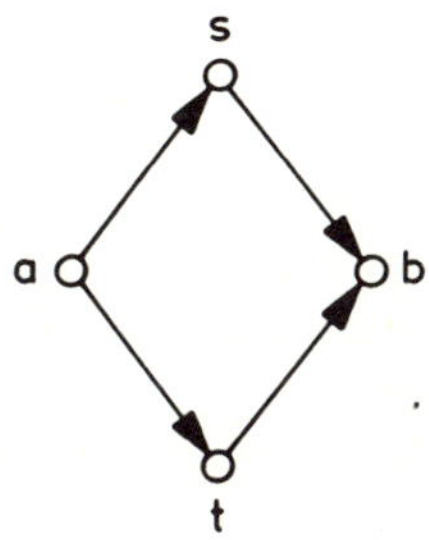

are equivalent in that $\langle a, s\rangle * \langle s, b\rangle = \langle a, t\rangle * \langle t, b\rangle = \langle a, b\rangle$. Therefore we can make

DEFINITION 5.5 Two complex events (processes) in a given event space are *equivalent* iff they have the same outcome [i.e. if they relate the same initial and final states].

Hence every complex event may be construed as an equivalence class of processes, namely the set of all processes with the same end points (states).

We wind up this section by bringing to the fore two notions that have occurred tacitly in the preceding and will be of capital importance in the sequel: those of process and of precedence.

DEFINITION 5.6 A complex event [i.e. one formed by the composition of two or more events] is called a *process*.

DEFINITION 5.7 Let e and e' be two events in a given event space [i.e., e, $e' \in E(x)$ for some thing x], such that e and e' compose to form a third event $e'' = e * e'$. Then e is said to *precede* e' relative to the reference frame involved in $E(x)$:

$$\text{If } e, e' \in E(x) \text{ then } e \prec e' =_{df} e * e' \in E(x).$$

This precedence relation is irreflexive because no proper event composes with itself; it is asymmetric: if two events compose in a given order, it does not follow that they also compose in the reverse order; and it is transitive: if one event composes with another, and the latter with a third event, then the first composes with the last. In short, $<$ is a *strict partial ordering* of $E(x)$ for any thing x. (Precedence is then formally the same as $<$ in a set of numbers and $\subset$ in a set of sets.) In short, we have

COROLLARY 5.5 For any given thing x and every state representation, $\langle E(x), < \rangle$ is a strictly partially ordered set.

That every event space is ordered by $<$ means that, provided two events in it are composable, one of them precedes the other. This order is not connected: there are events in every $E(x)$ that neither precede nor succeed each other, as shown by the very definition of the composition $*$ of events as a partial operation. Also, that order is *relative* or frame-dependent instead of absolute or frame-invariant, because $E(x)$ itself is frame-dependent. Thus if e precedes e' relative to a certain frame, there may exist another frame relative to which e' precedes e – unless of course e is necessary for the occurrence of e'. Hence the importance of mentioning the space to which the events of interest belong. More on this in Sec. 2.5.

2.3. *The Representation of Processes*

The method of tracing change step by step, i.e. of collecting pairs of states, is feasible only for finite state things. (Or rather, since there are no real things with finite state spaces, we should say that the method is restricted to finite state *models* of things.) If the state space is non-denumerable like the set of reals, then there is no single state following immediately upon a given state, i.e. there is no next state. In this case, given any two states a and b of a thing, such that b follows a, there are infinitely many intermediate states following a and preceding b. In such cases then we must look for a more powerful method for representing change. We shall presently expound such a method.

In Sec. 2.1 we introduced the concept of net change. In Sec. 2.2 we noted that one and the same net change can often be effected through different intermediate states: see Figure 5.5. The net change from state s to state s' in the state space $S(x)$ can be represented by the ordered pair $\langle s, s' \rangle$. But since the change along curve f is distinct from that along

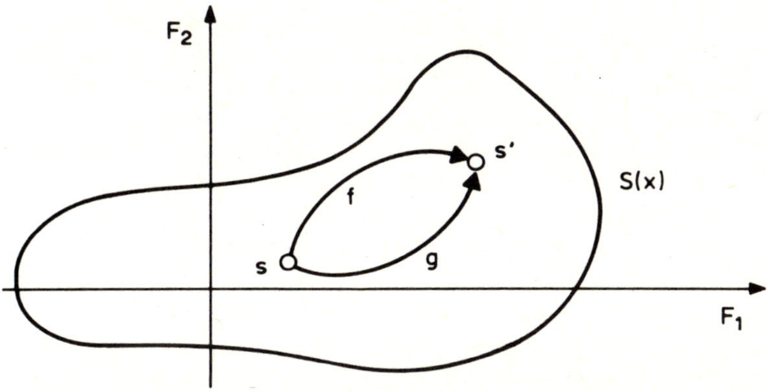

Fig. 5.5. Two different processes resulting in the same net change.

curve g, where $g \neq f$, we must represent the full events, or processes, by the ordered triples $\langle s, s', f \rangle$ and $\langle s, s', g \rangle$ respectively.

Now, the functions f and g occurring in our previous (imaginary) illustration are not supposed to be arbitrary: they must be lawful if we are to allow only lawful events and discard all the lawless (impossible) ones. In other words, f and g must be either laws possessed by the thing x concerned or transformations of the lawful state space $S_{\mathbf{L}}(x)$ compatible with those laws. (The functions need not be law functions: they may represent objective laws *cum* constraints and circumstances, such as initial conditions and boundary conditions, or they may be changes of representation leaving all the laws of the thing unchanged.) This suggests making

DEFINITION 5.8 Let $S_{\mathbf{L}}(x)$ be a lawful state space for a thing x. Then the family of *lawful transformations* of the state space into itself is the set of functions

$$G_{\mathbf{L}}(x) = \{g \text{ is a function } | g: S_{\mathbf{L}}(x) \to S_{\mathbf{L}}(x)$$
$$\& \ g \text{ is compatible with the laws of } x\}.$$

Clearly the identity transformation $i_S: S_{\mathbf{L}}(x) \to S_{\mathbf{L}}(x)$, where $i_S(s) = s$ for every $s \in S_{\mathbf{L}}(x)$, is a charter member of $G_{\mathbf{L}}(x)$. It represents no change and it can compose with any given nontrivial transformation.

Each such nontrivial transformation $g \in G_{\mathbb{L}}(x)$, where $g \neq i_S$, helps represent one feature of the total change undergone by x – e.g. its change of place, or of population, or of crowding. Since real changes are many-sided, they must be represented with the help of a number of different transformations composing to form the total change. Thus a thermoelectric event may be regarded as the composition of a thermal event and an electrical event – not however in the sense of 'composition' covered by the binary operation $*$ introduced in Sec. 2.2. (The acceptation of 'composition' employed in the present section is best understood by noting that $G_{\mathbb{L}}(x)$ is a proper subset of $S_{\mathbb{L}}(x)^{S_{\mathbb{L}}(x)}$, i.e. the family of all the functions on $S_{\mathbb{L}}(x)$ – not just those representing lawful changes. But the wider set has the monoid structure, since the composition of any two transformations of $S_{\mathbb{L}}(x)$ is a third transformation of the same set, and the latter includes the identity transformation.)

Now to the notion of interest:

PRINCIPLE 5.3 Let $S_{\mathbb{L}}(x)$ be a lawful state space for a thing x, and let $G_{\mathbb{L}}(x)$ be the family of lawful transformations of the state space into itself. Then a lawful event (or process) with end points s and s', where s, $s' \in S_{\mathbb{L}}(x)$, is representable as a triple $\langle s, s', g \rangle$, where $g \in G_{\mathbb{L}}(x)$ and $s' = g(s)$.

We call this the *functional representation* of change. Clearly, it subsumes the ordered pair representation, which obtains when the transformation g is fixed and the intermediate states are forgotten.

Consider the paradigmatic case of the classical model of a particle. The instantaneous dynamical state of such a (model) thing moving on a line is given by the instantaneous values of the position q and the momentum p of the particle, considered as independent variables. The nomological state space (called *phase space*) of such a thing is a subset of the cartesian plane $\mathbb{R}^2$. The transformation representing the motion can be calculated provided one knows or assumes (*a*) the equations of motion and (*b*) the forces and constraints acting on the particle. Assume, to fix ideas, that the force acting on the particle is elastic. In this case our thing is a linear oscillator and its laws are the canonical (or Hamilton) equations, which in this particular case reduce to

$$\dot{q} = p/m, \qquad \dot{p} = -kq,$$

where m is the mass of the particle and k the elastic constant. Then calling $q(t) = q$, $q(t+h) = q'$, where $h \in \mathbb{R}^+$, and similarly for the momentum, and expanding q' and p' in power series, one obtains

$$\begin{aligned} q' &= q \cos ah + a^{-1} p \sin ah \\ p' &= p \cos ah - aq \sin ah \end{aligned}, \quad \text{with } a = (k/m)^{1/2}.$$

The value q' of the position coordinate at time $t+h$ is then a function of the earlier values of both the position and the momentum, and similarly for the momentum value p'. A transformation representing this change (motion) is called a *canonical transformation*, or a transformation satisfying the canonical equations of motion. This transformation, for a given value of the time shift h, is a map $g: \mathbb{R}^2 \to \mathbb{R}^2$ such that $g(q, p) = \langle q', p' \rangle$ as above. (There are two functions involved but they may be regarded as just two components of g.) The transformation may be regarded as a rotation in the state space, where the cartesian axes are now $x = a^{-1}p$ and $y = q$, and the rotation angle $\alpha = ah$. This transformation, far from being arbitrary, is constrained by the laws of motion. The net effect of this constraint is that the ellipse $p^2/a^2 + q^2 = 1$ is transformed into the ellipse $p'^2/a^2 + q'^2 = 1$ rather than becoming a curve of a different type.

Unless the state space happens to be one-dimensional, change will be represented by more than one function defined on the state space. Thus in the preceding example the transformation function turned out to have two components, u and v, such that $q' = u(q, p)$ and $p' = v(q, p)$. But u and v are, as noted above, the two components of a single function g mapping $S(x)$ into itself and satisfying the equations of motion.

2.4. *The Space of Lawful Events*

In Sec. 2.2 we characterized the space of conceivable net events. We now turn to a characterization of the space of really possible (i.e. lawful) events, taking into account all the intermediate states between the end points defining the net events. Recall Principle 5.3 concerning the functional representation of a full process – not just its end points – as an ordered triple $\langle s, s', g \rangle$. The collection of all such triples for a fixed transformation g may be regarded as a set of ordered couples and a subset of the space $E(x)$ of conceivable events. It represents the totality of changes of type g happening in or to thing x. And the total space of events for a given thing is the collection of all such partial event spaces. More explicitly, we propose

DEFINITION 5.9 Let $S_{\mathbb{L}}(x)$ be a lawful state space for a thing x, and let $G_{\mathbb{L}}(x) \subseteq S_{\mathbb{L}}(x)^{S_{\mathbb{L}}(x)}$ be the family of lawful transformations of $S_{\mathbb{L}}(x)$. Then

(i) the g-lawful transformations (or set of *changes of kind* g) for x, or *space of g-lawful events* in x, is the set of ordered pairs of states

$$E_g(x) = \{\langle s, s'\rangle \in S_{\mathbb{L}}(x) \times S_{\mathbb{L}}(x) \mid s' = g(s) \ \& \ g \in G_{\mathbb{L}}(x)\};$$

(ii) the totality of lawful transformations of x, or *space of lawful events* (in all respects) in x, is the union of all the partial event spaces:

$$E_{\mathbb{L}}(x) = \bigcup_{g \in G_{\mathbb{L}}(x)} E_g(x).$$

(We could have defined $E_{\mathbb{L}}(x)$ as the family of all $E_g(x)$'s, but then a singular event belonging to any of the latter would not belong to the former.)

Let us now relax the condition that the transformations of the lawful state space be lawful themselves, and consider the space of conceivable events. We are now in a position to generalize the result obtained in Sec. 2.2 concerning net events. Consider the events $\langle s, s', g\rangle$ and $\langle s', s'', h\rangle$, where $s, s', s'' \in S_{\mathbb{L}}(x)$, and $g, h \in G(x) = S_{\mathbb{L}}(x)^{S_{\mathbb{L}}(x)}$. That is, we consider two conceivable processes occurring in one and the same thing. Because the transformations g and h compose, the events compose in the indicated order:

$$s \overset{g}{\mapsto} s' \overset{h}{\mapsto} s''$$

to produce the net change $\langle s, s'', h \circ g\rangle$, where '$\circ$' stands for function composition. In the particular case where $h = i_S$ (identity) we get

$$s \overset{g}{\mapsto} s' \overset{i_S}{\mapsto} s'$$

which reduces to $\langle s, s', g\rangle$. This shows that the space of conceivable events is a category. More precisely, we have proved in all generality.

THEOREM 5.1 Let $S_{\mathbb{L}}(x)$ be a lawful state space for a thing x, and $G(x) = S_{\mathbb{L}}(x)^{S_{\mathbb{L}}(x)}$ the totality of (lawful and unlawful) transformations of $S_{\mathbb{L}}(x)$. Moreover let '$\circ$' designate the composition of transformations (functions) in G. Then the triple $\mathscr{E} = \langle S_{\mathbb{L}}(x), G(x), \circ\rangle$, called the *conceivable event space* of x, is a category. The objects of this category are all the states $s \in S_{\mathbb{L}}(x)$ of the thing, the morphisms all the transformations $s \overset{g}{\mapsto} s'$ with $g \in G(x)$ and $s' = g(s)$, and the identities the functions $s \overset{i_S}{\mapsto} s$.

This result (obtained with the help of Arturo Sangalli) is not terribly important because it concerns the structure of the space of conceivable events not that of the lawful events. Nevertheless it exhibits the structure in which the lawful event space $\mathscr{E}_{\mathbb{L}}(x) = \langle S_{\mathbb{L}}(x), G_{\mathbb{L}}(x), \circ\rangle$ is embedded. Besides it teaches us how to compose full events rather than just net events. In fact we have learned that the composition of events proceeds as indicated by

DEFINITION 5.10 Let $S_{\mathbb{L}}(x)$ be a state space for a thing x, and $G_{\mathbb{L}}(x)$ the corresponding set of lawful transformations of $S_{\mathbb{L}}(x)$. Further, call $e = \langle s, s', g\rangle$ and $e' = \langle s'', s''', h\rangle$ two events in the lawful event space $E_{\mathbb{L}}(x)$ defined by $S_{\mathbb{L}}(x)$ and $G_{\mathbb{L}}(x)$. Then e and e' *compose* in the indicated order to form a third event e'' iff the end of the first coincides with the beginning of the second and if the two transformations compose to form a third lawful transformation. If the composition takes place then the first event is said to *precede* the second. In symbols:

(i) $\langle s, s', g\rangle * \langle s'', s''', h\rangle = \begin{cases} \langle s, s''', h \circ g\rangle \text{ iff } s' = s'' \;\&\; h \circ g \in G_{\mathbb{L}}(x) \\ \text{undefined otherwise} \end{cases}$

(ii) $e \prec e' =_{df} e * e' \in E_{\mathbb{L}}(x)$.

And now a few overdue illustrations.

Example 1 Let the thing be a conservative system with n components. Its classical state space is the $6n$ dimensional cartesian space spanned by the state function whose components are the $3n$ generalized position coordinates and the $3n$ momenta. The event space is included in the set of graphs of the canonical or contact transformations, which are those transformations of the state (phase) space that leave the laws of motion invariant. For details on the simplest case see the example in Sec. 2.3. *Example 2* The state space of a quantum-mechanical system is a Hilbert space – a certain set of functions. The operator $T = \exp(iHt/\hbar)$ defined in this space, i.e. $T: S \to S$, allows one to represent the changes in the states of the system. Thus if $Q(t_0)$ represents the value of a property Q of the system at time t_0, then its value at time t is $Q(t) = TQ(t_0)T^{-1}$. *Example 3* Some or even all of the components of the state function for a thing could be random variables. Then the corresponding state space would be a probability state space (Ch. 4, Sec. 4.2.) The laws of change, or at any rate some of them, would concern not the states in a direct manner but rather their probabilities or, rather, the probabilities of the various possible (lawful) state transitions. These probabilities are conditional. Consider the simple case of a state space with a finite number of points, every one of which has a definite probability. Each final state s_f can arise from (originate in) any other state s_i with a definite probability $Pr(s_f|s_i)$, which can of course be nought for some pairs of states. For a fixed i all these probabilities add up to unity. The total probability that state s_f be actualized is

$$Pr(s_f) = \sum_i Pr(s_i) \cdot Pr(s_f|s_i).$$

2.5. *Keeping Track of Changing States*

Whereas some properties of a thing, such as its composition, are frame-invariant or absolute, others – such as its energy – are frame-dependent or relative. Since every thing possesses energy, the states of every thing are relative: recall Ch. 3, Sec. 2.8. And because states are relative, so are events. True, events such as turning on a switch or kissing look absolute. But this is only because in thinking of them we focus on what seems typical, leaving aside a number of features that are frame-dependent, such as the location and speed of the thing(s) in which the events occur. A reasonably complete account of an event calls for the determination of a number of traits, most of which are bound to be relative rather than absolute. Hence

RULE 6.1 Refer every state function – ergo every state space and every event space – to some standard thing (reference frame) or other.

The explicit reference to standards serves not only as a reminder that change is relative: it can also be utilized to label states and keep track of them in a more precise way than we have done heretofore. This is in fact the point of Definition 3.14: it allows us to identify the thing states in terms of reference states, or states of a standard. Let us recall how this was done, and let us improve on that parametrization.

Let $S(f)$ be a state space of a reference frame f for a thing x with state space $S(x)$. Further, assume that the domain of the state function $\mathbb{F}$ that spans $S(x)$, far from being unrelated to the reference frame, equals the collection $S(f)$ of reference states. That is, set $\mathbb{F}\colon S(f) \to S(x)$, so that each reference state $t \in S(f)$ is paired off to a single thing state $s = \mathbb{F}(t) = \langle F_i(t) \mid 1 \leqslant i \leqslant n \rangle$, where n is the number of properties represented by $\mathbb{F}$. The state function $\mathbb{F}$ is now the list of properties of thing x *relative to* (or parametrized by) *the states of the reference frame f*. Given any reference state t, $\mathbb{F}$ picks the corresponding thing state $\mathbb{F}(t)$. And a substitution of an inequivalent frame f' for frame f will be accompanied by different values $\mathbb{F}(t')$ of the state function.

The states of a reference frame can be described with various degrees of precision. But if we are to use them for labeling or identifying thing states, then that precision must not be less than the precision employed in describing the thing states. Now, in science one usually seeks, and often attains, maximal precision, which consists in this case in assigning real values to the individual properties of things. That is, more often than not each value of the state function F_i for a given thing equals a real number r_i, so that the state of the thing at (relative to) $t \in S(f)$ is $\mathbb{F}(t) = \langle F_i(t) \mid 1 \leqslant i \leqslant n \rangle = \langle r_1, r_2, \ldots, r_n \rangle$. Such a high degree of accuracy calls for a similar accuracy in the description of the reference states. And such an accuracy is achieved by coordinatizing them, i.e. by assigning them coordinate values.

The coordinatization procedure is best visualized by construing $S(f)$ as a system of four cartesian axes, so that each reference state is assigned a quadruple of real numbers. That is, we single out those properties of a reference frame that can be mapped into a four-dimensional grid, leaving all other properties aside. (The points in the grid are eventually interpreted as points in spacetime.) Finally we refer the thing states not to the reference states but to their numerical labels – as when we say 'X was dreaming at the street corner at 7:00 p.m.'. That is, we pair each thing state off to a point in the reference grid.

The coordinatization of reference states is effected by a one to one correspondence $k: \mathbb{R}^4 \to S(f)$ between ordered quadruples of reals and reference states. The composition $\lambda = \mathbb{F} \circ k$ of the coordinatization function with the state function matches each point $r = \langle t, x, y, z \rangle$ in the reference grid with a thing state s, i.e. $r \overset{k}{\mapsto} t \overset{\mathbb{F}}{\mapsto} s$, as shown in Figure 5.6 and in the commutative diagram overleaf.

Now, the reference grid is partially ordered. Indeed, if r and s are quadruples of reals, then $r \leqslant s$ iff each component or coordinate of r is smaller than the corresponding coordinate of s. That is,

$$\langle r_1, r_2, r_3, r_4 \rangle \leqslant \langle s_1, s_2, s_3, s_4 \rangle \text{ iff } r_i \leqslant s_i \text{ for all } i = 1, 2, 3, 4.$$

Consequently $\langle S(f), \leqslant \rangle$ is a partially ordered set. This order is not connected: not every pair of reference states is related by $\leqslant$: see Figure 5.7. And the same kind of order is imposed, via $\mathbb{F}$, on the set $S(x)$ of thing

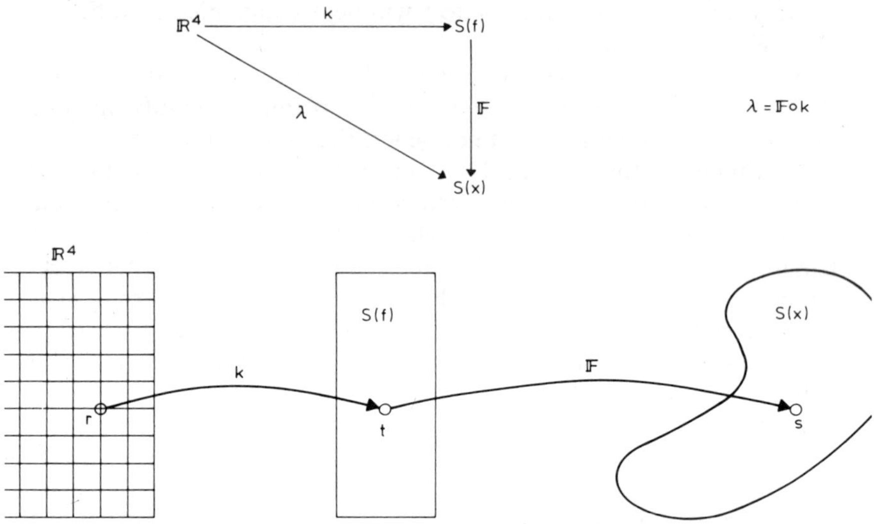

Fig. 5.6. Coordinatization of states. Each reference state $t \in S(f)$ is assigned at most one thing state $s \in S(x) \subseteq \mathbb{R}^n$ and is represented by a quadruple of reals $r \in \mathbb{R}^4$.

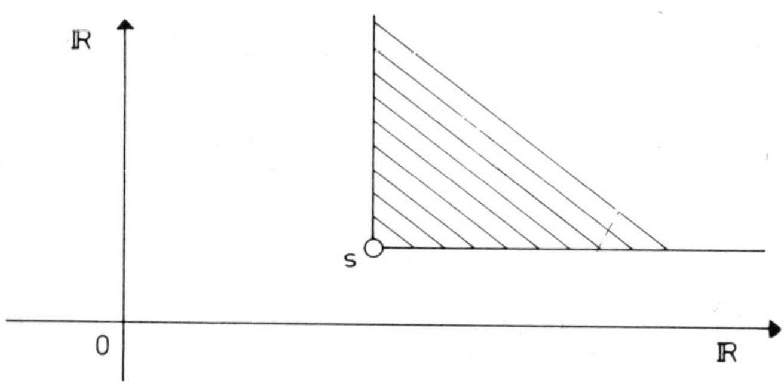

Fig. 5.7. State s precedes all others in the shaded region but none outside the latter.

states even if $\mathbb{F}$ does not take values in $\mathbb{R}^n$ – as when a mammal is said to pass through four stages of development.

The preceding considerations can be compressed into

DEFINITION 5.11 Let f be a reference frame for a thing x and let $S(f)$ and $S(x)$ be the respective state spaces. Further, assume that $S(f)$ is spanned by a four-component state function, while $S(x)$ is scanned by an n-component state function $\mathbb{F}$. Then a *coordinatization* of the states of both x and f is a $1-1$ correspondence $k: \mathbb{R}^4 \to S(f)$ such that

(i) k assigns each reference state $s \in S(f)$ a quadruple of real numbers r – as a consequence of which each thing state $s \in S(x)$ is identified as the n-tuple $s = (\mathbb{F} \circ k)(r)$;

(ii) k is order preserving, i.e. the partial order of $\mathbb{R}^4$ induces via k the partial ordering of reference states, which in turn orders the states of the thing: for all $t_1, t_2 \in S(f)$,

$$\text{if } r_1 \leqslant r_2 \text{ then } k(r_1) \leqslant k(r_2) \text{ \& } \mathbb{F}(t_1) \leqslant \mathbb{F}(t_2),$$
$$\text{where } t_i = k(r_i).$$

In other words, a state coordinatization is an order preserving function labeling and ordering the thing states via the reference states:

$$\langle \mathbb{R}^4, \leqslant \rangle \xrightarrow{k} \langle S(f), \leqslant \rangle \xrightarrow{\mathbb{F}} \langle S(x), \leqslant \rangle.$$

As the reference frame passes from one state to another, so does the thing: the successive states of a thing can then be matched to, hence identified by (not *with*), the successive states of the standard. And in turn, the ordering of states induces an ordering of events (in the same thing and relative to the same frame). Hence we can make

DEFINITION 5.12 Let s, s', s'', s''' be four states of a given thing relative to a certain reference frame, such that the first two form a net event $e = \langle s, s' \rangle$ and the last two another event $e' = \langle s'', s''' \rangle$. Then

(i) $e \leqslant e' =_{df} s \leqslant s'' \text{ \& } s' \leqslant s'''$;

(ii) $e \sim e' =_{df} e \leqslant e' \text{ \& } e' \leqslant e$.

The relation $\leqslant$ of event order is reflexive: two events are congruent iff their corresponding initial and final states are congruent, i.e.

$$e \sim e' =_{df} s \sim s'' \text{ \& } s' \sim s'''.$$

It is also clear that $\leqslant$ is antisymmetric and transitive. Hence $\leqslant$ is a partial order – unlike the relation introduced by Definition 5.7, which was only a strict partial order. Hence instead of Corollary 5.5 we have

COROLLARY 5.6 For every thing x and relative to any given reference frame, $\langle E(x), \leqslant \rangle$ is a partially ordered set.

Our new definition of event order is more comprehensive than Definition 5.7, as it includes the particular case when e composes with e' to form a third event $e'' = e * e'$ (in which case $s' = s''$), as well as the case when they do not compose. (See Figure 5.8.) The composition of events is possible only if the precedence is proper: $e * e' = e''$ iff $s' = s''$ & $e < e'$.

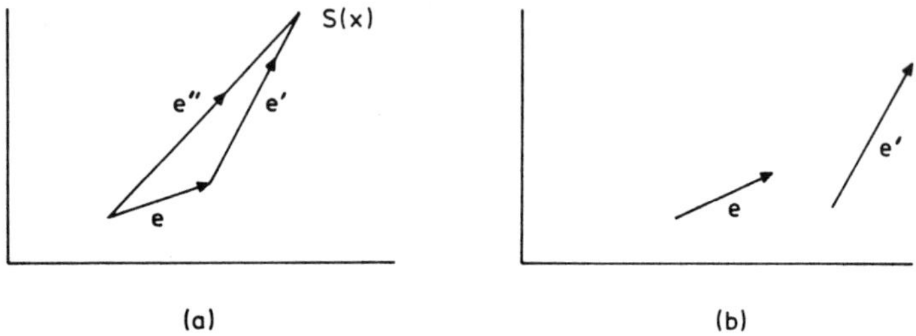

Fig. 5.8. Event order. (*a*) $e < e'$ and both compose to form e''. (*b*) $e < e'$ but they do not compose.

Consequently two events do not compose if neither precedes the other. Which is rather obvious, as the links in a chain of events are supposed to follow one another. Which in turn invites defining a *chain of events* as a denumerable and strictly partially ordered set of events $E^*(x)$ included in the event space of a thing x such that, for any $e, e' \in E^*(x)$, if $e < e'$ then the end state of e coincides with the beginning state of e', so that both events compose. More on this in Sec. 3.1.

Finally we shall proceed to formalize the notion of a variable collection, i.e. one whose membership changes as the reference states succeed one another – e.g. the population of a town. Obviously the concept of a set won't do because every set has a fixed membership. But instead of using a single set we may employ a family of sets indexed by a change parameter – call it t. Thus calling C_t a collection at t, we form the family $C = \{C_t | t \in T\}$, which may be called a variable collection. Or, equivalently, we may propose

DEFINITION 5.13 Let $t \in S(f)$ be the change parameter of frame f. Then any function

$$K: S(f) \to C, \text{ with } C = \{C_t | t \in S(f)\},$$

such that $K(t) = C_t$ for $t \in S(f)$, is called a *variable collection.*

This construal of a variable collection will be adopted henceforth, particularly in Vol. 4 of this treatise when dealing with populations of molecules, organisms, or persons – the membership of which, unlike that of a set, changes in time. It is particularly advantageous for analyzing properties of collections. Take for example the numerosity or population of a variable collection such as an ecosystem. It may be regarded as a function $p: C \to \mathbb{N}$ assigning each member of the family C a natural number $n \in \mathbb{N}$. This function composes with K to yield the function $f = p \circ K: S(f) \to \mathbb{N}$, which assigns to each reference state (or instant) $t \in S(f)$ a natural number, namely the population of the variable collection. (For discrete times the mathematical theory of sets through time is available.)

2.6. *Rate, Extent, and Change Potential*

A first notion we want to clarify is that of rate of change. Actually there is a large family of concepts of rate of change, of which we shall present only the two most common ones.

Suppose both the reference states and the thing states have been duly coordinatized (Sec. 2.5). Every reference state $t \in S(f)$ is then represented by a quadruple of reals: $t = \langle \alpha, \beta, \gamma, \delta \rangle$. Consequently the corresponding thing state is $s = \mathbb{F}(t) = \mathbb{F}(\alpha, \beta, \gamma, \delta)$. If $\mathbb{F}$ has a component F_i that is once differentiable with respect to one of the four reference parameters, say α, then the rate of change of F_i with respect to α is the partial derivative of F_i with respect to α. The thing does not change at all in the ith respect over the interval $[\alpha_1, \alpha_2]$ iff $\partial F_i / \partial \alpha = 0$ for all $\alpha \in [\alpha_1, \alpha_2]$. It changes swiftly in the same respect over the same interval if the corresponding rate of change is appreciable relative to F_i itself – otherwise it changes slowly. Needless to say, the reference parameter may but need not be the time. In sum, we can make

DEFINITION 5.14 Let the states of a thing x be coordinatized through those of a certain reference frame, so that each thing state is of the form $s = \mathbb{F}(\alpha, \beta, \gamma, \delta)$, where the arguments are the parameters identifying the reference states. Further, let the ith component of $\mathbb{F}$ be differentiable with respect to α. Then

(i) the *relative rate of change* of x in the ith respect and with respect to α is

$$V_i = \frac{1}{F_i} \frac{\partial F_i}{\partial \alpha};$$

(ii) the *relative extent of the change* of x in the ith respect, with respect to α and over the interval $[\alpha_1, \alpha_2]$, is

$$\Delta_i(\alpha_1, \alpha_2) = \frac{\ln F_i(\alpha_2) - \ln F_i(\alpha_1)}{\alpha_2 - \alpha_1} \quad \text{provided } F_i(\alpha_1), F_i(\alpha_2) > 0.$$

If the thing changes periodically with period $[\alpha_1, \alpha_2]$, the extent of its change over an integral multiple of the period is nil. This provides a definition of the concept of *cyclic change*.

The preceding definition is of no use in the case of a probability state space the points of which are not paired off to any reference frame. In this case we are interested in global changes that are frame invariant. The rate of change from state s_i to state s_j may be taken to be the conditional probability of s_j given that s_i has been actualized, i.e. $Pr(s_j|s_i)$. The rationale behind the choice is this. Think of a complex thing, n_i components of which are in state s_i, n_j in state s_j, and so on. Then $n_i Pr(s_j|s_i)$ will measure the most probable number of components jumping from state s_i to state s_j. The larger this product the greater the rate of change. The stochastic matrix $\|Q_{ij}\| = \|Pr(s_j|s_i)\|$ represents then the global rate of change of the thing. If the matrix is diagonal, i.e. if $Pr(s_j|s_i) = \delta_{ij} p_i$, where δ_{ij} is the Kronecker delta and p_i the probability of state s_i, then the thing does not change. The larger the change the larger the corresponding off-diagonal terms of $\|Q_{ij}\|$. And the extent of change, also called the *mobility*, equals the sum of the products of the off-diagonal terms by their symmetrical counterparts. We compress all this into

DEFINITION 5.15 Let $\langle S(x), Pr \rangle$, where $S(x)$ is a denumerable state space for a thing x, and Pr a probability measure on $S(x)$, be a possibility field for x. Then

(i) the *global rate of change* of x is the stochastic matrix

$$\|Q_{ij}\| = \|Pr(s_j|s_i)\| \quad \text{with} \quad s_i, s_j \in S(x),$$

where $Pr(s_j|s_i)$ is the probability of state s_j given that state s_i has been actualized;

(ii) the *extent of change* (or *mobility*) of x is

$$\Delta = \sum_{i \neq j} Q_{ij} \cdot Q_{ji} = \sum_{i \neq j} Pr(s_j|s_i) \cdot Pr(s_i|s_j).$$

The latter is a measure of the changeability of certain things in certain respects, namely those describable with the help of denumerable possibility fields. But it is not a measure of the overall changeability of an arbitrary thing. Just how changeable (in all respects) is an arbitrary thing? Equivalently: How big are event spaces? To answer this (imprecise) question consider the set of *all* properties of any real thing and a realistic theoretical model of such a thing – say, the relativistic theory of the electron, or of the gravitational field. Such a model, or functional schema, is based on a certain set on which an n-component state function $\mathbb{F}$ is defined, every component of which represents a (changeable) property of the thing. At least some of the components of the state function are bound to be continuous with respect to some of the "independent variables". Call F_i any such continuous state variable. As the thing changes, the values of F_i vary continuously (or at least piece-wise continuously) even if those of some other components of $\mathbb{F}$ change by jumps. (Think of a thing that gains or loses quanta of electric charge, or people, while losing or gaining momentum.) See Figure 5.9. Clearly, the (state) space spanned by $\mathbb{F}$ is nondenumerable.

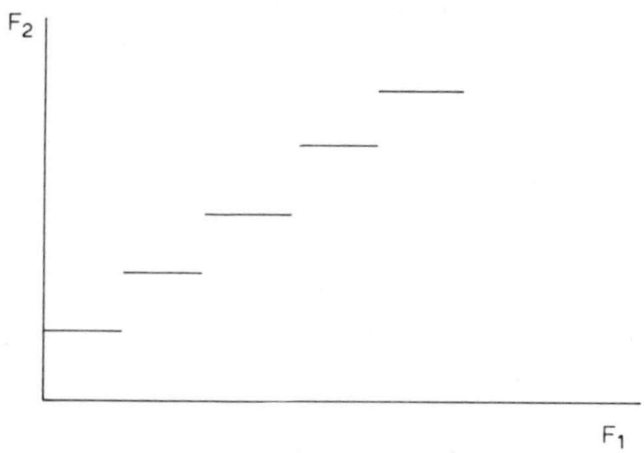

Fig. 5.9. A state space with a continuous axis (F_1) and a discontinuous one (F_2). The broken line represents the evolution of the thing. The totality of states is nondenumerable.

We assume that every thing has at least one property that changes continuously with respect to some variable or other – the "independent variable" being usually a time coordinate or a space coordinate:

POSTULATE 5.2 Every [realistic] state function has at least one continuous component.

Since a state space is the set of points scanned by the tip of the state function, we have

COROLLARY 5.7 Every [realistic] state space is nondenumerable.

And from the way events are built we also infer

COROLLARY 5.8 Every [realistic] event space is nondenumerable.

True, we often build models of things (model objects) whose state spaces are denumerable or even finite. But this is done for the sake of expediency rather than truth – i.e. because we are interested in just a few aspects of the whole thing. Thus, even though automata theory models certain machines as having only a finite number of states, the actual design and construction of such machines presupposes theories involving nondenumerable state spaces such as those analyzed in Ch. 3, Secs. 2.4, 2.5. Even if the universe were to consist of just two things in relative motion (e.g. one diatomic molecule), the world event space would be nondenumerable. In conclusion, realistic event spaces are as big as any subset of the real line can be: they all have the power of the continuum.

Because all realistic event spaces are equally bulky (Corollary 5.8), they do not supply a suitable measure of the change potential of real things. We must look elsewherc for such a measure. If we scan the factual sciences in search for a property shared by all things, regardless of kind and related to their changeability, we find but one universal property – namely, energy. Moreover energy has the property that it is greater the more complex the thing, hence the greater its potential for change. In fact, energy is strictly additive for noninteracting things, i.e. when the total state space of the compound equals the union of the partial state spaces (recall Ch. 3, Sec. 2.7). We therefore make

POSTULATE 5.3 There is exactly one property possessed by all things, that is additive for thing aggregates composed of detached parts. That is, if Θ is the class of things and F that of reference frames, then there exists one and only one real valued function $\varepsilon: \Theta \times F \to \mathbb{R}$ such that, for any x and y in Θ such that $x \rceil\!\!\lfloor y$,

$$\varepsilon(x \dotplus y, f) = \varepsilon(x, f) + \varepsilon(y, f) \text{ for all } f \in F.$$

This function ε is called *energy*, and its value at $\langle x, f \rangle$ – namely *the energy of x relative to f* – measures the *changeability of x relative to f*.

(*Energeia* was an ontological concept before being absorbed and transformed by modern physics. It is still sometimes regarded as a metaphysical notion though for the wrong reason, namely because it does not represent a perceptible property. The general concept of energy *is* a genuine ontological notion for a different reason, namely because it designates a universal property. For this reason it occurs everywhere in science, even though in some areas – particularly geography and history – it has entered only recently. In any event it is nice to see the concept of energy back in the philosophical fold. Because we now understand that energy is a property of things, there is no danger of it being reified, let alone be made into the only existent, as it was by the energetists of the turn of the century – e.g. Ostwald, 1902.)

It may be objected that Postulate 5.3 is not precise enough to define the energy concept. To this it may be rejoined, first, that science does not supply a definition of the general concept of energy. Second, to a physicist Postulate 5.3 is enough to *identify* ε as the energy function, for it is the only quantitative (and approximately additive) property all things have. (For things with interacting parts, the energy is slightly subadditive.) All our axiom states is that *all things have energy*. It does not supply a method for computing the energy of any thing in particular: this is a task for science. And no science can accomplish this task without some precise information (or conjecture) about the kind of thing, its structure, interactions, and circumstances.

One can conceive of certain properties that come in infinite degrees. For example, if the universe happens to be infinite in extension, then there are things that are infinitely distant from one another. On the other hand the energy of a finite part of the universe is normally taken to be finite. (True, both in classical and in quantum physics there are theories containing theorems – never axioms – according to which certain things, such as electrons and plane electromagnetic waves, have infinite energies. However, these formulas are regarded as blemishes of the theories not of nature, and several tricks are resorted to in order to avoid them.) We are therefore justified in adopting

POSTULATE 5.4 The energy of any thing composed of a finite number of things, and relative to any reference frame, is finite.

An important consequence of the finite energy (or potential for change) of every finite thing is that the pace and the extent of every change are bounded. (See Figure 5.10.) For example, the speed of a

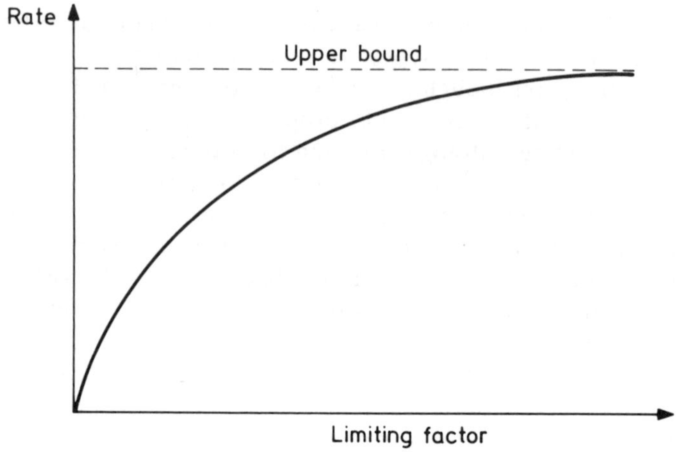

Fig. 5.10. Rate of process vs. intensity (or concentration) of a "limiting factor" such as energy.

body cannot exceed a certain limit precisely because it has a finite amount of energy. Likewise the limited energy resources of the biosphere – in particular of the soil – place a limit on human population. (For further examples and principles see Booij and Wolvekamp 1944.) All such particular statements concerning so-called "limiting factors" can be deduced from formulas regarding the energy of the particular system concerned. We compress them into

POSTULATE 5.5 The rate and the extent of every change in any thing composed of a finite number of things, are bounded.

(The restriction to finite things is due to the need for not excluding the possibility of the ceaseless expansion of the physical universe.)

An immediate consequence of the above postulate is that there is an end to every change. But this matter will be taken up in the next section.

Before closing this section we must ward off a possible misunderstanding concerning universal properties. It is often claimed that information constitutes an even more universal property than energy. However, not all things are information systems: only those constituted by an information source, an information channel, and an information receiver qualify as information systems, i.e. systems capable of generating, transmitting, and receiving information. Secondly, rather than being (relative to every reference frame) an intrinsic property, like energy, information is a mutual property in the sense that it is always an amount of information transmitted by an information carrier: change the channel (e.g. make it less noisy) and the amount of information will change. A cursory look at the basic assumptions and definitions of statistical information theory (e.g. Luce, 1960) will suffice to bear out our contention. Thirdly, while there can be energy exchanges without any information transmission – e.g. in atomic collisions – every transmission of information requires some energy. The fact that information theory pays no attention to the physical basis of an information process, hence to the energy exchanges underlying the transmission of information, does not annihilate that basis. We shall return to this matter in Sec. 4.4.

3. Process

3.1. *Serial Change*: *Types*

Heretofore we have studied change in general. In this section we shall analyze a particular kind of change, namely serial change or process. Not every old set of states or of events constitutes a process. For example an arbitrary collection of events occurring in different things that are comparatively isolated from one another does not constitute a process even if it is ordered in time. For a set of events to constitute a process it must satisfy two conditions:

(i) the events must involve or concern just one thing, however complex, and

(ii) the events must be ordered intrinsically.

In sum a process may be envisaged as a directed tree, in particular a chain, in some event space.

Before laying down general principles we may as well get hold of a stock of typical examples. We shall resort to these examples several times in the course of this work, so we had better display them ostensibly.

(*a*) *Chain*

The simplest process kind is the chain formed by a denumerable set of states, or of events, in a given thing. In this particular case we can formalize the notion of a process as a *sequence*, i.e. a function defined on the set $\mathbb{N}$ of natural numbers. More precisely, we can make

DEFINITION 5.16 Let $\mathbb{F}$ be a state function, and $S(x)$ the corresponding state space, for a thing x relative to a certain reference frame. Then x undergoes a *chain process* iff $\mathbb{F}: \mathbb{N} \rightarrow S(x)$, where $\mathbb{N}$ is the set of non-negative integers. In this case the process is represented by

$$\pi(x) = \langle \mathbb{F}(n) | n \in \mathbb{N} \rangle,$$

where $\mathbb{F}(n)$ is the nth *stage* of the process.

A finite segment of such a process is the restriction of the state function to a finite subset of natural numbers. The *length* of such a finite chain is of course the number of stages in it.

Example A population of unstable particles, or organisms, or any other things that come into and go out of being, will undergo a discontinuous process as far as the population number is concerned. If the rate of destruction is balanced by the rate of emergence, a steady state prevails in the given respect.

Discontinuous processes, though common and of great heuristic importance, are not universal. An even when they do take place it is worth while to try to analyze them further instead of taking the component events *en bloc* as was done above. This need not be done in the preliminary stages of a science – e.g. in learning theory at the present moment – but it is an ultimate desideratum. Finally because discontinuous processes do not exhaust the process genus, it makes no sense to ask about the number of events that have occurred during the history of the universe – not even during the last second of its existence.

(*b*) *Continuous process*

An obvious generalization of the denumerable chain is obtained by substituting an arbitrary directed set for the set of natural numbers in the previous definition. More exactly, we propose

DEFINITION 5.17 Let $\mathbb{F}$ be a state function, and $S(x)$ the corresponding state space, for a thing x relative to a certain reference frame. Then x undergoes a *serial change* iff $\mathbb{F}: T \rightarrow S(x)$, where T is a directed set [i.e. if $\mathbb{F}$ is a *net*]. In this case the process is represented by

$$\pi(x) = \langle \mathbb{F}(t) | t \in T \rangle.$$

Serial changes are often gapless or continuous, as when the state function is a continuous function of a parameter such as time. We need therefore

DEFINITION 5.18 Let $\pi(x) = \langle \mathbb{F}(t) | t \in T \rangle$ be a serial change in a thing x. Then $\pi(x)$ is *continuous* over the interval $U \subseteq T$ iff U is nondenumerable and $\mathbb{F}$ is a continuous function on U. And the process is *piece-wise continuous* over T iff T equals the union of finitely many nondenumerable sets, and $\mathbb{F}$ is continuous over every one of them.

Example 1 A reversible chemical reaction

$$A \underset{k_-}{\overset{k_+}{\rightleftharpoons}} B$$

in a closed reactor is described by the rate equations with positive coefficients

$$\frac{dA}{dt} = -k_+A + k_-B, \qquad \frac{dB}{dt} = k_+A - k_-B$$

subject to the closure or conservation condition $d(A+B)/dt = 0$. The state space of the whole thing is the $\mathbb{R}^+ \times \mathbb{R}^+$ plane and the evolution of the system is represented by an arc of a continuous curve in this plane.

Example 2 The trajectory of a particle that collides with other particles is broken but continuous even though the particle velocity changes abruptly at each collision. The simplest case is that in which the particle is initially at rest and obtains a constant velocity at $t = 0$ upon being pushed by another particle:

$$\dot{x}(t) = \begin{cases} 0 & \text{for} \quad t < 0 \qquad \therefore \quad x(t) = a = \text{const.} \\ 1 & \text{for} \quad t \geqslant 0 \qquad \therefore \quad x(t) = a + t. \end{cases}$$

It is sometimes claimed that continuity is more basic than discontinuity – and at other times that the converse is true. There seems to be no firm ground for either hypothesis, whence we shall adopt neither. We only seem to know that (*a*) both hypotheses have been ontologically and scientifically stimulating; (*b*) continuity hypotheses are empirically irrefutable or nearly so, and (*c*) while some properties (hence some processes) are continuous, others are not – at least according to contemporary science. For these reasons, instead of committing ourselves to either hypothesis we propose the more fertile heuristic

PRINCIPLE 5.4 Try to resolve every discontinuous change into a continuous process, and watch for discontinuities in every seemingly continuous process.

(*c*) *Path independence*

Whereas the outcome of some processes depends critically upon the precise trajectory in state space, others are *path independent* or nearly so. A particular case of great interest in all fields of inquiry, from physics to sociology, is that of equifinality, or same final state. The precise concept is elucidated by

DEFINITION 5.19 Two serial changes are *equifinal* iff their end states are the same.

Equifinality has been claimed to prove finality or goal-directedness, for it seems as if the system concerned – e.g. an organism – strives to attain a given goal by whatever means it can. However, equifinality is not evidence for teleology unless the existence of a purpose can be proved, for equifinal processes are common among physical things. For example, all closed thermodynamic systems approach a state of equilibrium.

Example Let the instantaneous value of a property F be given by one of the following functions on $\mathbb{R}^+$:

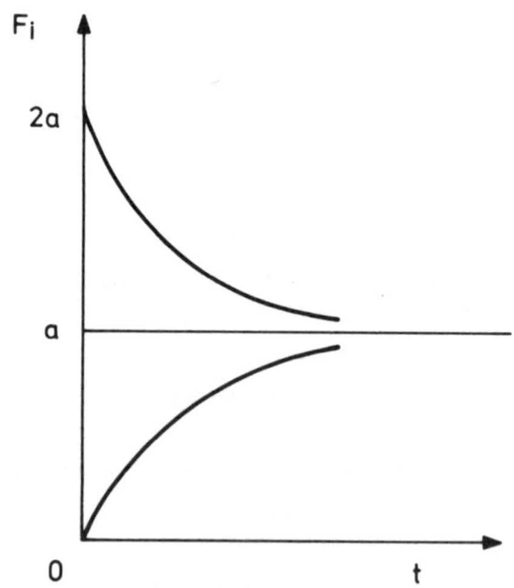

$$F_1(t) = a(1 + e^{-kt})$$
$$F_2(t) = a$$
$$F_3(t) = a(1 - e^{-kt})$$
with $a, k \in R^+$.

Fig. 5.11. Three different ways of reaching the same final state or asymptotic value $X_\infty = a$: starting from a, advancing to it, or declining to it.

(*d*) *Memory*

Suppose the state of a thing depends not only upon a parameter t (the phase of the process) but also upon some preceding state or states. In this case one says that the thing has a *memory* of some of its past states, or that it undergoes a *hereditary* process. Thus the elastic state of a piece of steel, the magnetic state of a magnet, and the genic state of an organism, all depend on their previous histories. In a number of cases such dependence can be expressed by a formula of the type

$$\mathbb{F}(t) = \int_0^t \mathbb{F}(\tau - t) \cdot M(\tau) \cdot d\tau,$$

where M is the memory function. (The process is memoriless iff $M = \delta$, or the Dirac delta.)

The general concept is clarified by

DEFINITION 5.20 Let $\mathbb{F}$ be a state function and $S(x)$ the corresponding state space, for a thing x relative to a certain reference frame. Further, call $t \in T$ the state parameter, where T is a directed set. Then x undergoes a *hereditary process* iff every one of its states $\mathbb{F}(t) \in S(x)$ is a functional of some other states $\mathbb{F}(\tau)$ with $\tau < t$.

There are two important species of hereditary process: (*a*) the memory is kept fairly intact throughout the process, and (*b*) the memory fades away until it disappears practically after a certain time called the *relaxation time*. Systems with weakly interacting components, such as non-ionized gases and extended empires, have short relaxation times. On the other hand organisms have long genetic and ontogenetic relaxation times. For example, malnutrition and deprivation of sociality in the early life of a mammal are separately sufficient to seriously impair some of its abilities for life.

(*e*) *Reversibility and irreversibility*

Some processes, particularly on the microphysical level – and probably only such – are reversible; all known macroprocesses are irreversible. The latter may be regarded as a generalization of a theorem proved in statistical mechanics. A system is called *ergodic* if it eventually scans its entire state space – or, more precisely, if, when left undisturbed (isolated), it occupies successively every state consistent with its total energy.

Until 1913 most authors asserted that, since individual particles move reversibly, every collection of particles must have the reversibility property and, moreover, must be ergodic. That year A. Rosenthal and M. Plancherel proved independently that, if the system satisfies the canonical equations (Ch. 3, Sec. 2.6), then it is not ergodic. That is, they proved that there are no classical-mechanical ergodic (or fully reversible) systems. Hence even if the universe were a huge aggregate of particles (which it is not) it could not retrace all of its steps. There is no eternal return.

Finally we note that in microphysics most, but by no means all, processes are assumed to be *stochastically reversible*, in the sense that the probability of an elementary process and that of its reverse are the same. This allows for momentary deviations from the general trend, i.e. for isolated events in which some states become overpopulated or underpopulated. However, certain microprocesses, notably radioactivity and, in general, transmutation processes, are irreversible.

All this invites

DEFINITION 5.21 Let $\pi(x) = \langle \mathbb{F}(t) | t \in T \rangle$ be a serial process in a thing x. Further, let $\bar{T}$ be the same set as T directed in the opposite way (e.g. $\langle \mathbb{R}, > \rangle$ instead of $\langle \mathbb{R}, < \rangle$). Then $\pi(x)$ is *reversible* iff $\bar{\pi}(x) = \langle \mathbb{F}(t) | t \in \bar{T} \rangle$ is just as possible (i.e. lawful) as $\pi(x)$.

If both $\pi(x)$ and its reverse $\bar{\pi}(x)$ can be assigned probabilities, one can think of defining the *degree of reversibility* of a process as the ratio of its probability and that of its reverse, or $r = \bar{p}/p$. For a fully reversible process, $r = 1$; for a fairly reversible process, $0 \ll r < 1$; for a nearly irreversible process, $0 < r \ll 1$; and for a fully irreversible process, $r = 0$. Such a concept might be useful in a number of fields, particularly in statistical mechanics.

It is common to declare a process to be reversible just in case its laws are invariant under time reversal. This is a mistake, because processes depend on circumstances as well as laws. For example, although the basic laws of radioactive disintegration (chiefly Schrödinger's) are T-invariant, the boundary conditions are such that the probability for a disintegration fragment to reenter the nucleus vanishes.

Finally, note that, while all reversible processes are cyclic, the converse is false. In fact most cycles are not reversible – i.e. they do not leave things as they were. For example, the water cycle and the nitrogen cycle on our planet are not reversible, as they are accompanied by energy degradation (entropy increase).

(*f*) *Random process*

A conspicuous kind of serial process, and one of great philosophical interest, is that of a random process. (For a definition of randomness see Ch. 4, Sec. 6.4.) A random or stochastic process is one described by a state function that is a random "variable", i.e. a function every value of which has a definite probability or probability density. In the simplest case the values s_i of the state function $\mathbb{F}$ form a discrete sequence $\langle s_i | i \in \mathbb{N} \rangle$, such as the times of arrival of customers at a counter. In this case the function f such that $f(s_i) = Pr(\mathbb{F}(t) = s_i) = p_i$, where Pr is a probability measure and p_i the probability that x happens to be in state s_i, is called the *probability distribution* of $\mathbb{F}$. (See e.g. Feller, 1968, Ch. IX.)

Many substantial properties, at all levels, must be represented by random "variables". Hence any real process – in contrast to an idealized model of it – is likely to have at least one random aspect. Therefore, instead of defining the concept of a real process (as is usual in the literature on stochastic processes) we prefer to adopt

DEFINITION 5.22 Let $\mathbb{F} = \langle F_1, F_2, ..., F_n \rangle$ be a state function for a thing x undergoing a serial change $\pi(x) = \langle \mathbb{F}(t) | t \in T \rangle$. This process is said to be *random in the ith respect* iff the component F_i of $\mathbb{F}$ is a random variable.

(*g*) *Stability*

A thing is said to be stable if its representative point in state space remains confined within a "small" region. More precisely, we can make

DEFINITION 5.23 Let $G_{\mathbb{L}}(x)$ be the set of lawful transformations of the state space $S_{\mathbb{L}}(x)$ of a thing x. Then x is *stable in the region* $A \subset S_{\mathbb{L}}(x)$ iff, for all $s \in S_{\mathbb{L}}(x)$, $g(s) \in A$ for every g in $G_{\mathbb{L}}(x)$.

In other words a thing is in a state of *dynamical equilibrium* if all the changes it undergoes send it to a fixed subset of its state space. If this subset reduces to a point then the thing is said to be in *static equilibrium* or not to change at all. In other words, a thing x is in static equilibrium at a state $a \in S_{\mathbb{L}}(x)$ iff a is a fixed point under every transformation of the state space into itself – i.e. if $g(a) = a$ for all g in $G_{\mathbb{L}}(x)$. In short, just as fixed subsets of the state space represent dynamical equilibrium, so fixed points represent states of static equilibrium.

States of static equilibrium can only be temporary. On the other hand dynamical equilibrium can be lasting. Figure 5.12 exhibits the states of

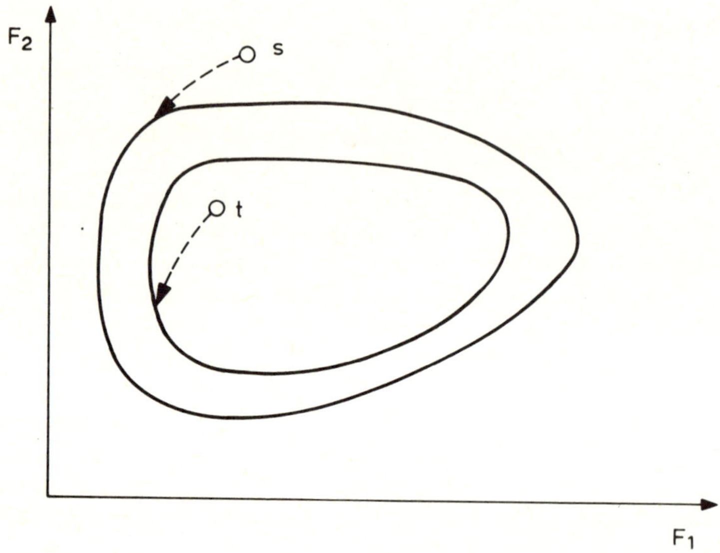

Fig. 5.12. Ecological equilibria are dynamical. If an external perturbation, such as farming or a draught, sends the representative point to *s*, or to *t*, the point eventually moves to an equilibrium orbit. (See e.g. Pielou (1969).)

dynamical equilibrium of an ecosystem consisting of two competing populations. Each curve represents a cyclic process – an oscillation in the population numbers. Each point lying on one of the closed curves represents a stable state of the system. If the populations are displaced – e.g. by some external disturbance – to values not too far from equilibrium, they tend spontaneously to return to an equilibrium point.

So much for special types of process. Let us hasten to warn that, in our view, no real process is of a pure type. This is because any (realistic) state function has a number of components of different kinds – some discrete, others continuous, some random, others nonrandom, and so on. In other words, call

$$\pi_k(x) = \langle F_k(t) | t \in T \rangle, \quad \text{with} \quad 1 < k \leqslant n,$$

the *k*th *component* of the serial process represented by $\pi(x) = \langle \mathbb{F}(t) | t \in T \rangle$. Then $\pi(x)$ will have at least one continuous component, at least one random component, and so on. In short, pure process types are ideal types: real processes are impure. We defer an exact formulation to the next subsection.

We first turn to some generalities that can be formulated in fairly exact terms.

3.2. *General Concepts and Principles*

In the previous subsection we discussed a few typical kinds of process. In the present one we shall treat processes in general, using some insights gained in the study of special process types. From that study it is apparent that a process may be regarded either as a *sequence of states* or as a *sequence of events*. In the former case we can describe a process in a thing x with the sole help of a state function $\mathbb{F}$ for it: our task boils down to following the wanderings of the tip of $\mathbb{F}$ – a task greatly facilitated by parametrizing the states $\mathbb{F}(t)$ relative to the states of a suitable frame. If on the other hand we construe a process as a sequence of events, then we must be given not only $\mathbb{F}$ but also the entire state space $S_{\mathbb{L}}(x)$ spanned by $\mathbb{F}$ as well as the totality $G_{\mathbb{L}}(x)$ of lawful transformations of the state space into itself. This alone allows us to form the individual events $\langle s, s', g\rangle$ and thereupon the whole event space $E_{\mathbb{L}}(x)$.

Each of these alternative construals has its advantages and disadvantages. The successive states approach is mathematically simpler, if only because the state space is n-dimensional whereas the corresponding event space is $2n$-dimensional. On the other hand the successive events approach, though mathematically clumsier, is philosophically more perspicuous. We can use either approach interchangeably and shall do so. In the present subsection we shall discuss some general principles in terms of events, but shall end up by introducing a notion of history based on the successive states approach.

First the general notions of process:

DEFINITION 5.24 Let $E_g(x)$ be the space of events of type g in thing x, for g in the set $G_{\mathbb{L}}(x)$ of lawful transformations of the state space chosen for x, and let $E(x)$ be the union of all $E_g(x)$'s, i.e. the total event space of x [Definition 5.9]. Further, let $<$ be the precedence relation [Definition 5.10]. Then

(i) a *process of type* g in thing x is a strictly partially ordered set of events of kind g:

$$\pi_g(x) = \langle E_g^*(x), <\rangle \quad \text{with} \quad E_g^*(x) \subseteq E_g(x) \subseteq E_{\mathbb{L}}(x);$$

(ii) a *process* in x is a strictly partially ordered set

$$\pi(x) = \langle E^*(x), <\rangle \quad \text{with} \quad E^*(x) = \bigcup_{g \in G_L(x)} E_g^*(x).$$

Since as a matter of fact every aspect g is related to some other aspect, we may safely surmise that there are no processes of a single type but only idealizations that single out now one feature of the thing of interest, now another. (Recall the end of the last subsection.) That is, we assume

POSTULATE 5.6 Every real process has more than one aspect.

And from the assumed lawfulness of the transformations in $G_L(x)$ it follows that every process involving a given thing satisfies all of the laws possessed by the latter:

COROLLARY 5.9 Every process is lawful.

Note that lawfulness is predicated of processes, i.e. intrinsically ordered sets of events, not of arbitrary sets of events. The latter, particularly if they occur in unrelated things, or in things with weakly coupled components, are not necessarily lawful even though every component is assumed to have a lawful behavior line. This is the ontological ground of the methodological rule employed tacitly by historians intent on discovering general historical trends or even laws – namely to focus on processes of a definite kind occurring in social systems instead of just looking for time series, which are bound to be erratic or lawless (see Braudel, 1969).

Remember however that chaos is not the same as randomness: whereas the former is lawless the latter possesses stochastic (probabilistic) laws such as those of quantum mechanics and genetics. Therefore we adopt an axiom employing Definition 5.22 of a process random in some respect:

POSTULATE 5.7 Every process is random in at least one respect.

Next we distinguish two important members of a process:

DEFINITION 5.25 Let $\pi(x) = \langle E^*(x), <\rangle$ be a process in a thing x. Then

(i) $\pi(x)$ *begins* at an event $b \in E^*(x)$ iff b is a lower bound of the set $E^*(x)$ [i.e. if every other member of $E^*(x)$ follows b];

(ii) $\pi(x)$ *ends* in an event $e \in E^*(x)$ iff e is an upper bound of the set $E^*(x)$ [i.e. if all the other members of $E^*(x)$ precede e].

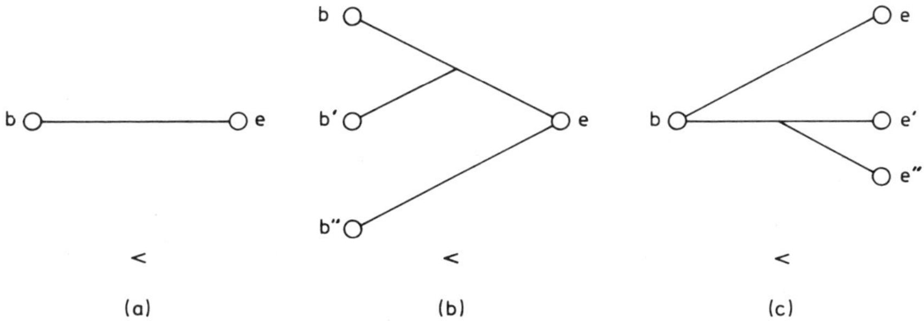

Fig. 5.13. Three kinds of process as to beginning and end. (*a*) Chain or linear process. (*b*) Converging or branching-in process. (*c*) Diverging or branching-out process.

A process may have no single beginner or single stop: see Figure 5.13. In other words, there may be multiple beginnings or multiple ends. We assume that every single process is bounded both below and above:

POSTULATE 5.8 Every process has beginning(s) and end(s). [I.e., if $\pi(x)$ is a process going on in a thing x, then $\pi(x)$ has at least one beginner and at least one end point.]

This hypothesis does not entail that there are absolute beginnings (i.e. processes starting out of the blue) and dead ends (i.e. processes that, on ending, do not start any other processes). Indeed the postulate is compatible either with a creationist and eschatologist cosmology or with the assumption that every process in a given thing, though bounded both below and above, is preceded and followed by some other processes either in the same thing or in other things. We adopt the latter hypothesis because it is the one supported by science. But first we need

DEFINITION 5.26 Let $\pi(x) = \langle E^*(x), \prec \rangle$ and $\pi'(x) = \langle E'^*(x), \prec \rangle$ be two processes in a thing x, such that $E^*(x) \cap E'^*(x) = \varnothing$. Then $\pi(x)$ *precedes* $\pi'(x)$ iff every event in the former precedes every event in the latter. Symbol: $\pi(x) \prec \pi'(x)$.

We are now ready for

POSTULATE 5.9 Every process is preceded and followed by other processes. More precisely, every thing x has parts y and z, not necessarily different from x, each of them equipped with its own event space, such that

$$\pi(y) < \pi'(x) < \pi''(z).$$

Since the world is the aggregation of all things, it follows

COROLLARY 5.10 The world has neither beginning nor end.

Remark 1 Like many another axiom of our ontology, the previous one is a sort of generalization of all we know. An ancient paradigm is of course the birth-growth-final disintegration process every metazoan undergoes. A simpler and more modern paradigm is that of the photon emitted by a star, travelling through interstellar space, and ending up trapped by a molecule or turning into a pair of electrons of opposite charges. *Remark 2* At one point some cosmologists wishing to save the steady state cosmology introduced the *ad hoc* hypothesis of the spontaneous creation of hydrogen atoms in empty space. This conjecture was not accepted because it clashes with the conservation laws (Bunge, 1962) – and anyway the steady state theory, which it tried to save, was eventually refuted by astronomical evidence. *Remark 3* Corollary 5.10 was first formulated by the Presocratic philosophers and was adopted by Aristotle. It would seem to contradict the "big bang" hypothesis of the "origin" of the universe, nowadays almost universally favored. However, this cosmological conjecture – which explains the expansion of the universe – does not illustrate creation *ex nihilo*, for it presupposes that the universe had been in existence – in a highly condensed state – before the expansion started. Nor does the old "thermal death" hypothesis (nowadays almost forgotten) illustrate absolute annihilation: all it asserts is the end of macrophysical motion once a closed system has attained thermal equilibrium.

So much for our general concepts and principles concerning process. Let us now adopt the successive states approach and introduce one more key notion that will prove indispensable in the sequel, namely that of history.

Recall that the states of a thing x can be coordinatized relative to the states $t \in S(f)$ of a reference frame f, by defining the state function as a mapping of $S(f)$ into $S(x)$, and by construing $S(f)$ as a four-dimensional grid conceptualizing a physical reference frame (Sec. 2.5). As the reference state t "moves" so does the tip $\mathbb{F}(t)$ of the thing state function. This invites

DEFINITION 5.27 Let $\mathbb{F}$ be a state function for a thing x relative to a reference frame f with states t in $S(f)$, and let the latter be coordinatized by a certain function $k: \mathbb{R}^4 \to S(f)$. Then the *history of x relative to f* is the set of ordered pairs

$$h(x) = \{\langle t, \mathbb{F}(t)\rangle \mid t \in S(f)\}.$$

This is an elucidation of the concept of a life line, behavior line, or trajectory. The history of a thing is regarded as the succession of its states but, instead of being represented by a line in the n-dimensional state space spanned by $\mathbb{F}$, it is represented as a curve in the $(n+4)$-dimensional space $\mathbb{R}^4 \times S_{\mathbb{L}}(x)$. (The dimensionality of this space can be reduced by 3 if we are interested only in the time coordinate, i.e. if t is construed as an instant of time rather than as a point in the spacetime region attached to the given frame.)

The above definition covers the most usual notions of history employed in the sciences. In particular it fits in with the notion of history occurring in one of the most fundamental theories in physics, namely quantum mechanics. Indeed in this theory the evolution operator for a thing is defined as

$$S(t_0, t) = \exp\left(-\frac{i}{\hbar}\int_{t_0}^{t} du\, H\right),$$

where H is the hamiltonian of the thing. The history of the latter is then

$$h(x) = \{\langle t, S(t_0, t)\psi(t_0)\rangle \mid u \in [t_0, t]\ \&\ \psi \in \text{Hilbert space of } x\}.$$

The broad concept of history adopted above may puzzle and even infuriate those scholars in the humanities who still regard man as supernatural. Until about the mid 19th century it had been tacitly assumed that all history is human history. Thus Hegel, for all his dynamicism, denied that nature has a history. That view changed with the revolutionary discoveries in geology and biology, to the point that Engels declared each science to be the history of something (Engels, 1878). Since then it is usually taken for granted that *every thing has a history* – i.e. that, for every thing x, $h(x)$ is bigger than a singleton.

Let us now apply the concept of history to a characterization of the notions of action and interaction.

4. Action and reaction

4.1. *Induced Change*

So far we have been concerned with change without regard for its sources. Now, a thing may change either by itself or under the influence of some other thing, i.e. spontaneously or by induction. We must therefore study the concepts of spontaneity and induction. To do this we shall avail ourselves of the concept of history (Definition 5.27 in Sec. 3.2) and of Leibniz' law (Ch. 2, Sec. 3.2).

Two different things of the same kind may be described with the help of the same functional schema $\langle M, \mathbb{F}\rangle$, hence of the same state space, though of course adding individuating circumstances in each case. In other words, the states of different kindred things may be represented in one and the same state space. Also: every state space, far from being owned by a single thing, may be assigned a reference class constituted by a more or less numerous equivalence class of things. Now, Leibniz' law (Postulate 2.5 or Corollary 2.1) states that different things have different properties. Hence even if a given state function describes adequately two things, they cannot occupy exactly the same states – for, if they did, they would be one. Consequently we have the following rephrasing of Leibniz' law:

COROLLARY 5.11 Different things have different histories.

Surely a thing can occupy now exactly the same state another thing was in a moment ago, but no two things can occupy exactly the same state at the same time – or, more generally, for the same value of the change parameter. (See Figure 5.14.)

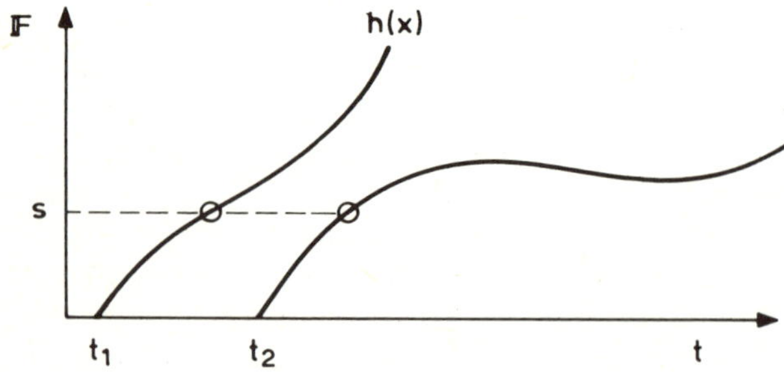

Fig. 5.14. Different things can occupy the same state provided they do it at different times.

Corollary 5.11 is our present version of Leibniz' *principium individuationis* and it supplies the clue to what we are after. Indeed, if a thing is under the action of another thing, then its history must differ from the history when free from such action. (See Figure 5.15.) (This is what scientific experiment, as different from observation, is all about, namely the deliberate switching on and off of actions or influences to compare the forced behavior of a thing to its free behavior and thus to assess the importance of the variable that is being manipulated.)

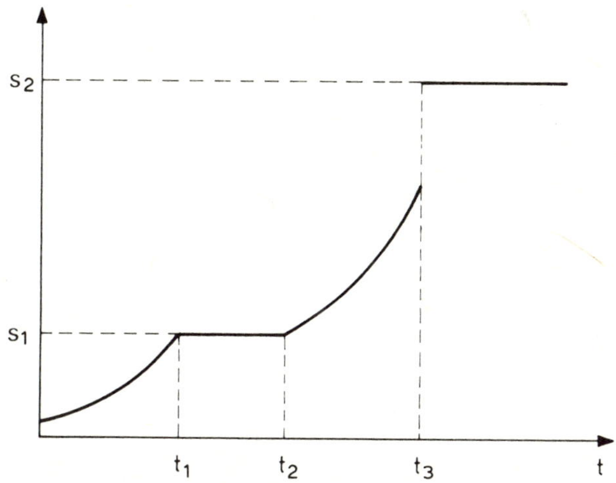

Fig. 5.15. Imaginary example of an action. The agent, by acting during the $[t_1, t_2]$ interval, forces the patient to remain in state s_1. The external action resumes at time t_3, forcing the patient to jump to state s_2.

In the absence of any actions of the agent x upon the patient y, the history of their physical sum or juxtaposition $x \dotplus y$ equals the union of their individual behavior lines. In other words, we have

DEFINITION 5.28 Thing x changes by itself, or *spontaneously*, iff $h(x)$ is not a singleton and, for any other thing y,

$$h(x \dotplus y) = h(x) \cup h(y).$$

Clearly, if one of the things acts upon the other, then while the behavior line of the agent remains unaffected, that of the patient becomes, as it were, the history of a different thing, namely the thing-acted-on-by-the-agent. (See Figure 5.16.) More exactly, we make

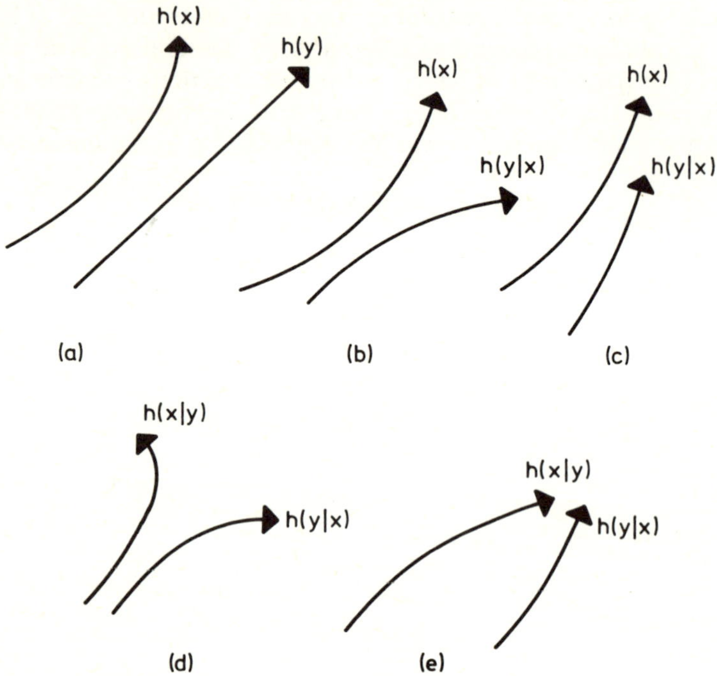

Fig. 5.16. Induced change. (*a*) No action: spontaneous change of both *x* and *y*. (*b*) Pushing action of *x* on *y*. (*c*) Pulling action of *x* on *y*. (*d*) Pushing interaction. (*e*) Pulling interaction.

DEFINITION 5.29 Let *x* and *y* be two different things with state functions $\mathbb{F}$ and $\mathbb{G}$ respectively relative to a common reference frame *f*, and let

$$h(x) = \{\langle t, \mathbb{F}(t)\rangle | t \in S(f)\}, \qquad h(y) = \{\langle t, \mathbb{G}(t)\rangle | t \in S(f)\}$$

be their respective histories. Further, let $\mathbb{H} = g(\mathbb{F}, \mathbb{G}) \neq \mathbb{G}$ be a third state function, depending on both $\mathbb{F}$ and $\mathbb{G}$, and call

$$h(y|x) = \{\langle t, \mathbb{H}(t)\rangle | t \in S(f)\}$$

the corresponding history. Then *x acts on y*, or $x \rhd y$ for short, iff, for some state function $\mathbb{H}$ determining the trajectory $h(y|x)$, $h(y|x) \neq h(y)$.

We may now rephrase Definition 5.28 of spontaneity: A thing *changes spontaneously* iff it changes and if there is no other thing that acts upon it. And, from the hypothesis of Definition 5.29 (in particular, that $x \neq y$), it follows that *nothing acts upon itself.*

The definition of reciprocal or mutual action is now obvious:

DEFINITION 5.30 Two different things x and y *interact* iff each acts upon the other. In symbols:

$$x \rhd\!\lhd y =_{df} x \rhd y \;\&\; y \rhd x.$$

Experimental scientists assume tacitly (a) that all things are studiable, (b) that there are no totally isolated things – i.e. that every thing acts on, or is acted upon by, some other thing, and (c) such actions are the basis of empirical knowledge. Indeed if a presumed thing does not act on any of our observation means then we cannot observe it; and if some putative thing is not acted upon by any of our experimental means then we cannot experiment on it. In either case we won't bother with "it". In short, the following axiom belongs to the ontology inherent in science, so we adopt it:

POSTULATE 5.10 Every thing acts on, and is acted upon by, other things.

This assumption should not be mistaken for the far stronger, and false, hypothesis that any two things interact – let alone with the same intensity. We do not react back to the light activity of the sun, even less to that of an extinct star. In other words, the universal interaction principle, first suggested by the Stoics and much favored during the acme of Newton's gravitation theory, is false.

The effects of an interaction may be lasting, even in the case of microthings. Thus if two microthings interact for a while and then cease to do so – e.g. by becoming widely separated in space – they will continue to be correlated. Hence observing one of them may yield information about the other – just as observing the behavior of a divorced person may shed light on his/her former mate. In the jargon of quantum mechanics, such things are said to be *non-separable*, or to exhibit an EPR correlation. (This is what the Einstein–Podolsky–Rosen paradox seems to be all about. More in Sec. 4.2.)

We need a qualitative concept of the size of an action or an interaction. This is given by the amount of the distortion effected by the agent(s) on the behavior line of the patient. In other words, we make

DEFINITION 5.31 Let x and y be two things, and let $h(x)$ be the history of x, and $h(x \mid y)$ the history of x when acted upon by y, and similarly for x. Then

(i) the *total action* or *effect* of x on y equals the difference between the forced trajectory and the free trajectory of y:

$$A(x, y) = h(y|x) \cap \overline{h(y)};$$

(ii) the *total reaction* of y upon x is the difference between the free trajectory and the forced trajectory of x:

$$A(y, x) = h(x|y) \cap \overline{h(x)};$$

(iii) the *total interaction* between x and y is the union of action and reaction:

$$I(x, y) = A(x, y) \cup A(y, x).$$

If $A(x, y) = \varnothing$ we can say that x exerts the *null action* on y.

Since the action of x on y is a set of events in y, this set cannot be the same as the reaction of y on x, which is a set of events in x. Therefore we have

COROLLARY 5.12 No action equals the corresponding reaction. I.e., for any things x and y, $A(x, y) \neq A(y, x)$.

This result is not inconsistent with the third law of Newtonian particle mechanics. Indeed this law does not state that actions and reactions are the same but that their strengths (the corresponding forces, or else torques) are equal. Anyway the law does not hold in other theories, such as electrodynamics.

Now, actions may exist in certain respects but not in others. That is, the action of thing x upon thing y may affect only some of the components of the state function of y. Hence we may analyze the total action into partial actions, namely thus:

$$A(x, y) = \bigcup_{i=1}^{m} A_i(x, y),$$

where m is the number of mutually independent actions that x exerts on y. Similarly for the reaction of y on x and for their interaction.

This distinction allows us to supplement Postulate 5.8 with

POSTULATE 5.11 Every thing changes spontaneously in some respects, inductively in others.

Finally, the concept of action permits us to elucidate the important distinction made in Ch. 2, Sec. 5.1, between a relation that affects its relata and one that does not. (See Figure 5.17.) The difference is rendered precise by

DEFINITION 5.32 Two different things are *bonded* (or *linked* or *coupled*) together iff at least one of them acts upon the other. In symbols: If x and y are things, then

$$Bxy =_{df} x \rhd y \quad \text{or} \quad y \rhd x.$$

Note that we are not including the condition that the bond be attractive. Two enemies are just as bonded as two friends.

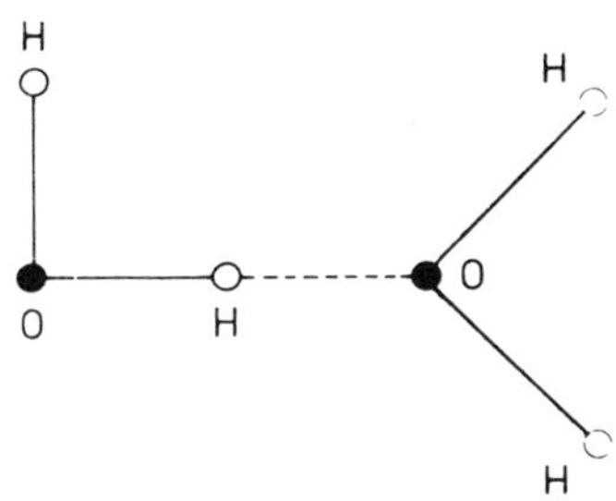

Fig. 5.17. Three different relations between two water molecules. (*a*) The valence bonds between the atomic components of each molecule (full lines); (*b*) the hydrogen bond between both molecules (dotted line); (*c*) the relation of being to the left (or its converse). Both (*a*) and (*b*) are bonds, links, or couplings, whereas (*c*) is a nonbonding relation.

The collection of bonds among members of an arbitrary set T of things deserves a special name:

DEFINITION 5.33 The *bondage* of a set T of things is the set $\mathbb{B}_T$ of all bonds among members of T.

In principle $\mathbb{B}_T$ includes couplings of all n-arities – dyadic, triadic, and so on. In practice the dyadic or binary bonds are by far the most important. If only binary bonds are taken into account, the above definition reduces to

$$\mathbb{B}_T = \{B_i \in \text{Binary relations on } T \mid (\exists x)(\exists y)$$
$$(x, y \in T \;\&\; B_i xy \;\&\; 1 \leq i \leq m\},$$

where $m \geq 1$ is the number of kinds of binary bond [or of action] holding among members of T.

It will be recalled that we assumed that there are finitely many general substantial properties (Postulate 2.3). Therefore we have

COROLLARY 5.13 The bondage of any set of things is finite.

The class of all nonbonding (or "mere") relations among T's will be designated by $\bar{\mathbb{B}}_T$. And the union of $\mathbb{B}_T$ and $\bar{\mathbb{B}}_T$ equals of course the set of all relations on T. This concept will become crucial in our treatment of the concept of a concrete system in Vol. 4, Ch. 7, so it invites

DEFINITION 5.34 Let T be a set of things and $[T]$ its aggregation [physical sum]. Then the *structure* of $[T]$ equals the totality of relations, both bonding and nonbonding, on T:

$$\mathscr{S}([T]) = \mathbb{B}_T \cup \bar{\mathbb{B}}_T.$$

We shall use the concept of a bond to characterize systems and reference frames.

4.2. *Aggregates and Systems*

We shall now employ the concepts of bond and history to clarify the distinction between a heap or aggregate, on the one hand, and a whole or system, on the other. Obvious examples of systems are atoms and molecules. On the other hand the nucleons and electrons composing an atom are not systems since they have no separable components. Another example of a nonsystem is the aggregate formed by a thing and a reference frame. Since reference frames are supposed to record without either influencing or being influenced, the history of a thing-frame compound must reflect that independence. How this is done will be seen in Sec. 4.3 once we have properly introduced the notion of a system.

An analysis of the histories of the members of a set of things should help us discover whether or not the aggregation of the set constitutes a system, i.e. whether or not its components are bonded. Indeed, the history of an aggregate of noninteracting things is uniquely determined by the histories of such components: it is just their union. Not so in the case of a system: here the history of every component is determined at least partly by the states other components are in, so that the history of the whole does not equal the sum of the individual histories. An example or two will help bring this point home.

Think of the histories of a moth and a candle before and after the former spirals into the latter. If a more distinguished example is needed, think of a system composed of two electrons close enough to interact appreciably, as is the case with those of a helium atom. This system is described by a Schrödinger equation (classically by two coupled equations of motion) jointly with the Pauli exclusion principle. The latter selects those state functions that are odd (antisymmetric) in the coordinates of the electrons. (That is, the additional law is: $\psi(x, y) = -\psi(y, x)$, where x and y are the position coordinates of the electrons. It follows that $\psi(x, x) = 0$, i.e. the two entities cannot coexist at the same point – unless they have different spins.) The Pauli principle expresses a global or systemic property, one that the individual components fail to possess, whence it cannot be represented in the partial state spaces. Therefore the construction of the state space of the system must proceed from scratch rather than on the sole basis of the state spaces of the individual electrons. (Fortunately the Pauli principle does not hold for the entire universe but only for certain parts of it, namely the subsystems formed by half integral spin components. Otherwise there would exist no fairly isolated systems in reality. (Cf. Margenau (1966).) What holds for state spaces holds a fortiori for histories.

We summarize and generalize the preceding remarks into

DEFINITION 5.35 Let X be a thing composed of parts X_i for $1 \leq i \leq n$. Then X is an *aggregate* (or *conglomerate* or *heap*) iff, in every representation of X (i.e. for every choice of state function), its history $h(X)$ equals the union of the partial histories $h(X_i)$. Otherwise X is a *system*.

The cautious phrase 'in every representation', occurring in the preceding convention, is due to this. It is often possible to choose a representation of a thing wherein the mutual actions of its components are "absorbed" in the latter. In other words one can often model the real system as a conceptual aggregate of noninteracting components. But this trick does not work for every representation.

Recalling the Definitions 5.31 of action and 5.33 of bondage, we obtain

COROLLARY 5.14 Let X be a thing with composition $\mathscr{C}(X)$. Then X is a *system* iff the bondage of $\mathscr{C}(X)$ is not empty, i.e. $\mathbb{B}_{\mathscr{C}(X)} \neq \varnothing$.

Now, the universe or world is that thing which is composed of all things (Postulate 3.3); and by Postulate 5.10 every thing – hence every component of the world – is bonded to some other thing. Hence

COROLLARY 5.15 The world or universe is a system.

By Definition 5.35 it follows in turn that the history of the world is not equal to the union of the histories of its parts.

What happens if a system breaks down, e.g. if its components fly apart and cease to interact? It would seem that, when this happens, the history of the aggregate does equal the union of the histories of the former components, which are now free from one another. This is not necessarily so. In fact the famous Einstein–Podolsky–Rosen paradox of quantum mechanics may be interpreted as follows. If two quantal things interact at any given time then, even after they have ceased interacting, they are still correlated in the sense that the history of each is affected by what happens to the other. (Equivalently: the state vector of the whole does not equal the direct product of the state vectors of the parts.) Thus interaction at some time is sufficient to induce a correlation that is reflected in the nondecomposability of the total state space, and a fortiori in the nonseparability of the partial histories. If this interpretation is correct, then the above definitions may be kept but a new assumption must be added: *Once a (micro)system always a (micro)system.* However, this is still debatable, so we shall not incorporate it into our system.

4.3. *Reference Frame*

Another important application of the notion of an aggregate is to the characterization of a reference frame. This concept is of paramount importance not only in physics but also in our ontology, for in real cases states are frame dependent. (The paradigm is of course the state of motion of a body: its track in state space is frame dependent. So much so that it won't move at all relative to a frame rigidly attached to itself.)

One condition for a thing to qualify as a reference frame for another is that the two not influence each other, so that their respective states be neatly separated. (To indulge in anthropomorphism: a reference frame must be an impartial spectator. But this is just a didactic prop: we must not mistake frames for observers.) Another condition is that the states of the reference frame for a given thing be utilizable for parametrizing the states of the thing of interest, in the sense that, for every state t of the reference frame, the thing is in a given state $s = \mathbb{F}(t)$, where $\mathbb{F}$ is the n-component state function for the thing. (See Figure 5.18) These two

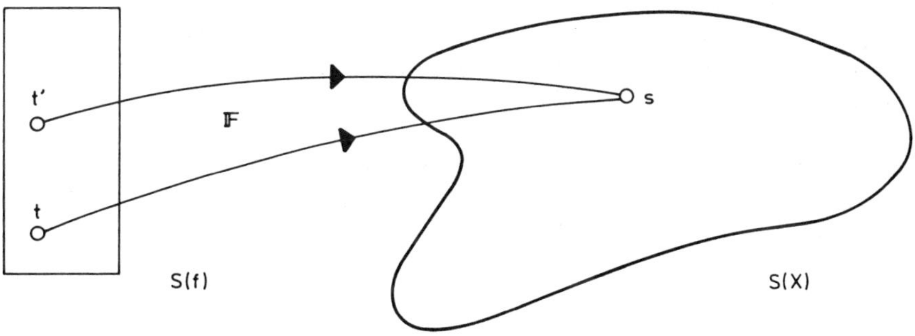

Fig. 5.18. The mapping $\mathbb{F}$: $S(f) \to S(X)$ of reference states onto thing states. $\mathbb{F}$ need not be 1–1: two different reference states may correspond to a single thing state, as in periodic motion. But $\mathbb{F}$ must be onto: each thing state must be paired off to at least one reference state.

conditions – independence and parametrizability – will be taken to be jointly necessary and sufficient to characterize the general concept of a reference frame. (Each scientific theory may define its own specific concept of a reference frame by adding its own additional conditions – usually that certain laws be satisfied relative to the frame. See Bunge (1967b)). In sum, we make

DEFINITION 5.36 Let X be a thing represented by the functional schema $X_m = \langle M, \mathbb{F} \rangle$ and let f be another thing with states in $S(f)$. Then f is a *reference frame* for X iff

(i) $f \dotplus X$ is an aggregate, so that $h(f \dotplus X) = h(f) \cup h(X)$, and

(ii) the domain of the state function of X equals the state space of f, i.e. $\mathbb{F}$: $S(f) \to S(X)$, whence $M = S(f)$.

Remark 1 We have defined a reference frame as a concrete thing of a certain kind. Because every reference frame can be *represented* by a coordinate patch in a manifold, the misconception has spread that reference frames are identical with their conceptual representatives. That this is mistaken is realized upon recalling that frames are assigned substantial properties, such as velocities, which would be impossible if they were concepts. *Remark 2* A reference frame need not be a perfectly rigid body. This is so firstly because there is no such thing as a rigid body in the real world, secondly because a frame must have moveable parts if it is to function as a (natural or artificial) clock. A frame can be a field (Dehnen, 1970).

The concept of a reference frame allows us to elucidate the notion of a thing's being in a given state relative to some frame. In fact we can introduce

DEFINITION 5.37 Let f be a reference frame for a thing X represented by the functional schema $X_m = \langle S(f), \mathbb{F}\rangle$. Then

(i) the *state of* X *at* $t \in S(f)$ *relative to* f is

$$s = \mathbb{F}(t) = \langle F_i(t) | 1 \leqslant i \leqslant n\rangle;$$

(ii) the *state space of* X *relative to* f is

$$S(X) = \{s = \mathbb{F}(t) | t \in S(f)\} = \mathbb{F}(S(f)).$$

Because there is an unlimited number of reference frames, and the choice of functional schema to represent a thing is partly conventional, states can be parametrized in a number of ways. Some such representations are equivalent – as when the frames are inertial in a certain sense – while others are not. This does not entail that actually states are absolute (frame independent) and that we should blame our own cognitive limitations for the relativity of states. Except in the highly artificial case when all the components of a state function are assumed to be invariant under all possible frame substitutions, *the states themselves are relative.* This is enough to regard with suspicion all the philosophical theories that employ absolute state descriptions such as "Thing X is in state s" or, worse, "The state of the world at a given instant is such and such". (Possible worlds metaphysics and inductive logic abound with such curiosities.) But relativity must not be mistaken for subjectivity: that every state of a thing is relative to a frame does not entail that it is the private state of a subject – unless of course one manages to confuse thing states with reference states or states of a standard of reference, and frames with subjects or observers. (For criticisms of such confusions see Bunge, 1967b, 1973b.)

So much for technicalities. Let us now turn to a few rather nontechnical yet important matters of principle.

5. Panta rhei

5.1. *Fact*

Having now at hand exact definitions of a state (Ch. 3, Sec. 2) and of a change of state, or event (Sec. 2), we can make full sense of Definition 4.3 of a fact. We can now say that *a (real) fact is either the being of a thing in a given state, or an event occurring in a thing.* The following comments spring to mind.

Firstly, the above definition of a fact applies also the case of the conversion of one thing (e.g. a caterpillar) into another (e.g. a butterfly), provided we construe both things as a single thing in two successive stages of development. In fact, as suggested in Sec. 1.2, all we have to do is to feign that we are dealing with one and the same thing whose representative point moves now in one state space (e.g. the caterpillar state space), now in another (e.g. the butterfly state space), the two state spaces being correctly assembled.

Secondly, constructs – such as concepts and propositions – do not qualify as facts since they are neither states of things nor changes in things. Hence it is mistaken to call a true factual proposition *a fact.* At most, a proposition might be regarded as a certain equivalence class of mental states – but a class is a concept not a thing. Hence it is mistaken to identify the two.

Thirdly, our construal of facts is objective. We do not define a fact as "anything that may be observed" because (*a*) microphysics, megaphysics and the physiology of perception teach us that most facts remain beyond human observation (though not therefore beyond comprehension), and (*b*) not all observations are reliable: some correspond to no external facts, whence reports on them, unless critical, are false.

Fourthly, the same applies, *mutatis mutandis,* to the semantic definition of a fact as "the kind of thing that makes a proposition true or false" (Russell, 1918, in Marsh, Ed., 1956, p. 183). And this for the following reasons: (*a*) unless one wishes to construct a subjectivist ontology (as was the case with Russell (1914), Whitehead (1919), Carnap (1928), and Goodman (1951), one will refrain from defining ontological concepts in semantical, epistemological, or psychological terms; (*b*) most factual propositions are at best partially true rather than wholly true; (*c*) the semantic definition of a fact commits one to asserting that there are negative facts (those verifying negative statements) and general facts (those confirming general statements). The weird metaphysics of negative and general facts is avoided if, contrary to the early Wittgenstein and the Russell of logical atomism, one drops the assumption that propositions and facts stand in a 1–1 correspondence. The correspondence cannot be such because all facts are "positive" and singular. (Recall that a fact is either the being of a thing in a state, or the occurrence of a change of state of a thing.) The nonoccupancy of a state and the nonoccurrence of an event are not facts. Thus not having been inoculated against smallpox is not an event, hence it does not count as a cause of smallpox. And since we do not countenance negative events, we avoid talking of "logical events" such as "*e* or not-*e*", and so we do not have to worry about their causes. We take this to be a distinctive advantage over the probabilistic theory of causality (Suppes, 1970), where any element of a probability space qualifies as an event and where the causal relation is assumed to hold among possible events, even negative ones, rather than among actual events. In short it is false that whatever makes a proposition true or false counts as a fact.

Fifthly and lastly, facts, whether possible or actual, are objective but – pace Wittgenstein (1922) – they do not constitute the world. The world is not the totality of facts but the totality of things, i.e. concrete individuals endowed with all their properties, some of which consist in their being able to change in definite (though possibly randomly) lawful ways. This is how both physical cosmology and our ontology construe the term 'world'. Wittgenstein's construal requires either dispensing with things altogether – an awkward situation for science – or attempting to define them in terms of facts – which attempt has not even been broached. More on this in Sec. 6.

5.2. *Dynamicism*

Ours is a world of things – but of changing things not quiescent ones. Indeed, according to Postulate 5.11 in Sec. 4.1, every thing has a nontrivial history. In other words *every thing changes*. This is our version of the Heraclitean *panta rhei*. Note that we are not yet taking a stand concerning the stronger ontological hypothesis that every thing changes *incessantly*: we do not have yet a suitable concept of time. For Postulate 5.11 to hold it is necessary and sufficient that every thing changes at least once – i.e. that the history of every thing consists of at least two states or, equivalently, of one event.

We take change – not its concept – for granted rather than either deny or justify it. What has got to be justified in each case is not change but any assumption, or pretence, to the effect that a certain thing fails to change in some respect. We do this kind of thing all the time, by saying that certain changes, though real, are small, or that they consist in nothing but relative motion, etc. Moreover we explain permanence (in certain respects) in terms of insulation, or isolation, or a balance of forces, or what not. Statics is a particular case of dynamics, electrostatics of electrodynamics, and so on. The most powerful and deep theories of contemporary science are theories of change not theories of being: permanence is a particular, exceptional, temporary case of change. To deny that all things are restless, to deny that universal change is objective – as Weyl (1949), Costa de Beauregard (1963) and Grünbaum (1967) – have done is an exercise in sophistry.

The dynamicist view inherent in contemporary science, and adopted by our ontology, contrasts not only with Parmenidean staticism but also with the Aristotelian and Thomist hypothesis that rest, rather than motion, is the "natural" state of things. Our version of dynamicism goes also one step beyond Democritus', for whom atoms themselves were subject to no changes. Both high and low energy physics suggest that not even the "ultimate" components of things – "elementary particles" and fields – are quiescent. There is no inert matter, no perpetually unchanging stuff that can be disturbed only by external influences but is otherwise passive. As Bolzano saw, such external influences and the resulting relative motions could not be explained "if no change could take place in the interior of the simple substances themselves"; hence "the capacity for change through mutual influence" must be attributed to all substances, whether complex or simple (Bolzano, 1851, §§ 50, 51). The "immutable pole" exalted by some poets and philosophers is not a thing or a set of things but the set of basic objective laws.

Our view is definitely dynamicist but not excessively so; we assume that every thing changes in some respect or other but refrain from asserting that it changes in every respect, let alone incessantly. Science does not seem to support such an exaggeration: it shows that some traits remain invariant throughout certain changes. (The constants of the motion of a physical system are among such invariants.) Moreover change in some respects is possible only on account of permanence in others. (Thus, life processes are impossible unless homeostasis is maintained.) Likewise whatever is permanent (invariant, unchanging) is so relative to (under) some specific set (groupoid, semigroup, or group) of transformations. Consequently the invariants of any one such structure may not be the same as those of a different set of transformations.

What holds for every single thing – namely that it is changeable – is true of the thing composed of all things, namely the universe. That is, the world as a whole is in flux – though not as a single well integrated system or cosmos. The universal flux is a huge bundle of processes some of which intertwine (influence each other) while most probably do not. Hence it is mistaken to speak of "the world series of events" – as if it were something like an oversized world baseball series. A fortiori it is mistaken to affirm or to deny that such a series has a beginning or an end. There just *is* no world series of events. Because the precedence relation among events is local (bound to some reference frame or other), the totality of events has no overall order. All we can assert – and have in fact asserted in Postulate 5.9 – is that every single process is preceded and succeeded by other processes. Hence although the world as a whole is not engaged in a relay race, every bit of it does participate in some local relay race or other.

5.3. *Interconnectedness*

Postulate 5.10 in Sec. 4.1 states that every thing influences, or is influenced by, some other things. In other words, nothing – except for the universe as a whole – is totally isolated. If this is so then the world is an interconnected whole, i.e. a system rather than an aggregate. (Recall Definition 3.12 in Ch. 3, Sec. 2.7.) However, it is not a very tightly knit one.

The hypothesis that the world is a system is weaker, but presumably nearer the truth, than the organicist cosmology of Plato, the Stoics, or Hegel, according to whom every thing is bonded to everything else, so that the world as a whole is an "organic whole". If this other hypothesis were true then it would be impossible to study any particular thing without taking cognizance of every other thing, and it would be impossible to act on any particular thing without disturbing the entire universe.

Both contemporary science and our ontology steer a middle course between the extremes of organicism ("Everything hangs together forming a solid block") and atomism ("Everything goes its own way"). No part of the universe is isolated (or entirely free from bonds) but every thing is isolated in some respects from some other things. This *partial interconnectedness* of the parts of the world renders their study possible, for the study of every thing is partial and rests on the possibility of making contact with the thing.

Moreover, Postulate 5.10 supplies the existence criterion commonly used in the sciences, to wit: *Whatever exists (really, physically) is influenced by, or influences, some other things.* To repeat:

CRITERION 5.1 For any object x other than the world, x *exists really* iff there is a thing y other than x such that y acts on x or x acts on y.

If the thing y that acts on or is acted upon by the putative thing x, is controllable by an observer, then x is a subject of experimental research. But such possibility of effective control is sufficient, not necessary, to conjecture existence.

In short, our ontology includes the thesis of the limited or partial interconnectedness of all the parts of the universe.

5.4. *Three Misconceptions*

We have defined change with reference to concrete objects or things. Every collection of events is a set of changes in some thing or other. We have found no use for the fiction that there are changes that fail to consist in modifications of the states of some thing or other. Anyone claiming that there are such changes in reality should exhibit empirical evidence to this effect and proceed to build a theory of such thingless (or immaterial) changes.

Nevertheless once in a while philosophers and even scientists are found claiming that there are such thingless changes. Let us examine briefly the claims for three such items. One of them is the information process, sometimes regarded as not being based on the transport of matter or the propagation of fields. The reason for this misconception is that statistical information theory is a black box or phenomenological theory so general that it disregards the precise kind of signal carrying the information, the information transmission mechanism, and the kind(s) and quantity of energy involved. For this reason information theory is a nonphysical theory: it is instead a theory qualifying as a piece of scientific metaphysics, since it deals in an extremely general way with a genus of concrete things. But, of course, information is a property of certain physical (or chemical or biological) processes, namely signals, which are processes of a kind. No signal, no information transmission. And signals, let us repeat, are chains of events occurring in concrete things: the transmission of information consists in events propagating across space and carrying energy. Remove such real processes and only parapsychological anecdotes remain.

Another alleged example of a nonphysical yet real change is the parallelistic thesis that psychical or mental ("behavioral") events are not to be conceived of as changes in the nervous system of the animal concerned. If this is a point in methodology, it may have been sound in the beginnings of experimental psychology. But it has now become obnoxious for it blocks research in physiological psychology, and it is absurd when taken metaphysically, i.e. as restating the moth-eaten dogma that the soul is above matter. There is no evidence whatever for brainless ideation, whereas on the other hand there is definite evidence that every mental event is an event in the brain of some animal. We defer the detailed investigation of this matter to Vol. 4, Ch. 10.

A third point of incidence of immaterialism is the standard quantum mechanical theory of measurement (von Neumann, 1932). According to it the most striking difference between quantum physics and classical physics is that, while the latter treats measurement processes as purely physical processes, according to quantum mechanics they are not because of the intervention of the observer's mind and his decisions. So much so, it is alleged, that only natural processes obey the Schrödinger equation (but then they are unobservable) whereas measurement processes obey a separate theory centered on a different postulate, namely the projection postulate. This interpretation is not accepted by the orthodox followers of Bohr, according to whom *all* quantum events (which they call 'phenomena'), not just those controlled by measurement devices, involve observation acts. According to this alternative view there would be no spontaneous (unprovoked and unobserved) microevents ("quantum phenomena"). Realists reject of course Bohr's thesis as well as von Neumann's (Bunge, 1973b). One argument for the realist stand is that, while it is true that every measurement process is designed, executed and interpreted by some person (eventually with the help of automatic devices), it is a strictly physical process so far as the measured and the measuring things are concerned. So much so that (*a*) some measurement processes can be completely automated and (*b*) none of the existing measurement theories includes any mental variables. But even if the larger system including the observer and his staff had to be considered, this would not render measurement a nonphysical process – unless of course the mind were construed as an immaterial substance in the style of Aristotle and of the public (not the secret) Descartes. But modern physics need not ally itself with an obsolete metaphysics.

6. Concluding remarks

A basic polarity in traditional metaphysics is that of being and becoming: event is opposed to thing, process to stuff, change to structure. This opposition makes no sense in our system, where every change is the transformation of some thing or other, and every thing is in flux. This view, incompatible with much of traditional metaphysics, is consistent with science: the latter provides no ground for hypothesizing the existence of thingless events any more than it suggests that there might be changeless things. Surely the formulas of theoretical science do not always contain thing variables in an explicit way, but the context – what is sometimes called "the prose identifying the variables" – usually makes it clear that the formulas refer to things of some kind or other. The precise description of every event requires mentioning the thing or things subject to the change in question. This is why any correct axiomatization of a scientific theory starts by making assumptions concerning the kinds of thing the theory refers to, and ends up by characterizing events as changes in some of the properties of such referents. (See e.g. Bunge, 1967b, 1973b.) The converse course, namely that of starting with events and ending up by defining things, has been suggested as a program (e.g. by Russell (1914) and Whitehead (1929)) but never actually carried out. Let us take a glimpse at it.

Some process philosophers – anxious to flog the dead horse of Parmenidean ontology – have gone to the point of postulating that the being of an entity consists in its becoming – that "the actual world is a process" (Whitehead, 1929, pp. 29 and *passim*). Likewise the physicist Fokker characterized reality as a stream of events (Fokker, 1965, p. 1).

A few other physicists have reinvented this idea and even denied the existence of things (Bohm, 1970; Finkelstein, 1973; Stapp, 1976). This idea, that events and processes are the stuff the world is made of, has been encouraged by a misleading terminology introduced by some workers in relativistic physics, who call any point in spacetime an *event*, whether or not the point happens to be occupied by some thing. Further, they call *world* the set of all such "events" – even for a hollow universe. In this way the world is regarded as a set of events and the things that change are lost sight of.

In point of fact not the world but just a *world line* or tube or history – the region of spacetime scanned by some thing – qualifies as a sequence of events or process. (See Figure 5.19.) What can be said is that any region of spacetime is the seat of *possible* events occurring to things (particles, fields, etc.) that might exist within that region. But this is a far cry from identifying an event with a quadruple of real numbers.

In any case we need not spend more time with the quaint version of process metaphysics that dispenses with the thing concept, because it is logically untenable. Indeed, it is circular to define a *thing* as the

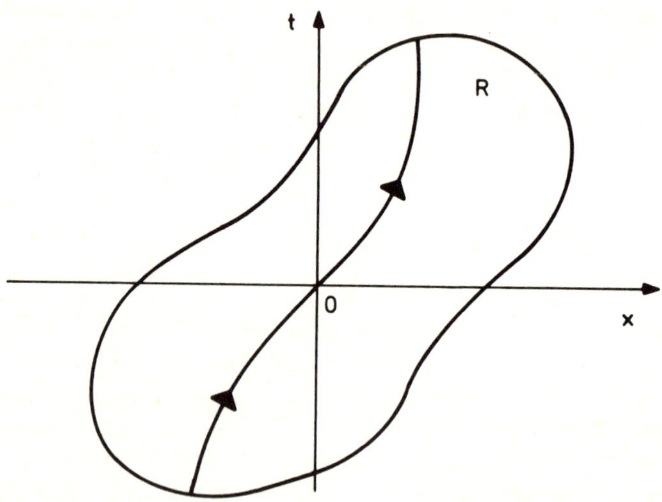

Fig. 5.19. The spatiotemporal track of a point particle (or world line of the latter) is a process in spacetime. Unless there are actually other things in the spatiotemporal region *R*, all the points other than those constituting the world line of the particle are uneventful and therefore do not deserve being called 'events'.

collection of events occurring in the *thing*. This is why our study of change was preceded by an examination of the concept of a thing.

The dual of process metaphysics is of course being metaphysics. One version of it is structuralism, currently fashionable in France (cf. Lévi-Strauss, 1958). Just as traditional metaphysics opposed being to becoming (or thing to process), and even extolled one of them at the expense of the other, so structuralism tends to oppose structure to change and to divorce the former from things. Just as the extreme dynamicists speak of thingless events, so structuralists speak of structures in themselves and concentrate their attention on structural permanence amidst flux. This too is mistaken: except for pure mathematics, where there are structures in themselves – e.g. Boolean algebras – every structure is the structure *of* some (complex) thing – atom, molecule, cell, organ, organism, community, ecosystem, or what have you. (Recall Definition 5.34 of the structure of an aggregation of things.) We do not say of a certain physical object that it *is* a rotation group but that a certain property of it (e.g. the energy) has rotation symmetry, or is invariant under rotations (members of the rotation group). Likewise we do not say of a community that it *is* a social structure but instead that it *has* such a structure. Moreover given the changeability of all things it is unlikely that there be anything with a permanent structure. In short, structure is a property – and a changeable one.

In conclusion, neither processism nor structuralism are viable alternatives to our metaphysics of changing things. Becoming is not pure flux but consists of things' being in successive states, or of successive things. (Remember Aristotle's *Physics*, Bk. III, Ch. 1, Sec. 1: "there is no such thing as motion *over and above* the things".) And structures are not just fickle but also lacking in independent existence: they are structures of things, features of concrete objects. In sum change and structure are so many distinct but interlocked features of changing things endowed with some structure or other. Therefore the traditional metaphysical poles, namely process metaphysics and being metaphysics, have each a grain of truth and both are overcome by our ontology.

We come to the end of our study of change in general. We have formed the main building blocks for constructing the concepts of time and space. Once these have become available we shall be able to give a more detailed account of change – in particular qualitative change, which will concern us in Vol. 4, Ch. 8. We proceed then to a study of extension and duration.

CHAPTER 6

SPACETIME

Science takes space and time for granted. Although not every scientific theory makes detailed assumptions about space and time, these two are usually regarded as constituting an independently existing framework allowing one to locate things and date events. So much so that facts are said to be or to happen *in* space and time even when we do not care for their precise spatiotemporal characteristics. So much are space and time taken for granted in science that most descriptions of things are effected in terms of space or time – i.e. space and time coordinates are used as independent variables. Moreover space and time are usually regarded as external to things and their changes: they are taken to constitute a fixed scenario.

True, on the relativistic theory of gravitation things influence the structure of spacetime: they distort its metric, hence its intrinsic curvature. But they do not create the basic manifold. (Indeed, setting the matter tensor equal to zero in the gravitational field equations one is left with equations describing a riemannian space.) In other words, even the general theory of relativity is consistent with – although it does not necessitate – a modified absolute theory of spacetime, i.e. one according to which spacetime has an independent existence instead of being a mesh of relations among facts. (The revision required by general relativity is this: the metric is not a priori, or purely geometric, but depends upon the momentary distribution of matter and fields. Hence spacetime is homogeneous just in case that distribution is itself homogeneous – i.e. actually never.) In sum science tells us what the structure of spacetime is but not what spacetime is.

This procedure, though legitimate in science, is unsatisfactory in philosophy. Here we cannot take space and time for granted because philosophy takes for granted nothing but the existence of the whole universe. In philosophy we must ask 'What are space and time?': we must attempt to understand how they come about, what their roots in things and events are. For, in the absence of things, there should be no spatial relations; and in the absence of change there should be no temporal relations. Indeed, it takes at least two things to make sense of 'here', 'there', 'to the left', and the like. And it takes at least two different states of a thing to make sense of 'before', 'after', 'meanwhile', and their kin.

Therefore instead of assuming that space and time are absolute (autonomous, self-existing) or even mildly absolute (influenced but not created by the furniture of the world), we should try and construct them out of facts. Philosophers are not interested in the empirical question 'Where and when did x happen?' They are interested instead in the question 'Do facts give rise to space and time, and if so how?' It behoves them to characterize the gross – algebraic and topological – structure of space and time, if possible in terms of factual items. Once this task has been accomplished, the philosopher should step aside and let the scientist investigate the fine structure – in particular the metric – of spacetime. It would be presumptuous of him to legislate that spacetime is – or is not – euclidean: this is a matter for the scientist to investigate. The philosopher will not try to compete with the scientist unless he is intent on being beaten by him.

From a methodological point of view the situation is this: there are three kinds of geometry – mathematical, physical, and ontological. *Mathematical geometry*, or geometry for short, is the collection of theories defining spaces of all kinds, where by 'space' is meant an arbitrary set endowed with a minimal structure, namely a topology. This structure can be modest or rich, and it may or may not depend upon the nature of the elements of the set. For example let S be a two member set $S = \{a, b\}$ – never mind what these elements are. Form the power set of S, i.e. $\tau = 2^S = \{\varnothing, \{a\}, \{b\}, \{a, b\}\}$. Then S is said to be *topologized* by τ. Equivalently, the ordered couple $\langle S, \tau \rangle$ is called a *topological space* – or space for short. A theory of a mathematical space is said to be "concrete" if the nature of the elements of the base set S is specified – e.g. if the members of S are n-tuples of real numbers. Otherwise, i.e. if the "points" of the space are faceless (unspecified individuals), then the theory is said to be abstract, or to define an abstract space.

A *physical geometry*, on the other hand, is a theory purporting to represent things and events in physical space (or in spacetime). It consists of a mathematical geometry of the "concrete" type (i.e. interpreted in mathematical terms) enriched with semantic assumptions ("correspondence rules") specifying the referents of the theory (e.g. light rays) and what properties of such referents (e.g. their mutual separation) those concepts represent. For example, in such a geometry a (physical) triangle could be determined, say, by three intersecting light rays, and the length of its sides could be equalled to the times of flight of the photons constituting such rays. That is, the objects of a physical geometry are physical objects and as such subject to measurement at least in principle.

Finally we shall call *philosophical geometry*, or *chronotopics*, the set of theories explaining the deep structure, i.e. the factual support, of spacetime. (A philosophical geometry is a part of both philosophy – in particular ontology – and the general foundations of physics. It should not be mistaken for the philosophy of geometry, which is a metatheory concerned with the nature of geometric objects, the kinds of geometric argument, the relation of mathematical geometry to factual science and to reality, and so on.) Chronotopics makes use of both ontology and pure mathematics, and it intends to underpin physical geometry. It attempts to lay bare the ontic substratum of spatiotemporal relations in order to answer the philosophical questions that neither mathematical nor physical geometry answer, namely 'What are space and time?'. The subject of the present chapter is precisely chronotopics. And our answer to the question formulated a wink ago will be this: Space and time are what chronotopics say they are. Or, in a less sibylline way: *Spacetime is the collection of facts together with their separations* – where "separation" is construed in terms of facts only. Precisely how it is construed will be seen shortly. But before embarking on the construction of the theory we must examine the kinds of theory we might build.

1. Conflicting views

1.1. *The Three Main Views*

There are three main views on the nature of space and time: we call them the container, the prime stuff, and the relational views. (For a history of these ideas see Jammer, 1954.) Here they are in a nutshell:

(i) *The container view*: space and time constitute the fixed stage where things play out their comedy. Physical objects exist in space and time, which in turn are not physical objects. Moreover a physical object may be defined as anything occupying a region of space and a stretch of time. The container exists by itself (is absolute), hence it would continue to exist if all things contained were to change radically in kind or even were to pass out of existence. Not being a physical thing or a relation among physical objects, the supreme container cannot be described in physical terms: space and time have got to be described in purely mathematical terms without even the assistance of concept-fact bridges or semantic assumptions. Thus instead of stating that a certain geometry represents physical space we must say that the two are identical. Such a view is compatible with both the objectivist conception invented by the Greek atomists, and refined by Newton, and the subjectivist conception of space and time as the intuitive scaffolding necessary for human experience (Kant).

The container view of space and time is strongly suggested by the way science uses these concepts most of the time. Thus one calculates or measures the place or time *at* which something happens, and one computes or measures spatial and temporal intervals regardless of the kind of fact involved. However, the fixed framework view has been shaken by the relativistic theory of gravitation, where the very structure of spacetime is influenced (though not determined) by the things contained. (As you turn the pages of this book you modify, albeit by tiny amounts, the distances between neighboring objects as well as the trajectories of the light rays reaching your eyes.) Besides, the doctrine is philosophically unsatisfactory because it does not really answer – except metaphorically – the question 'What are space and time?'. It is particularly unacceptable to any ontology which, like ours, has no room for objects that are neither things nor properties of things nor relations among things. Therefore we cannot accept the container view even though we may feign to do so in everyday life.

(ii) *The prime stuff view*: spacetime is the elementary substance of which every physical object is made. Whatever is part of the world is a chunk of spacetime, and every substantial property is a property of a chunk of spacetime. Hence everything physical ought to be explained in spatiotemporal terms – be it Anaximander's *ápeiron*, Clifford's space hills, or Wheeler's wormholes. Physics becomes geometry and moreover a geometry in no need of physical interpretation, as it generates its own interpretation. Geometric monism is substituted for the spacetime-matter dualism of the container view: space and time are not just experientially and epistemically prior (as for Kant) but also ontologically primary (as for Alexander). Indeed, every physical object is regarded as a local warping of space (or of spacetime). Thus an electrically charged particle is said to be identical to a wormhole in the basic spacetime manifold. (See Wheeler, 1962.) In short, on this view space and time are just as absolute as on the container view, in so far as their properties depend on nothing. The difference is that, according to the prime stuff view, things are not in spacetime but instead the latter constitutes the former.

This fascinating theory is unfortunately semantically and ontologically too simple to be true. Semantically because the mathematical formalism is declared to be in no need of semantic hypotheses ("correspondence rules"). Thus instead of saying that a certain figure in spacetime *represents* a black hole one simply *calls* that surface a black hole: definitions take the place of semantic assumptions, hence the border line between formal science and factual science disappears – perhaps also that between constructs and things. And the theory is ontologically too simple because it attempts to reduce the entire variety of kinds of stuff to extension and shape (in four dimensions) alone. This Cartesian reduction to *figures et mouvements* does not account for the varieties of elementary particles and fields.

(iii) *The relational view*: space and time are not self-existing objects but a network of relations among factual items – things and their changes. What is assigned mathematical (topological, affine, or metrical) properties is not spacetime itself but the set of things – atoms, fields, etc. – and their changes. No changing things, no spacetime. Of course things may be said to have spatiotemporal properties – but the latter boil down to relations among things or events. The concepts of space and time used in science should result (and also be made more precise) upon one's abstracting from the facts, as when "Point x is included in sphere s with center at the origin of coordinate system k" is abstracted from "Thing X is inside ball S centered at the origin of the material frame of reference f (represented by coordinate system k)". Likewise "Instant t_1 precedes instant t_2 in the four-dimensional coordinate system k" should be construed as an abstraction from "Event e_1 precedes event e_2 in reference frame f (represented by the four-dimensional coordinate system k)".

The relational view has been expounded, with various degrees of clarity, by thinkers as diverse as Aristotle, Lucretius, Augustine, Leibniz, Lambert, Lobachevsky, Riemann, Engels, Mach, Whitehead, Robb, van Dantzig, Fokker, and Penrose. It was best summarized by Leibniz in oft quoted words: space is an *order of possible coexistents* and time an *order of successives* (Leibniz, 1956, II, p. 1083). Hence talk of space or time is elliptic talk of facts (Mach, 1872, 1883). When saying

'The position of thing *X*' we actually mean "The position of thing *X* relative to thing *Y*", where *Y* is either a reference frame or just the environment of *X*. And when saying 'The state of thing *X* at time *t*' we actually mean "The state of thing *X* when the chosen reference frame (or the environment) is in state *t*". However, *pace* Mach it does not follow that the concepts of space and time are dispensable: elucidating is not necessarily explaining away or eliminating. We cannot afford such a luxury because mention of space and time landmarks amounts to a public and global, hence objective and economical, reference to the state of the environment – of any environment, of any kind. Moreover there is no better way of getting quantitative precision than to fix the spatiotemporal coordinates of a thing or an event. But this is a methodological point: what matters to ontology is that space and time are not self-existing (absolute) objects of uncertain ontological status (neither things nor thing properties). The relational view is that *spacetime is the basic structure of the totality of possible facts*. But, unlike the prime stuff view, which eventually developed into a full fledged theory (geometrodynamics), the relational view has heretofore remained at the heuristic stage. The subsequent sections will see us through a hypothetical-deductive system spelling out the basic intuitions of the relationist thinkers. But before piling up bricks let us take a look at architectural styles.

1.2. *Approaches to Chronotopics Building*

There are two epistemological approaches to the problem of building a relational chronotopics: the subjectivist and the objectivist ones. In the former one takes the cognitive subject and his experiences as the point of departure, whereas in the second case one proceeds from a consideration of the things themselves. The subjectivist approach has been tried by Whitehead, Nicod and Basri among others. Let us glance at these attempts.

Whitehead (1919) assumes that space is euclidean and tries to define in experiential terms all the concepts that euclidean geometry takes as primitive. (There is an echo of the Berkeley–Mach program in Whitehead's attempting "the deduction of scientific concepts from the simplest elements of our perceptual knowledge": Whitehead, 1919, p. vii.) Thus Whitehead's is not really a radical construction but a reconstruction with a strong a priori component, namely a particular mathematical conponent, and an equally strong philosophical bias, namely radical empiricism. Because of that a priori element Whitehead's theory of space and time is incompatible (by design) with the two theories of relativity and therefore hopelessly obsolete at birth and beyond repair. And because of its pronounced subjectivism the theory is inconsistent with the epistemological outlook of science, which is realist and the one adopted in this work. (Recall our Rule 7 for philosophizing: Introduction, Sec. 4.) Lucas' more recent attempt (Lucas, 1973) is subject to similar criticisms.

Nicod's (1923) venture was more radical though technically far less accomplished than Whitehead's and just as far removed from the science of his time. It was more radical because Nicod started from (imaginary) experiences and attempted to build geometry out of them without imposing any a priori structure on space. (One is reminded of Condillac and his statue.) In other words Nicod – following leads from Mach and Russell, and independently of Carnap's similar attempt – tried to build physical geometry out of (common sense) psychology. As a matter of fact he delivered nothing of the sort: he did not even produce a theory of any of the perceptual spaces such as the visual or the acoustic space. (Neither he nor Whitehead realized the differences between physical space and the perceptual spaces.) Like Whitehead's, Nicod's was a scientific and philosophic failure. (For detailed criticisms see McGilvary, 1951 and Vuillemin, 1971.)

Finally the most recent work along the Mach-early Russell–Whitehead–Carnap line, namely Basri's (1966), is just as subjectivistic but on the other hand it has the merit of being admirably rigorous and of respecting the formulas of relativistic physics. However, it does not serve as a foundation for the latter because it violates the spirit of the two relativities, which are field theories (Basri admits only particles) and moreover observer-invariant ones (Basri's is observer-bound). And although this theory professes to be concerned with human observers and their sensations and operations, it makes no contribution to the physiology or the psychology of space and time perception. In short, since the theory is subservient to an unscientific philosophy – namely empiricism – it serves neither physics nor psychology.

The subjectivistic approach cannot yield what we want in scientific ontology, namely an objective chronotopics compatible with physics but soft enough that the physicist may shape it according to his theories and data. And, whether subjectivist or merely conventionalist, the a priori approach to the geometry of physical space is unphysical because it assigns space a structure independent of the actual distribution of things. We must therefore try an objectivist approach, i.e. one starting with factual items as the prime stuff and postulating their spatiotemporal relations quite apart from considerations of perception and even measurement. This does not mean that the construction should proceed a priori, i.e. in advance of any experience. Quite the contrary, we shall be guided by the finding of scientific experience, that physical space is a three-dimensional connected generalized continuum. For the sake of simplicity we shall build first a geometry, then a chronology, and finally a chronotopics or philosophical theory of spacetime.

2. Space

2.1. *Interposition*

We shall build our relational theory of space using only materials elaborated in the previous chapters. In fact we need only the following background assumptions and definitions:

(i) there are concrete objects or things, none of them is a construct, and all of them exist independently of the cognitive subject;

(ii) all things can juxtapose or aggregate: if $x, y \in \Theta$, then $x \dotplus y \in \Theta$ or, equivalently, $[\{x, y\}] \in \Theta$;

(iii) the world is the aggregation of all things – i.e. $\square = [\Theta]$ – and whatever is a part of the world is a thing – i.e. if $x \sqsubset \square$ then $x \in \Theta$;

(iv) all things are, or are composed of, basic things or members of $B \subset \Theta$: for any $x \in \Theta$ there is a unique subset B_x of B such that $x = [B_x]$; and basics are elementary, i.e. have no parts: if $x, y \subset B$ and $x \sqsubset y$, then $x = y$.

(v) every thing is in some state or other relative to a given standard thing or reference frame, and the collection of really possible (lawful) states of a thing x is called its lawful state space $S_{\mathbb{L}}(x)$;

(vi) the evolution or history of every thing x is representable as a trajectory in its lawful state space: if $\mathbb{F}$ is the state function of x, and $S(f)$ the state space of reference frame f, then $h(x) = \{\langle t, \mathbb{F}(t)\rangle \mid t \in S(f)\}$;

(vii) a thing x acts upon another thing y iff the former modifies the history of the latter: $x \rhd y =_{df} h(x \dotplus y) = h(x) \cup h(y \mid x)$ & $h(y \mid x) \neq h(y)$;

(viii) the total effect or action of thing x upon thing y is $A(x, y) = h(y \mid x) \cap \overline{h(y)}$;

(ix) every change is a change in the state of some thing, and the net change of thing x from state s to state s' is representable as the ordered pair $e = \langle s, s' \rangle \in S_{\mathbb{L}}(x) \times S_{\mathbb{L}}(x)$;

(x) if $e = \langle s, s' \rangle$ and $e' = \langle s'', s''' \rangle$ are events in a thing x relative to a frame f, then e precedes e' if e composes with e' in the indicated order to form a third event in x; i.e. $e < e' =_{df} e * e' \in E_{\mathbb{L}}(x)$, where $E_{\mathbb{L}}(x)$ is the proper lawful event space of x, i.e. the set of really possible changes of x.

Our first geometrical notion is that of interposition, which we proceed to build in terms of the ontological notions occurring in the previous list. Our definition will be explicit, namely via

POSTULATE 6.1 Let R be a ternary relation among things and abbreviate "R holds among x, y, z in the given order" to $x|y|z$, where $x, y, z \in \Theta$. Then R is the *interposition* or *betweenness* relation iff, for any $u, v, x, y, z \in \Theta$,

(i) $x|y|z \Rightarrow x \neq y \neq z \neq x$ or $x = y = z$, i.e. R holds either among different things or trivially for a single thing;

(ii) $x|y|z \Rightarrow z|y|x$, i.e. R is symmetric in the outer variables;

(iii) $x|y|z$ & $u|x|y$ & $y|z|v \Rightarrow u|y|v$, i.e. whatever interposes between two given things lies also between two outer things;

(iv) $x \neq y$ & $x \sqsubset y \Rightarrow \neg(\exists z)(z \in \Theta$ & $x|z|y)$, i.e. nothing interposes between the part and the whole;

(v) $y \in B \Rightarrow (\exists x)(\exists z)(x, z \in B - \{y\}$ & $x|y|z)$, i.e. any basic thing can be "surrounded" by two other basics;

(vi) $\neg(x \sqsubset y)$ & $\neg(y \sqsubset x) \Rightarrow (\exists z)(z \in B$ & $x|z|y)$, i.e., there are basics interposed between any two detached things: the world is dense, it constitutes a *plenum;*

(vii) $x|y|z$ & $x \rhd z \Rightarrow (\exists u)[u \in A(x, y)$ & $(v)(v \in A(x, z) \Rightarrow u < v)]$, i.e. if y interposes between x and z, and x acts on z, then some of the effects of x on y precede all of the effects of x upon z. (See Figure 6.1.)

The first three clauses seem intuitive. The fourth is a sort of interpretation of Hilbert's second axiom group for elementary geometry – in our notation $\neg(x|y|y)$ (Hilbert, 1899). The fifth is a sort of dual of Hilbert's 4th component of the second axiom group – namely that there is a thing lying between any two distinct things. The sixth is countenanced by both gravitation theory and quantum electrodynamics: according to these theories no region of space is totally devoid of some entity or other. (This is of course the plenistic hypothesis defended by Aristotle, Descartes and Einstein among others.) And the last clause exhibits,

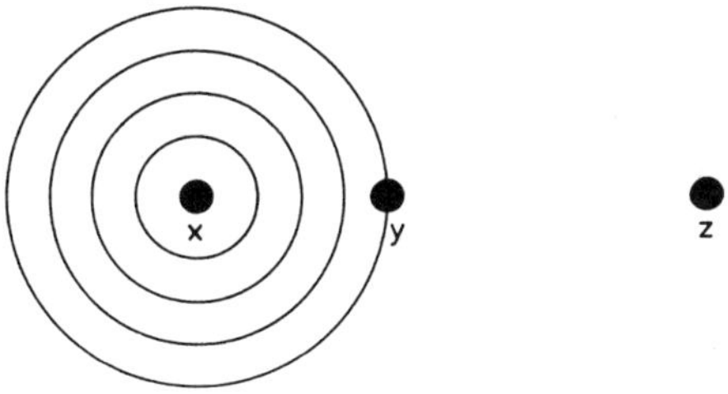

Fig. 6.1. Illustration of clause (vii) of Postulate 6.1. A spherical wave emitted by x reaches y before it does z.

perhaps even more forcefully than all the others, the material basis of the interposition relation. It would be objected to by the Leibnizians, who would argue that there would be space even in a world composed of mutually independent entities (monads), none of which acted upon any other. Maybe, but ours happens not to be such a world (Postulate 5.10). Clause (vii) suggests that the interposition relation might not even be properly definable for a changeless universe.

2.2. *A Philosopher's Space*

A basic concept in any geometry is that of separation. There are several notions of separation. One of them is the topological notion of separation between sets (Wallace 1941). We cannot use it because we wish to clarify the notion of separation between concrete things not between sets, which are constructs. Other concepts of separation are metrical or quasimetrical. Any of these consists in a real valued function d on an abstract set S obeying well known conditions. The members x and y of S are said to be separate iff $d(x, y) \neq 0$. We cannot use these notions either because it is not our purpose to compete with physicists by assigning precise quantitative measures to the separations among things. We need a more basic, qualitative notion of separation. We shall get it with the help of the concept of interposition.

We define the separation between two things as the set of things that interpose or lie between the given things. More precisely, we make

DEFINITION 6.1 Let $B \subset \Theta$ be the set of basic things. The function $\sigma: \Theta \times \Theta \to 2^B$ such that $\sigma(x, y) = \{z \in B \mid x|z|y\}$ for $x, y \in \Theta$ is called the *thing separation*.

Because the null thing is part of every thing, the separation between an arbitrary thing and the null thing is nought, i.e. $\sigma(x, \square) = \varnothing$. And because every thing is part of the world, there is no separation either between an arbitrary thing and the world: $\sigma(x, ▯) = \varnothing$. These two consequences follow from Postulate 6.1 and Definition 6.1, as does

THEOREM 6.1 For any two things $x, y \in \Theta$,
(i) $\sigma(x, x) = \{x\}$;
(ii) $\sigma(x, y) = \sigma(y, x)$.

Proof The first part follows from clause (i) of Postulate 6.1. The second, from clause (ii).

Note the similarities and dissimilarities between σ and the distance (or quasidistance) functions d mentioned at the beginning of this section. Property (ii) is analogous to the corresponding properties of the d's. On the other hand property (i) is not, nor is the frame invariance of σ, which contrasts with the frame dependence of the d's.

Further, if a thing interposes between two things which in turn interpose between two other things, then the former interposes between the latter. In terms of the separation function:

THEOREM 6.2 For any things $u, v, x, y, z \in \Theta$: If $y \in \sigma(x, z)$, $x \in \sigma(u, y)$ and $z \in \sigma(y, v)$, then $y \in \sigma(u, v)$.

Proof By Postulate 6.1(iii) and Definition 6.1.

We could derive further theorems but we shall not need them for our purposes. What we do need is another property of the separation function, which the previous assumptions and definitions fail to entail. We must therefore postulate it (see Figure 6.2). More exactly, we assume

POSTULATE 6.2 Let x_1, x_2, y, z_1 and z_2 be basic things. If $y \in \sigma(x_1, z_1) \cap \sigma(x_2, z_2)$ and $x_1 \neq z_1$, $x_2 \neq z_2$, then there are basic things x_3 and z_3 such that $x_3 \neq z_3$ and

$$y \in \sigma(x_3, z_3) \subset \sigma(x_1, z_1) \cap \sigma(x_2, z_2).$$

We now have all we need to build an important notion:

DEFINITION 6.2 The set B of basic things, together with the separation function σ, is called the thing space, abbreviated $\vartheta = \langle B, \sigma \rangle$.

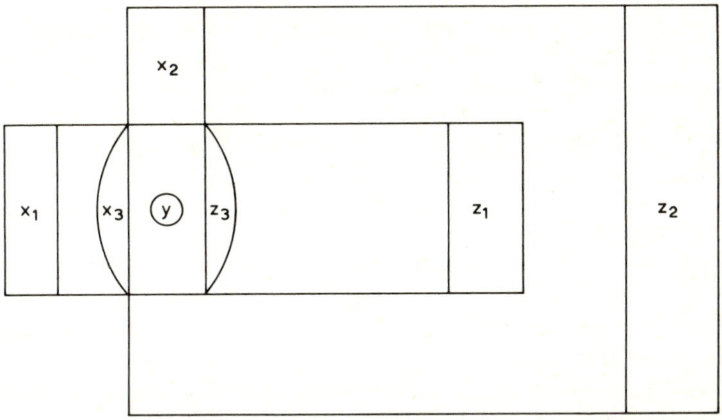

Fig. 6.2. Postulate 6.2.

In other words, the thing space is nothing but *the collection of spaced things*, or the set of things related by their mutual separations. This space deserves being regarded as a real (or ordinary or physical) space because its coordinates, B and σ, are real. Indeed, the basics have been assumed to be real and moreover the fundamental constituents of all concrete objects; and the separation function is a *relatio realis* in the sense of Ch. 2, Sec. 5.1, Definition 2.17, for σ obtains between real things.

The space ϑ is all the philosopher needs, for it has been built out of a few extremely general (ontological) concepts. We may call ϑ a *philosopher's space*, and the corresponding theory (summed up in the preceding postulates, definitions and theorems), a *philosophical geometry*, and moreover one both relational and objectivist. However, so far we have no evidence that ϑ is indeed the physical space: we have not proved that our philosophical geometry can be specified to yield a geometry utilizable by physics. This task will be taken up in the next subsection, adapted from Bunge and García Máynez (1977).

2.3. *The Physicist's Space*

The thing space introduced by Definition 6.2 would seem to have too poor a structure to qualify as physical space or even as a mathematical space. However, this is not so, as will be shown in the present subsection.

To begin with $\langle B, \sigma \rangle$ has a definite topological structure, as shown by

THEOREM 6.3 The family of sets of basic things

$$\tau = \{X \in 2^B | \text{For all } y \in X \text{ there are } x \text{ and } z \text{ in } B - \{y\} \text{ such that } y \in \sigma(x, z) \subset X\}$$

is a topology for B.

Proof By Postulate 6.1(v), $B \in \tau$. Besides, it is obvious that $\varnothing \in \tau$ and that every union of members of τ belongs to τ. Finally, by Postulate 6.2 the intersection of any two elements of τ is also in τ. Hence τ is indeed a topology for B.

We shall assume then that the set B of basic things, equipped with the topology τ, is or rather represents physical space:

POSTULATE 6.3 The topological space $\Sigma = \langle B, \tau \rangle$ represents physical (ordinary, real) space.

However, we have so far little justification for assuming this hypothesis aside from the fact that B is constituted by things and has a topology. For one thing we have assigned Σ no definite dimensionality, whereas we feel certain that physical space is three-dimensional. For another we want Σ to be connected (in one piece). These and other shortcomings will be remedied by foisting certain properties on Σ. That is, we shall *force* Postulate 6.3 to become true.

To begin with we define, in the usual way, closures in τ:

DEFINITION 6.3 If $A \in \tau$, then

$$Cl\,A =_{df} \{x \in B | \text{For all } V \text{ in } \tau \text{ containing } x,\ V \cap A \neq \varnothing\}.$$

For these closures to be of any help they must be large enough. This is ensured by a mild enough assumption:

POSTULATE 6.4 Every pair of separate basics is in the τ-closure of their separation. I.e.,

$$\text{For any } x, y \in B, \text{ if } \sigma(x, y) \neq \varnothing, \text{ then } x, y \in Cl\,\sigma(x, y).$$

By taking $x = y$ it follows immediately that every finite subset of B is closed. Now, a space with this property is called a T_1-space. In other words, we have

COROLLARY 6.1 The physical space $\Sigma = \langle B, \tau \rangle$ is a T_1-space.

We need one more assumption to convert Σ into a Hausdorff (or T_2) space. Actually Σ should have further properties: the physicist wants Σ to be a connected and metrizable space locally homeomorphic to $\mathbb{R}^3$. In other words Σ should be a Euclidean three-manifold.

Now, it has been known for a long time that it is possible to express every theorem in Euclidean three-dimensional geometry with the sole help of the concepts of sphere and of inclusion. In fact these are the sole extralogical undefined notions in the set of postulates for elementary geometry proposed by Huntington (1913). For example, a *point* is defined in this geometry as a sphere such that there is no other sphere within the given sphere. And a *segment* is defined as follows: Let a and b be any given points, i.e. minimal spheres. If x is a point such that every sphere which contains both a and b also contains x, then x is said to belong to the *segment* (ab) or (ba). (See Figure 6.3.)

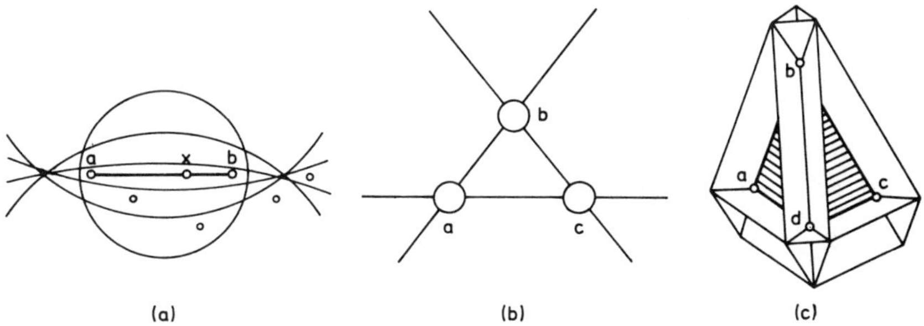

Fig. 6.3. Building space with spheres: (a) a segment, (b) a triangle, (c) a tetrahedron. From Huntington (1913), figs. 1, 3, 9.

We shall adopt Huntington's postulates but, instead of taking the notion of sphere as primitive, we shall define it in terms of some of our previous concepts. This will endow Huntington's postulates with a physical (or rather protophysical or ontological) interpretation and authorize us therefore to claim that they characterize physical space (in the small).

Our first task is then to define the concept of a sphere. We will define a sphere lying between two given things as the closure of a separation between those things. More precisely we make

DEFINITION 6.4 For any pair $x, z \in B$ of distinct basic things, the *spheres* lying between x and z are

$$Sxz = \{Cl\sigma(u, v) | u, v \in B \;\&\; \varnothing \neq Cl\sigma(u, v) \subset \sigma(x, z)\}.$$

Next we assume that these spheres satisfy Huntington's postulates:

POSTULATE 6.5 For any given pair $x, z \in B$ of distinct basics, if the spheres Sxz are nonempty, the structure $\langle Sxz, \subset \rangle$ satisfies (is a model of) the Huntington (1913) axioms.

Hence

THEOREM 6.4 The nonempty separation $\sigma(x, z)$ between two basic things $x, z \in B$ is homeomorphic to three-dimensional Euclidean space.

Proof The minimal spheres are the single points of $\sigma(x, z)$. By Huntington's (1913) Theorem 47 this collection can be topologized in such a way that it is homeomorphic to $\mathbb{R}^3$. Now, a basis for this topology consists of all the τ-interiors of elements of Sxz. Hence this latter topology is identical with the relative topology of $\sigma(x, z)$.

We are now justified in framing

DEFINITION 6.5 Every member $Cl\sigma(u, v)$ of the family of spheres Sxz lying between the things x and z is said to be a *Euclidean ball.*

Remember now that physical space is a T_1-space (Corollary 6.1). Since Postulate 6.5 supplies regularity we obtain

COROLLARY 6.2 $\Sigma = \langle B, \tau \rangle$ is a regular three-manifold without boundary.

We have come near the goal. We shall attain it by adding one last assumption, namely that any two basic things can be bridged by a string of spheres (Figure 6.4). More precisely, we assume

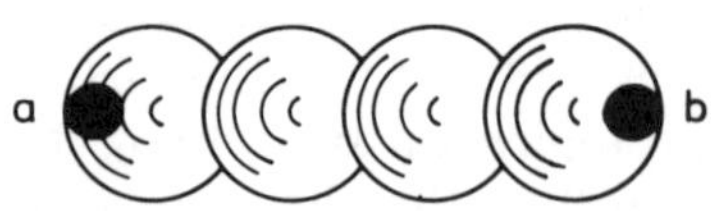

Fig. 6.4. The chain of partially overlapping spheres bridging things a and b (black dots).

POSTULATE 6.6 The set B of basics contains two sequences, $x_1, x_2, \ldots$ and $z_1, z_2, \ldots$, such that $x_i \neq z_i$ for each $i \in \mathbb{N}$ and, for every pair a, b in B, there is a simple chain

$$\langle C_i | 1 \leqslant i \leqslant n \rangle, \text{[i.e. } C_i \cap C_{i+1} \neq \varnothing, \ 1 \leqslant i \leqslant n-1]$$

such that $a \in C_1$ and $b \in C_n$, and every C_i is of the form $\sigma(x_j, z_j)$.

We can now prove that Σ is connected, i.e. in one piece or gapless. Moreover we shall prove in the same breath that Σ is second countable, i.e. that it has a countable basis. (A basis for τ is a subset β of τ such that each member of τ is a union of members of β, so that the latter are the building blocks of τ.) In fact we can also prove that Σ is metrizable. All of this is stated by

THEOREM 6.5 $\Sigma = \langle B, \tau \rangle$ is a connected, second countable and metrizable three-manifold without boundary.

Proof Being homeomorphic to $\mathbb{R}^3$ (by Theorem 6.4), every $\sigma(x_i, z_i)$ is connected and Lindelöf. Hence by Postulate 6.6 $\langle B, \tau \rangle$ is connected and Lindelöf. Now, every regular Lindelöf space is paracompact and every paracompact locally metrizable space is metrizable. Therefore $\langle B, \tau \rangle$ is connected, Lindelöf and metrizable. Finally $\langle B, \tau \rangle$ is second countable because, in metrizable spaces, the properties of being Lindelöf and of being second countable are equivalent. [For these notions see e.g. Gaal (1964).]

In other words, physical space is a three dimensional connected manifold. And, because Σ is metrizable, the physicist may assign it an adequate metric. We shall of course abstain from doing so because our goal was to build a concept of physical space broad enough that it can be used in the axiomatic foundations of any of the current physical theories. (Besides, the theories of relativity have taught us that any attempt to assign space a metric, independently of time, is bound to fail: only spacetime can be assigned the proper metric.)

This completes our justification of Postulate 6.3, that $\Sigma = \langle B, \tau \rangle$ represents physical space. But why should we have required Σ to have precisely the properties postulated or deduced in the present subsection? Why could we not hypothesize an entirely different set of properties of physical space? The answer to these questions depends upon the relation of one's philosophy with science, hence indirectly with reality.

In an a priori philosophy, physical space can be assigned any structure whatsoever. For example one could postulate – as Whitehead (1919) and Lucas (1973) have done – that physical space is globally euclidean. But in a philosophy that seeks to be continuous with science no such freedom exists. In any science-oriented philosophy one requires the geometry of ordinary space to agree with physics. Moreover one lets physics have the upper hand because physicists, not philosophers, are competent to decide on matters of the fine structure of physical space.

Now, physics happens to assume that physical space is a three-dimensional differentiable manifold. (See e.g. Trautman, 1965.) This is necessary to write down the basic equations of contemporary physics, though usually insufficient to solve them. (We disregard the speculative deviant theories. If any of them were shown to be true then we would have to change our geometry.) Surely most theories require some additional structure. For example classical mechanics and non-relativistic quantum mechanics require ordinary space to be globally euclidean, classical electrodynamics requires it to be an affine space, and the relativistic theory of gravitation assumes that space is riemannian. But the assumption common to all of these specifications is that physical space is a three-dimensional manifold. This being the minimal geometrical assumption, it should be a necessary and sufficient constraint for philosophy – until further notice. (Recall Rule 10 of the method of scientific ontology: Introduction, Sec. 4.)

The geometry described in the present subsection is then compatible with the mainstream of contemporary physics. A new physics might call for a new protophysical (or ontological) geometry. Two radical changes of the sort have been suggested a few times. One is that the spatial generalized continuum may have to be replaced by a discontinuous or atomic space with a fundamental length built into it. (This hypothesis is suspect because it usually has an operationalist motivation, namely the impossibility of measuring distances below a certain value, say 10^{-11} cm.) Another possibility is that our current metrics may be sorts of statistical averages of stochastic metrics describing a fluctuating spacetime. (This other hypothesis is far more plausible in view of the zero-point fluctuations of the electromagnetic field.) However none of these ideas seem to have been carried beyond the hand-waving stage. The fact is that the theories actually employed by physicists make no assumptions about physical space contradicting our results.

In sum, we may declare our ontological geometry true because contemporary physics says so. But at the same time we should be prepared to see it revised or even revolutionized by new developments in physics.

2.4. *Bulk and Shape*

Now that we have grounds for modelling physical space as a manifold we can use the latter to determine the bulk and shape of things. First we exactify the notion of the *place occupied by a thing*:

DEFINITION 6.6 Let the three-manifold $M^3 = \Sigma = \langle B, \tau \rangle$ represent physical space. Then the function $\beta\colon \Theta \to 2^{M^3}$ from things to regions of space is called the *bulk*, and its value $\beta(x)$ at $x \in \Theta$ the *bulk of* x, iff

(i) β is injective;

(ii) for every thing x other than the null thing, $\varnothing \subset \beta(x) \subseteq M^3$;

(iii) for all $x, y \in \Theta$, $\beta(x \dotplus y) = \beta(x) \cup \beta(y)$.

The first clause makes room for things, such as photons, that can share the same region of space. The second states that every real thing has a nonempty bulk – even if, as in the case of the mythical point particle, the bulk consists of a single point of space. The third, that β is additive. The three clauses together elucidate the notion of *room* or *space occupied by a thing*.

None of the above conditions presupposes or implies that all things have sharp boundaries: so far as Definition 6.6 is concerned, a thing can have "colonies", i.e. pieces outside its main bulk. Consequently nothing is implied about the measure of the bulk of a thing. Questions of measure are specific scientific questions answered by computing or measuring volumes not bulks. (And volumes, unlike bulks, depend on the interactions among the components, as well as on the reference frame.) On the other hand Definition 6.6 allows one to compare bulks. Indeed if x and y are things then we can say that x is *bulkier* than y just in case $\beta(x) \supseteq \beta(y)$.

The concept of bulk allows us to define that of shape:

DEFINITION 6.7 Let $x \in \Theta$ be a thing with bulk $\beta(x) \subset M^3$. Then x has *shape* f iff the boundary $\partial\beta(x)$ of $\beta(x)$ is homeomorphic to a piece-wise smooth function $f\colon \mathbb{R}^2 \to \mathbb{R}$.

Remark 1 The values of the shape function f depend upon the choice of reference frame, hence of coordinate system. Hence shape is frame-dependent rather than frame-invariant. (Recall the flattening of a sphere into an ellipsoid relative to a moving frame.) *Remark 2* Shape is not a universal property of things. For example electrons have no shape of their own – or, if preferred, they acquire the shape of whatever macrophysical body contains them. Nor do porous bodies and gaseous bodies have a definite shape of their own. Therefore we cannot regard shape as a basic property of all things. Hence we cannot adopt Descartes' program of explaining the world in terms of *figures et mouvements* only. *Remark 3* Shape, hardly a property of basic things, emerges rather definitely at the macromolecular level and becomes the more definite, the bulkier the thing. It is therefore a derivative property. Moreover it emerges from nongeometric characteristics. Thus the helicoidal configuration of a DNA molecule results from chemical forces such as the hydrogen bonds between an NH group and a carbonyl group, and it is influenced by the environment of the molecule – to the point that the pattern disintegrates at high temperatures. Likewise the shape of a macrobody is determined jointly by the inner stresses and the external forces. In general, shape or geometric pattern, when it exists at all, is an outcome of the interplay of internal forces and environmental constraints. *Remark 4* Although shape is a secondary property, once acquired it conditions the acquisition or loss of further properties, which are called *steric properties*. Suffice it to recall that the specific activity of enzymes depends upon their shape.

2.5. *Concluding Remarks*

We have obtained a solution to the problem we set out to solve, namely that of building a theory of physical space based on the concept of a changing thing. This does not mean that changing things are more fundamental than space: only the corresponding categories are ordered. Ours is then a full fledged relational theory – not just a program – and moreover one based on general ontological ideas and couched in strictly objective terms, as enjoined by Rule 7 of our philosophical methodology. (See Introduction, Sec. 4.)

However, ours is not the only possible relational theory of space. For example there is Basri's (1966) theory. But we rejected it in Sec. 1 for being a particle theory and therefore inconsistent with field physics, and also for being subjectivist, hence inconsistent with the epistemology inherent in the scientific approach. There is also Penrose's (1971) equally rigorous and moreover objectivist relational theory of space. But, because it is based on the concept of angular momentum, it is not general enough to serve as a foundation for all physical theories. On the other hand our own theory is based on extremely general ideas.

Our theory has one seemingly unsatisfactory feature, namely that it postulates the three-dimensionality of space instead of explaining it. (The hypothesis is contained in Postulate 6.5.) This assumption may seem arbitrary. Actually it is not, for it is suggested by the actual behavior of real things: if things were different then physical space might not be three-dimensional. In other words, the three-dimensionality of space is rooted in the laws of things. What would the dimensionality of physical space be if things possessed ("obeyed") laws different from those we know? This question can receive a partial answer, namely along the following lines. (For a detailed investigation see Penney (1965).) Take for instance the pervasive processes of propagation of waves of various kinds and write out any wave equations in spherical coordinates in an n-dimensional manifold. Any such equation will contain a radial term with a factor $(n-1)$. So, if experiment were to show that wave propagations proceed according to say, $n = 4$, then we would have to conclude that physical space is four-dimensional. But of course this is not the case. In short experiment points, in however devious a fashion, to the three-dimensionality of ordinary space. (Note by the way that, on a nonrelational theory of space, one would ask instead the question: 'What would things look like if space were not three-dimensional?' But this question cannot be answered except dogmatically.) Hence in postulating that space is three-dimensional our ontology bows to experience, just as Peirce wanted.

What would become of space if all things were to go out of being? The answer is provided by the separation function (Definition 6.1). Since σ is defined only for pairs of things, and its values are sets of things, unless there are (changing) things there is no physical space, hence neither a relational philosophical geometry nor a physical geometry. That is, a hollow universe would be spaceless. Likewise a block universe, such as Parmenides' undifferentiated One, would be spaceless.

This corollary of our philosophical geometry has some bearing on the interpretation of the relativistic theory of gravitation. What happens if the matter tensor in the gravitational field equations is set equal to zero everywhere? Sure enough the solutions characterize a riemannian manifold: there would seem to be space even in the absence of things (particles, fields, etc.). Is this a physically possible solution? Not on our geometry: here, if there are no things, none of the relations among points in the manifold is a real relation. The relations involved in a riemannian geometry (e.g. the distance relation) become real to the extent that they are satisfiable by things. In short, the homogeneous equations of gravitation, corresponding to a hollow universe, are physically meaningless. (For the notion of factual content see Vol. 1, Ch. 5, Sec. 3 of this *Treatise*.) It is nice to see that ontology can be of some use to science.

Ordinary space, then, is just as real as any other real relation. (Actually space is not a relation but a set of related things or, if preferred, a relational structure: recall Definition 6.2.) But, not being a thing, physical space has no causal efficacy. In other words, the spatial relations are nonbonding relations rather than bonds or couplings. (For the concepts of bonding and nonbonding relations see Ch. 5, Sec. 4.1.) That is, just as things do not act upon space (since space is not a thing), so space does not react back on things. This is one point our theory shares with Newton's absolute view.

To the practising physicist, who takes space for granted, space is basic and things are somehow embedded in or superimposed on space. To the philosopher upholding a relational doctrine of space, neither is the more basic. In particular, on our theory things come with their mutual separations, hence neither is prior to the other: there is neither a thingless space nor a spaceless thing. This is of course in sharp contrast to the container view of space, according to which space, though not a thing, exists by itself and independently of the things it contains. According to our theory neither space nor things exist by themselves. Only mutually spaced things exist.

Moreover, the separations or spacings among things may alter with changes in the things themselves. Hence real space is as much in flux as are things. Real space is then a dynamic structure of the collection of things. Ordinary space may be pictured as an elastic net, or fluctuating lattice, the nodes of which are things. But temporality belongs in the next section.

3. Duration

3.1. *Intuitive Idea*

A duration is the duration of some event or process: a changeless universe would be timeless. Just as space is the spacing of things (Sec. 2), so time is the pace of events. (Aristotle, *Physics* IV, 11, 220a 25, called time *arithmós kinéseos*, the measure of motion.) And just as spatial distance is the separation among things, so temporal interval is the separation between different states. This is the intuitive germ of the relational theory of time.

To get an idea of this theory consider a world consisting of a single thing the states of which can be mapped on the set of natural numbers: it can be a clock or a heart. The history of such a heart is, roughly, the sequence of its heart-beats; and the life span of the heart is the total number of its heart-beats. Since there are not other hearts (or clocks) in our imaginary universe, there is no question of slowing down or quickening the pace of the heart-beats – not even of missing a heart-beat now and then. The heart-beats generate a uniform or well tempered time – i.e. same intervals between beats. The time coordinate attached to this clock is a function $t: B \to \mathbb{N}$, where B is the set of heart-beats and $\mathbb{N}$ that of natural numbers, such that $t(b) = n \in \mathbb{N}$ for $b \in B$. We interpret '$t(b)$' as the time at which b occurs. This function determines the time lapse function

$$\tau: B \times B \to \mathbb{N} \quad \text{such that } \tau(b, b') = |t(b) - t(b')| = |n - n'|$$

for $b, b' \in B$. (Clearly, the structure $\langle B, \tau \rangle$ is a metric space.) If the heart stops beating altogether then both t and τ lose their support and time ceases to "exist", i.e. the temporal relations cease to be real (to hold among real events). If we now multiply the number of time keepers it becomes possible to compare their paces and eventually to pick the most regular among them. But this is a technical point of interest to physics and astronomy rather than ontology. So much for the intuitive background.

3.2. *Before and After*

On a relational theory of time we cannot accept a purely mathematical characterization of time – e.g. as the ultimate independent variable, or as the parameter of a one-parameter continuous group of transformations. In other words there is no such thing as mathematical time in our ontology: there are only mathematical representations of time. (See however Whitrow (1961) for the concept of mathematical time.) Mathematics, unlike mathematicians, is alien to time: constructs are neither changeable nor mutable (Corollary 5.1 in Ch. 5, Sec. 1.2); hence they are neither in time nor out of it. Nor would a definition of time in terms of the bare individuals of Ch. 1 do, because no state space and a fortiori no event space can be defined for them. We need the concept of a changing thing, hence that of state, if we are to define duration as a sort of distance between states.

Our starting point is then the concept of a lawful state space for a concrete object, or thing, of any kind. (Recall Ch. 3, Sec. 2.5.) In Ch. 5, Sec. 2.5, we saw that the states of a thing can be ordered by reference to those of a standard thing or reference frame, such as a clock mounted on a ruler. That was an *extrinsic* order, i.e. one induced on the thing states *ab extrinseco* by the natural or intrinsic ordering of the reference states. But actually any thing can serve as a clock even though only a few things are good clocks. In other words, the states of every thing are ordered. Because this order is intrinsic and independent of the kind of thing, we are justified in identifying it with temporal order. More explicitly, we assume the *intrinsic* time order of thing states:

POSTULATE 6.7 For every basic thing $x \in B$ and every lawful state space $S_{\mathbb{L}}(x)$ for x, there is exactly one ordering relation $\leqslant$, connected in $S_{\mathbb{L}}(x)$, such that, for any $s, s' \in S_{\mathbb{L}}(x)$,

$$s \leqslant s' \quad \text{iff} \quad s' = g(s), \quad \text{where} \quad g: S_{\mathbb{L}}(x) \rightarrow S_{\mathbb{L}}(x)$$

is a really possible (lawful) transformation of $S_{\mathbb{L}}(x)$, i.e. one representing a lawful change of x.

This assumption acquires a precise ontological sense by virtue of its semantic partner:

POSTULATE 6.8 The set $S_{\mathbb{L}}(x)$ of states of any basic $x \in B$ is *ordered temporally* by $\leqslant$. Furthermore for any $s, s' \in S_{\mathbb{L}}(x)$

(i) s *precedes temporally* s' iff $s \leqslant s'$;

(ii) s is *simultaneous with* s' iff $s \leqslant s'$ and $s' \leqslant s$ (i.e. $s \sim s'$).

The basic temporal relations have been defined only for the states of a single thing. So far we do not have any concept of temporal relations between states of different (in particular distant) things. It will also be noted that the preceding elucidations depend upon the notion of lawful change but not upon that of causation. Our postulates allow for temporal relations in a stochastic universe, i.e. one in which every single law is stochastic. Causality is sufficient for temporality and it may be necessary to *ascertain* which events precede which, but it is not necessary for the relations of before and after to *obtain*. The relation between precedence and causation is this:

For all events e and e′, if e causes e′ then e precedes e′.

But the converse is false. Hence precedence in time is not definable in terms of causation, nor is the converse move possible. (See Bunge, 1959.) Therefore our theory of time, though relational, is not causal – unlike the theories of Robb (1914, 1921, 1936) and Reichenbach (1928, 1956). Ours is a nomological rather than causal theory of time.

The equivalence relation $\sim$ introduced by Postulate 6.8(ii) effects a partition of $S_{\mathbb{L}}(x)$ into the states which are simultaneous and those which are not. The latter are related either by the *before* relation $<$ or by the *after* relation $>$. However, if we were to consider a rather bulky thing rather than a basic component (or member of B), there might appear pairs of states which are neither simultaneous nor related by $<$ or by $>$. Such would be the states of components of a thing not connectible by light waves. This is one reason for defining temporal relations on B rather than Θ.

We can now clarify the notions of present, past, and future of a basic thing:

DEFINITION 6.8 Let $S_{\mathbb{L}}(x)$ be a state space for a basic thing $x \in B$, and call $s_0 \in S_{\mathbb{L}}(x)$, a distinguished state of x, the *origin* or *zero* state. Then

(i) the *present* of $x =_{df} \{s \in S_{\mathbb{L}}(x) | s \sim s_0\}$;
(ii) the *past* of $x =_{df} \{s \in S_{\mathbb{L}}(x) | s < s_0\}$;
(iii) the *future* of $x =_{df} \{s \in S_{\mathbb{L}}(x) | s > s_0\}$.

Obviously, we are free to choose the origin s_0 and may call it *now* if we want to – provided we do not insist that this word designates the egocentric "now". Indeed, s_0 may be chosen to be the beginner of a process in the thing concerned, and all interested observers may agree on this choice for the sake of convenience. This, *pace* Hugo Bergmann (1929) and Grünbaum (1967), does not entail that becoming is unreal, hence temporal order conventional. Likewise the relativity of the splitting of states into present, past and future, to a given thing (in the above the very thing whose states are being partitioned) implies no subjectivity. Relativity and parochialism – as shown by the special choices of thing and of initial state – are one thing, subjectivity another.

The relativity of temporal orderings to things cannot be remedied: the two theories of relativity – the special and the general – have taught us that every thing has its proper time, so that there is no universal time. (There would be one if the universe were to pulsate as a whole – but this it does not.) In other words *there is no absolute or thing-free time*. This is a generalization of the statement in relativistic physics, that time is relative to some reference frame or other. We feel justified in making that generalization because, as noted before, almost anything can be regarded as a clock since all things are assumed to tick in some way or other. Of course some things tick more regularly than others and therefore make better clocks. But this is a matter for metrologists, astronomers, and atomic physicists, not for ontologists. From the point of view of ontological principles any thing the states of which are ordered in a certain way qualifies as a clock, whether or not it is actually used as such. The order in question is the strict partial order, i.e. asymmetric and transitive, so that no two states are simultaneous. Therefore we adopt

DEFINITION 6.9 A thing $f \in \Theta$ is a *potential clock* iff every one of its state spaces $S_{\mathbb{L}}(f)$ is strictly partially ordered.

DEFINITION 6.10 Let f be a potential clock with state space $S_{\mathbb{L}}(f) = T_f$. Then

(i) T_f is called the *time span* of f;
(ii) any interval of T_f is called an *f-moment*; and
(iii) any member of T_f is called an *f-instant*.

Each thing, and in particular each clock, ticks its own way. However, things are not monads: recall Postulate 5.10 in Ch. 5, Sec. 4.1. That is, every thing is connected with other (though by no means with all) things. And some such connections or couplings are of the physical information kind, so they can establish relations among the local temporal orderings. (For example, if two bodies move with constant velocity relative to one another then their time coordinates are related by a Lorentz transformation.) This gives us a revised notion of temporal universality: time orders are universal not because they are the same for all things – which they are not – but because they are mutually translatable within bounds. The kind of translation is specified by the following convention and the subsequent axiom.

DEFINITION 6.11 Let $f \in \Theta$ be a potential clock [Definition 6.9] with state space $S_{\mathbb{L}}(f)$ and $x \in \Theta$ any thing connectible with f, with state space $S_{\mathbb{L}}(x)$. Then

(i) an *f-time assignment* to the states of x is an injection

$$\varphi_f\colon S_{\mathbb{L}}(f) \to S_{\mathbb{L}}(x);$$

(ii) an f-time assignment φ_f is *faithful* iff, for any $t, t' \in S_{\mathbb{L}}(f)$,

$$t \leqslant t' \Rightarrow \varphi_f(t) \leqslant \varphi_f(t').$$

The next axiom ensures that time assignments are always possible, so that every potential clock can be "utilized" by any thing connectible with it:

POSTULATE 6.9 Let $x \in \Theta$ be any thing. Then there exists a potential clock $f \in \Theta$ and a faithful time assignment φ_f: $S_{\mathbb{L}}(f) \to S_{\mathbb{L}}(x)$.

In other words, the states of any thing can be ordered by the (reference) states of a potential clock connectible with the given thing.

So much for time order and its universal translatability (universal within bounds). We proceed now to build the concept of duration.

3.3. *Duration*

Relativistic physics teaches that durations are frame-dependent. Any given process in a thing has as many durations as there are (inequivalent) potential clocks connectible with the given thing. (And the shortest of all such durations is the one relative to the thing itself or to a frame attached to it. That is, the shortest duration is the so-called *proper* duration, i.e. the one relative to the thing undergoing the process of interest.) Therefore our ontological theory of time must take into account the relativity of duration.

We assign each event or process in a thing a collection of contiguous reference states, i.e. an interval of the time span of a clock. (See Figure 6.5.) The assignment is made by

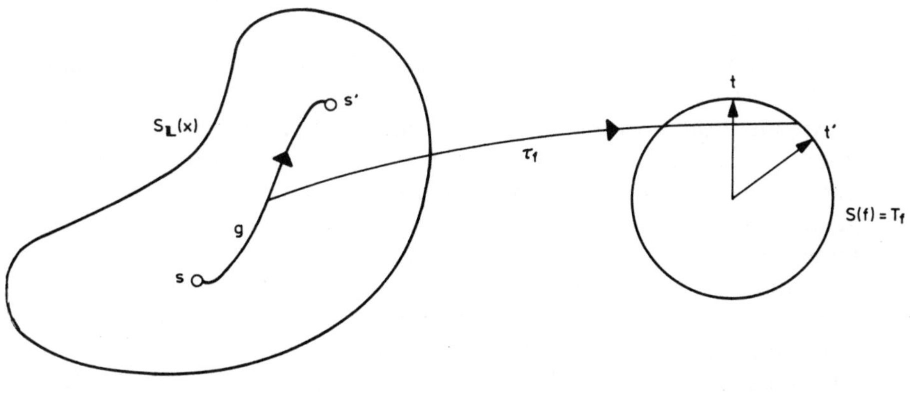

Fig. 6.5. The process $\langle s, s', g\rangle \in E_{\mathbb{L}}(x) \subseteq S_{\mathbb{L}}(x) \times S_{\mathbb{L}}(x)$ in thing x is assigned the interval $[t, t'] \subset T_f$ of reference states.

DEFINITION 6.12 Let T_f be the time span of a clock f and let $E_{\mathbb{L}}(x) \subseteq \subseteq S_{\mathbb{L}}(x) \times S_{\mathbb{L}}(x)$ be an event space for a thing x connectible with f. The *duration relative to f*, or *f-duration*, is the function

$$\tau_f: E_{\mathbb{L}}(x) \to 2^{T_f}$$

from events to intervals of reference states such that, for any events $e = \langle s, s', g \rangle$, $e' = \langle s'', s''', h \rangle \in E_{\mathbb{L}}(x)$, where g and h are lawful transformations of $S_{\mathbb{L}}(x)$,

(i) if $s \sim s'$ then $\tau_f(e) = \varnothing$;

(ii) if $s < s'$ or $s' > s$ then $\tau_f(e) \supseteq \varnothing$;

(iii) if e and e' compose to form $e'' = \langle s, s''', g \circ h \rangle \in E_{\mathbb{L}}(x)$, then

$$\tau_f(e) \cup \tau_f(e') = \tau_f(e'')$$

$\tau_f(e)$ is called the *duration* of e relative to f.

The duration function enables us to define in a precise (yet still nonquantitative) manner a number of notions in wide usage:

DEFINITION 6.13 Let τ_f be the duration function associated with a clock-thing pair $\langle f, x \rangle$. Then for any two events $e, e' \in E_{\mathbb{L}}(x)$ occurring in thing x,

(i) e is *instantaneous* (or a *point event*) iff $\tau_f(e) = \varnothing$;

(ii) e is *temporally extended* (or *takes time*) iff e is not instantaneous;

(iii) *e lasts just as long as e'* relative to $f =_{df} \tau_f(e) = \tau_f(e')$;

(iv) *e lasts longer than e'* relative to $f =_{df} \tau_f(e) \supset \tau_f(e')$;

(v) e and e' are *temporally contiguous* relative to $f =_{df} \tau_f(e) \cap \cap \tau_f(e') \neq \varnothing$;

(vi) e and e' are *temporally detached* iff they are not temporally contiguous.

An identical transition $s \mapsto s$, or nonevent, is of course trivially instantaneous. But it may also happen that $\tau_f(s, s', g) = \varnothing$ even if $g(s) \neq s$. That is, there might exist instantaneous proper events, i.e. genuine changes taking no time. Whether such changes exist we do not know and we need not know for building our theory of time, because the values of the duration function are intervals between $\varnothing$ and T. Hence we do not face the difficulty met with by alternative theories of time – such as Russell's and Whitehead's – based on the assumption that all events are analyzable into point (instantaneous) events. (This is a difficulty because, according to the quantum theories, there are temporally extended elementary events.)

We have now all we need to state in a precise fashion the gist of the relational theory of time:

DEFINITION 6.14 Let x be a thing and $E_{\mathbb{L}}(x)$ an event space for x. Further, let τ_x be the duration function relative to x. [I.e., x is now its own reference frame or clock.] Then the ordered couple $\langle E_{\mathbb{L}}(x), \tau_x\rangle$ is called the *local time* in (or relative to) x.

If $E_{\mathbb{L}}(x)$ were to shrink to nought, i.e. if x were immutable, the duration function would lose its support and there would be no time left. In other words we infer

COROLLARY 6.3 There is no time where there are no changing things.

Hence we are justified in calling a thing *timeless*, or out of time, just in case it is unchanging. But by Postulate 5.11 in Ch. 5, Sec. 4.1 there are no immutable things. Hence there are no timeless things. Therefore if anything looks timeless then it has been improperly investigated.

We emphasize that, since $E_{\mathbb{L}}(x)$ is the set of lawful events in thing x, our time concept hinges on the notions of thing, change, and law. In a chaotic (i.e. lawless) thing there would be no time. (Remember the distinction between stochastic orderliness or regularity, on the one hand, and chaos on the other: cf. Ch. 4, Sec. 6.3.) If we can discern temporal relations even in an arbitrary set of events, such as those composing a slice of our experiential stream, it is only relative to some set of regular events such as those of our biological clock.

Note also that Definition 6.14 allows for as many times as there are things. The assumption that all these times are the same, i.e. that the pace of events is the same relative to all things, is the hypothesis of *universal time*. We do not adopt this assumption because we wish our metaphysics to be compatible with physics, and the latter admits only local times. However, a relational theory of time including the universal time hypothesis would still be rela*tional* though not rela*tivistic*.

So far our concepts of time are qualitative. We can easily metrize the concept of duration provided we borrow from the foundations of physics the concept of a chronometric scale-*cum*-unit system. Indeed, the precise measure of a duration depends upon the way the successive stages of a process are assigned numbers, i.e. on the chronometric system. Moreover since time units – like all units – are conventional, they can be generated *ad libitum* to form a whole set U_t. Reckoned on a certain time unit $u \in U_t$, the duration of an event $e = \langle s, s', g\rangle$ in a thing x, relative to a clock f, can be taken to be the value, at $\langle e, u\rangle$, of a certain function T_f to be defined presently, i.e. $t = T_f(e, u)$. This function is characterized by

DEFINITION 6.15 Let f be a clock and x a thing, possibly the same as f, with event space $E_{\mathbb{L}}(x)$. Further, let τ_f be the duration function for the given pair $\langle f, x\rangle$. Then the corresponding *metric duration* is the mapping

$$t_f\colon E_{\mathbb{L}}(x) \times U_t \to \mathbb{R}$$

satisfying the conditions that, for any fixed unit $u \in U_t$ and any two events $e = \langle s, s', g\rangle$, $e' = \langle s'', s''', h\rangle \in E_{\mathbb{L}}(x)$,

(i) $t_f(e, u) \geqslant 0$ iff $\tau_f(e) \supseteq \varnothing$;

(ii) if e and e' compose to form $e'' = \langle s, s''', g \circ h\rangle \in E_{\mathbb{L}}(x)$, then

$$t_f(e, u) + t_f(e', u) = t_f(e'', u).$$

Clearly, metric durations are frame-dependent, which is in keeping with relativistic physics. Note also that clause (i) does not exclude negative durations but states only that, if state s precedes state s', then the duration of this event is non-negative.

Clearly, for every $u \in U_t$, the structure $\langle E_{\mathbb{L}}(x), t_f\rangle$ is a metric space. This justifies our calling $t_f(e, u)$ the *metric duration* of event e, relative to frame f, in units u.

We derive a couple of immediate consequences. First

COROLLARY 6.4 The duration of a point event is nil: for any frame f,

If $\tau_f(e) = \varnothing$ then $t_f(e, u) = 0$ for any $u \in U_t$.

Proof By clause (ii) of Definition 6.15 applied to the case of identical change or improper event. In fact in this case $2t_f(e, u) = t_f(e, u)$, whence $t_f(e, u) = 0$.

COROLLARY 6.5 The metric duration is an oriented interval [or directed set]: for any frame f, any unit $u \in U_t$ and all states $s, s' \in S_{\mathbb{L}}(x)$ forming the conceivable events $e = \langle s, s', g\rangle$ and $e' = \langle s', s, g^{-1}\rangle$,

$$t_f(\langle s, s', g\rangle, u) = -t_f(\langle s', s, g^{-1}\rangle, u).$$

Proof By clause (ii) and Corollary 6.4 upon setting $s''' = s$.

The last consequence expresses the so-called *anisotropy* or *asymmetry* of time, about which so much nonsense has been written. There is no mystery in it and it has nothing to do with irreversible processes.

Indeed, Corollary 6.5 is just a consequence of a convention, namely Definition 6.15, which assigns positive metric durations to changes occurring in the natural order and negative durations to their conceptual inversions. (We might as well have chosen the backwards reckoning.) More on this subject in Sec. 5.1.

4. Spacetime

4.1. *Spacetime, the Basic Network of Events*

Since Einstein's 1905 revolution, space and time have ceased to be regarded as independent: they are now treated as two aspects of a single four-manifold called *spacetime*. The separate relational theories of space and time should accordingly be replaced by a unified theory of spacetime. According to our relational point of view this formal unification is not a mere mathematical trick but formulates the idea that spacetime is generated by changing things. Before proceeding to the mathematical development let us give an intuitive idea.

Imagine a miniature universe composed of just two particles initially superposed at the origin (see Figure 6.6). All of a sudden one of the

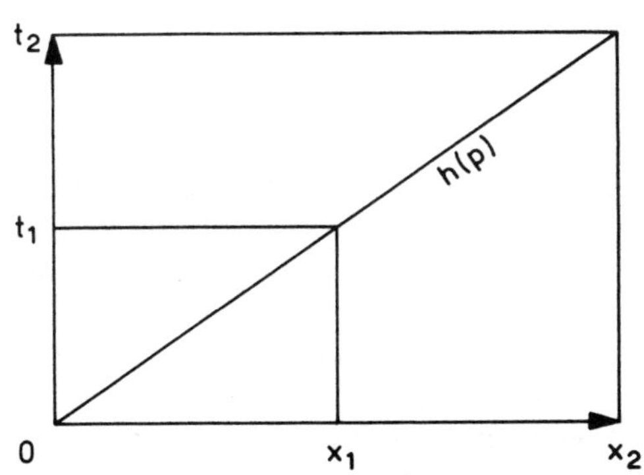

Fig. 6.6. The edge of spacetime advances as particle p moves away from the particle located at 0.

particles, call it p, starts moving while the other remains at 0. (Actually since there is no third thing that can function as a reference frame they just move relative to one another.) As p moves the edge of spacetime advances. When p reaches position x_1, spacetime covers the interior rectangle. This region expands up to the outer rectangle as p reaches position x_2. If p stops here there is no further displacement of the spacetime edge, and the outer rectangle remains frozen: the system has a past but no future.

We shall construct our view of spacetime on the basis of the concepts of event and of spatiotemporal separation between events occurring in things that may be different. Spacetime will then be defined as a certain set of events together with their spatiotemporal separations. However, this new concept of separation will be based on the concepts of thing separation (Definition 6.1) and of event duration (Definition 6.12). This strategy will allow us to advance rather quickly.

We start by defining the spatial distance between events as the separation between the things where they occur:

DEFINITION 6.16 Let $x, y \in B$ be two basic things, possibly identical, with event spaces $E_{\mathbb{L}}(x)$ and $E_{\mathbb{L}}(y)$ respectively relative to a common frame f. Moreover assume that all three things are connectible. Then the *spatial separation* between the events in x and those in y equals the separation between the corresponding things. In symbols,

$$\delta_{sf}: E_{\mathbb{L}}(x) \times E_{\mathbb{L}}(y) \to 2^B$$

such that

$$\delta_{sf}(e, e') = \sigma(x, y) \quad \text{for} \quad e \in E_{\mathbb{L}}(x) \quad \text{and} \quad e' \in E_{\mathbb{L}}(y).$$

If the things are not connectible – e.g. by signals of some kind – then the inter-event separation is not defined. And if the things coincide then the events are not spatially separated. Indeed, by Theorem 6.1(i), if $x = y$ then $\delta_{sf}(e, e') = \sigma(x, x) = \{x\}$. Note finally that the index f occurring in the event separation function is missing in the thing separation function. This is because, whereas events are relative to frames, their carriers (things) are not. Think of an emitter moving towards a stationary receiver – moving, that is, relative to a certain reference frame. The wavelength of the emitted signal differs from that of the received signal (Doppler effect). On the other hand the set of things that can interpose between the two bodies is frame-invariant.

Next we define the temporal distance between two events as the lapse between the end state of the first and the beginning state of the second. (We are presupposing here Postulate 5.8 in Ch. 5, Sec. 3.2, i.e. that every change has a beginning and an end.) More exactly, we make

DEFINITION 6.17 Let $x, y \in B$ be two basic things, possibly identical, with event spaces $E_{\mathbb{L}}(x)$ and $E_{\mathbb{L}}(y)$ respectively relative to a common frame f with states in $S(f)$. Moreover suppose that all three things are mutually connectible. Then the *temporal separation* between the events in x and those in y is the function

$$\delta_{tf}: E_{\mathbb{L}}(x) \times E_{\mathbb{L}}(y) \to 2^{S(f)}$$

such that, if $e \in E_{\mathbb{L}}(x)$ and $e' \in E_{\mathbb{L}}(y)$,

$$\delta_{tf}(e, e') = \tau_f(\text{end state of } e, \text{ beginning state of } e').$$

Now, the two separation functions δ_{sf} and δ_{tf} can be regarded as so many projections of a third function $\Delta_f = \delta_{sf} \times \delta_{tf}$ determined by the former in accordance with the following diagram:

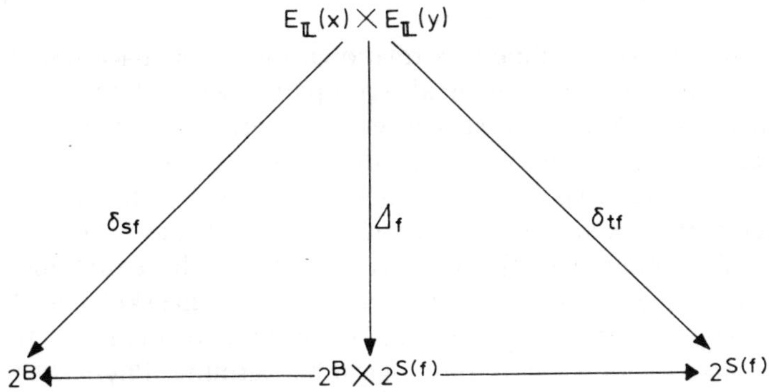

More explicitly, we propose

DEFINITION 6.18 Let $x, y \in B$ be two basic things with event spaces $E_{\mathbb{L}}(x)$ and $E_{\mathbb{L}}(y)$ respectively relative to a common frame f with states in $S(f)$. Moreover assume that all three things are mutually connectible. Then the function

$$\Delta_f = \delta_{sf} \times \delta_{tf}: E_{\mathbb{L}}(x) \times E_{\mathbb{L}}(y) \to 2^B \times 2^{S(f)}$$

such that

$$\Delta_f(e, e') = \langle \delta_{sf}(e, e'), \delta_{tf}(e, e') \rangle \quad \text{for} \quad e \in E_{\mathbb{L}}(x), e' \in E_{\mathbb{L}}(y),$$

is called the *spatiotemporal separation* (or *interval*).

We have now all we need to define a philosopher's relational concept of (local) spacetime:

DEFINITION 6.19 Let $B_0 \subset B$ be a set of basic things and call f a common reference frame. Furthermore suppose that all these things, i.e. $B_0 \cup \{f\}$, are mutually connectible. Finally, call E_f the event space, relative to f of the aggregation $[B_0]$ of basic things, and Δ_f the corresponding spatiotemporal separation. Then the ordered pair $\langle E_f, \Delta_f \rangle$ is called the *f-spacetime* (or *local spacetime of frame f*).

In physics a far more sophisticated structure is defined, namely a connected four-manifold M_f^4. We shall not construct M_f^4 out of the event f-spacetime. But this can be done along lines similar to those followed in Sec. 2 for building the physicist's space. In other words it is possible to map the events connectible with a frame f, and their mutual separations, on to a local spacetime chart M_f^4. And once such a chart for an arbitrary frame f has been defined, one can compose the atlas formed by all such local charts – i.e. the global or cosmic spacetime M^4.

So much for our outline of a relational theory of spacetime. Such a theory is not only relational but also compatible with relativistic physics, in that (*a*) it assumes the structure of spacetime to depend upon its furniture, and (*b*) it does not postulate a global structure. However, the theory is not *relativistic*: it does not include any of the special laws characterizing the various relativistic theories, such as for example the frame independence of the velocity of light, or the equations of the gravitational field. The relational theory of spacetime sketched above is just a component of the background of any general-relativistic theory – if one cares to add such an ontological background. Physicists usually don't: they are in the habit of postulating the four-manifold without inquiring into its roots in events. (Surely they call *events* the points of the manifold, but this is not done consistently, for such a definition presupposes a bijection between events and points in spacetime, and actually there is no such bijection, i.e. they do not mean that there is something going on at every point in spacetime.) Conversely, a relational theory of space and time may not be consistent with relativistic physics. Thus Robb's original theory (Robb, 1914), based on a set of point-like events and an absolute (frame-independent) relation of precedence, was relational but at variance with relativistic physics. In sum the theory of spacetime outlined in this subsection is relational and compatible with relativity but not relativistic. Only relativity is relativistic; and it is also relational provided it is explicitly based on a relational theory of spacetime.

4.2. *Position in Spacetime*

It is likely that events, however elementary, are not pointlike but take up nonvanishing regions of spacetime. That is, the event space E_f associated with a frame f ought to be mapped on the power set of the four-manifold M_f^4 rather than on M_f^4 itself. However, we may pretend, for the sake of simplicity, that events are pointlike, so that each event $e \in E_f$ can be assigned a point p in the corresponding local chart M_f^4. We shall make this simplifying assumption here because we are concerned only with clarifying certain ideas. The refinement is left to the reader.

We assume then that the event spacetime E_f can be mapped on a chart M_f^4. Now, by definition of a manifold, the latter can in turn be mapped locally on a set of ordered quadruples of real numbers. Consequently by composing the two maps we can assign numbers to events. This can be called a spatiotemporal coordinatization of events. The gist of it is shown in Figure 6.7, where 'N_f' designates an open subset of M_f^4.

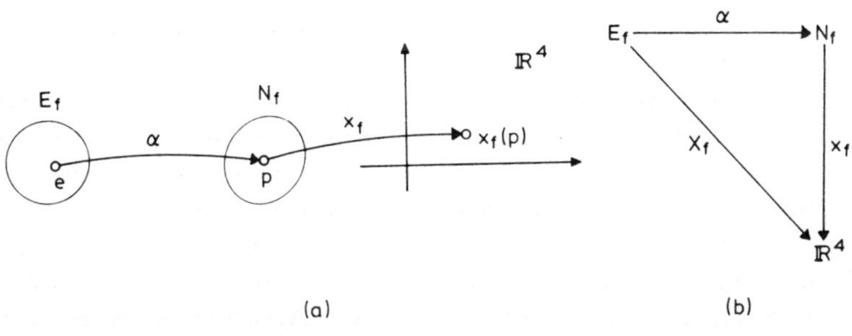

Fig. 6.7. (*a*) The mapping of the image N_f of E_f onto numbers and (*b*) the composition of the maps α and x_f to form the map X_f. The latter is a physical coordinate while x_f is a geometric coordinate. Whereas α is injective, x_f is bijective. Hence although every event is paired off to a quadruple of reals, the converse is false: there might be points in spacetime where nothing happens. (This is impossible according to our ontology, but scientific theories are partial and most of them do not assume that something is going on at every point in spacetime.)

In other words, for every reference frame f there is a coordinate system $\langle M_f^4, x_f \rangle$ representing f, where x_f, the local coordinate function, is a homeomorphism from an open subset N_f of M_f^4 into the cartesian space $\mathbb{R}^4$. This homeomorphism assigns each *point* $p \in N_f$ a quadruple of real numbers called the *coordinate values* of p relative to f:

$$x_f(p) = \langle x_1(p), x_2(p), x_3(p), x_4(p)\rangle, \quad p \in N_f \subset M_f^4.$$

(The choice of either of the four coordinate functions as the time coordinate is conventional, but one of them has got to be interpreted as the time coordinate.)

On the other hand the value of the composition of x_f and α, i.e.

$$X_f(e) = (x_f \circ \alpha)(e), \quad \text{where} \quad e \in E_f,$$

equals the spatiotemporal location of the *event* e relative to f. Note the difference – seldom recognized – between the geometrical coordinate functions x_f and the physical coordinate functions X_f. (The plural is apposite because for each f there are four coordinate functions of each kind.) The former locate *constructs*, the latter locate *events* represented by those constructs. Because α is not onto, X_f is not onto either, i.e. there are points and quadruples of reals that represent no events at all. But because α and x_f (hence also X_f) are functions, *every event is somewhere in spacetime* (*relative to a frame*). And because every event is a change in some thing, it follows that *every thing is at some place or other* (*relative to a frame*).

Nowadays we take it for granted that, even if space (and time) were absolute – i.e. self-existent –, position (in space and time) is not. But the idea that places are relative to frames has not been easy to come by. In all archaic and ancient cosmologies place was regarded as absolute and unique, every thing was supposed to occupy a definite place in space, and every event to occur at a moment of time likewise independent of any frame. These ideas are found not only in Aristotle but as recently as in Strawson (1959, pp. 22, 25–26). The credit for exactifying the notion of place relative to a coordinate system usually goes to Descartes. Galilei and his followers introduced the notion of a physical reference frame, which they dimly saw as representable by, though not identical with, a coordinate system. By the same token they showed that things and events have no absolute places – or, equivalently, that every single thing or event has as many places as there are reference frames, i.e. in principle infinitely many. These developments, stemming from Descartes and Galilei, disposed then of the principle that every thing and every event have or occupy a unique hence absolute place. In our own century Einstein crowned that development by demolishing the axiom that every fact happens relative to any reference frame (not just relative to the connectible frames) and consequently that it happens at a single absolute time irrespective of any frames.

Other developments that completed the demise of the *One fact-one place in space and time* principle were the following. First, Newton hypothesized that gravity can act everywhere, whence it has no unique or natural place. Second, Faraday showed that electricity and magnetism can likewise extend almost anywhere. Third, different bosons (e.g. photons) have been shown to be able to occupy the same place. Fourth, quantum mechanical things – whether particles or fields – seem to occupy no clear cut spatial regions, i.e. they have no sharp boundaries. These discoveries have refuted the old principles concerning place and have even diminished the importance of the category of place. Parallel developments in epistemology have devalued the notions of here and there, now and then. These have proved to be egocentric particulars (Russell, 1942) rather than objective items. That is, they are meaningful only in relation to a very particular reference frame, namely the self. Remove all subjects and no trace is left of here and there, now and then. Spacetime, on the other hand, is supposed to outlive the hecatomb for, *pace* Kant, it is not phenomenal.

Place in spacetime is then viewed as being relative to some reference frame or other. Being relative, location in spacetime is not an intrinsic property of things and events. And, being neither intrinsic nor causally effective, place in spacetime is not essential *per se*. In other words, coordinate values are rather arbitrary labels. What may be of importance is not the place that the mapping α assigns to an event nor consequently the relative positions of points and regions in the spacetime manifold, but rather the connections or bonds among things. In particular, it is irrelevant whether thing a is near thing b unless such nearness makes it possible for a to act upon b or conversely.

To sum up. Location in spacetime is both relative to a frame and objective. It is epistemologically important because it enables us to individuate or identify things and events that, aside from their distinct spatiotemporal location, might be equal (i.e. "only numerically distinct", as it used to be said). But spatiotemporal location is arbitrary to the extent that the choice of reference frame is arbitrary within broad bounds. Hence it is ontologically of secondary importance. So much so that the basic laws are invariant under changes in spatiotemporal position – which is a way of saying that coordinates are artifacts. But this is a matter for the next section.

4.3. *Change in Spacetime*

Change in general can be described without the assistance of any concepts of space and time: see Ch. 5. But specific kinds of change, such as change of place (relative to some frame) require of course some concepts of space and time. The very first thing to do in order to describe the motion (Aristotle's local change) of a thing is to model the latter either as a piece of space or as a chunk of spacetime (see Figure 6.8).

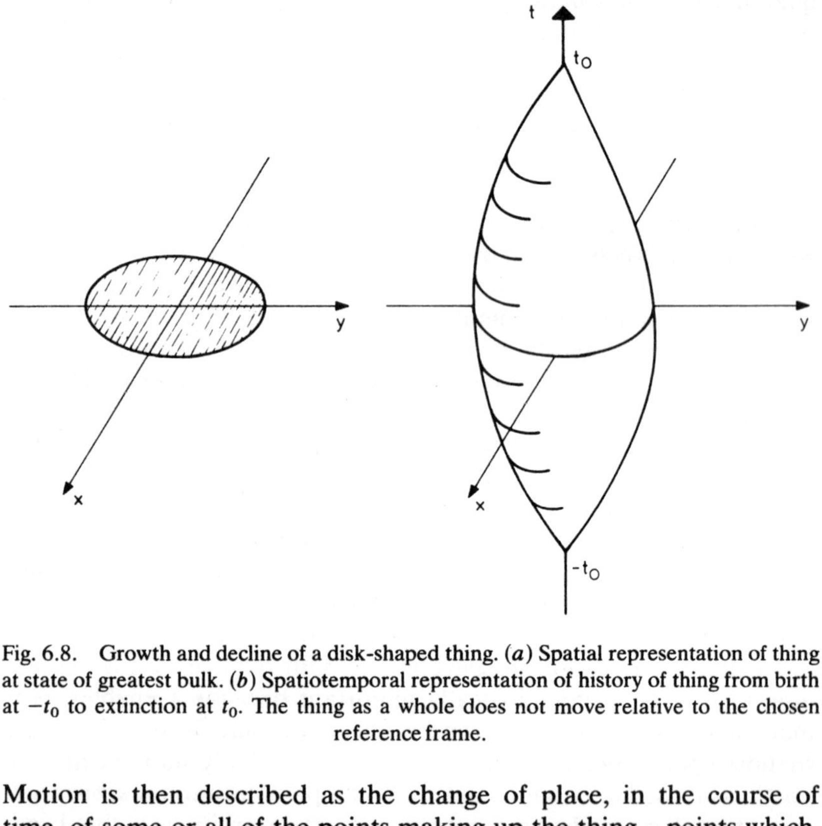

Fig. 6.8. Growth and decline of a disk-shaped thing. (*a*) Spatial representation of thing at state of greatest bulk. (*b*) Spatiotemporal representation of history of thing from birth at $-t_0$ to extinction at t_0. The thing as a whole does not move relative to the chosen reference frame.

Motion is then described as the change of place, in the course of time, of some or all of the points making up the thing – points which, unlike those of empty space, are located with the help of physical coordinate functions X_f (Sec. 4.2).

In building a model of a moving thing we start from Definition 3.6, in Ch. 3, Sec. 1.4, of a functional schema. A functional schema of a thing X

is a model thing X_m consisting in an ordered couple $X_m = \langle M, \mathbb{F} \rangle$, where M is the domain on which the state function $\mathbb{F}$ is defined. The latter is usually frame-dependent and it consists in an n-tuple of functions, usually real valued ones. We obtain a mechanical model if we equate M with space, or with spacetime, and some of the components of $\mathbb{F}$ with the physical coordinates of the thing (relative to the given frame), in the sense of Sec. 4.2. The values of these coordinate functions differ from zero only at the places (in space or in spacetime) at which the thing is. More precisely, we can make

DEFINITION 6.20 Let $X_m = \langle M, \mathbb{F} \rangle$ be a functional model of a thing X relative to a reference frame f [that occurs in the very construction of $\mathbb{F}$]. This model is a *mechanical model* just in case

(i) $M = N_f \subset M_f^4 =$ the spacetime chart attached to f;

(ii) some components of the state function $\mathbb{F}$ are physical coordinates serving to locate the parts of the thing:

$$X_f\colon N_f \to \mathbb{R}^4 \quad \text{or} \quad X_f\colon N_f^3 \times T_f \to \mathbb{R}^4$$

where N_f^3 is a three-dimensional projection of N_f.

In most theories the second choice of position coordinates is available, so that a general value of the position vector would be

$$X_f(x_f, t) = \langle x, y, z \rangle \in \mathbb{R}^3.$$

Note that the physical coordinate depends upon the geometric coordinate. The values of the two functions coincide only where the thing is. It is only in very special theories, such as classical particle mechanics, that X_f depends not on x_f but only on t.

A special kind of mechanical model is that of kinematical model: here all of the state functions are physical coordinates. That is, we have

DEFINITION 6.21 Let $X_m = \langle N_f, \mathbb{F} \rangle$, with $N_f \subset M_f^4$ be a mechanical model of a thing X. This model is a *kinematical model* iff $\mathbb{F} = X_f$.

The concept of motion, or mechanical change, may be regarded as a specification of the general concept of history introduced by Definition 5.27 in Ch. 5, Sec. 3.2, as shown by

DEFINITION 6.22 Let $X_m = \langle N_f, \mathbb{F} \rangle$, with $N_f \subset M_f^4$, be a mechanical model for a thing X relative to a frame f, and call X_f the position coordinate functions in $\mathbb{F}$. Further, call T_f the time projection of the state space $S(f)$ of frame f. Then a *motion* (or *mechanical change*) *of X relative to f* is any nonempty subset of its kinematical history

$$\mu_f(X) = \{\langle t, X_f(t, x_f)\rangle \in \mathbb{R}^5 | t \in T_f\}.$$

Spacetime is so important because it is an inescapable part of the state space of any moving thing. (The two seldom coincide, for change of place often brings about changes in other properties, or conversely. It is only in extremely idealized models, such as that of the point particle, that the history coincides with the kinematical history.) Moreover whenever there is change of any kind there is also motion, if not of the whole thing at least of some of its parts. (Think of a community established in an old town: although the town does not move relative to the earth, every member of the community moves about a great deal.) This is so important and sweeping a generalization that it deserves being conferred the title of

POSTULATE 6.10 The history of any basic thing, relative to an arbitrary frame, has a nonempty projection onto the spacetime attached to the frame.

This postulate must not be mistaken for the hypothesis that all change is reducible to motion. This other thesis, which characterizes all mechanistic ontologies, is plainly false: just think of the propagation of an electromagnetic wave or of an industrialization process. Surely the things concerned, or at least some parts of them, move in the course of such processes; but they also change in other respects. This remark too deserves being generalized and the generalization exalted to the axiom rank:

POSTULATE 6.11 Whatever changes does so in more than one respect.

The implication for epistemology is clear: no theory concerning a single type of change suffices to account for the multifarious changes of real things.

5. Spatiotemporal properties

5.1. *Does Spacetime have any Properties?*

Spacetime is often said to have mathematical properties. According to our view only mathematical objects can have mathematical properties and, since spacetime is not such an object, it can have no mathematical properties. Not spacetime itself but rather every one of its conceptualizations, in particular the four-manifold M^4, has mathematical properties, such as that of being the carrier of continuous functions. That the distinction between spacetime and its mathematical representations is seldom made in the scientific literature should be no excuse for the philosopher's failure to draw it.

How about physical properties: does spacetime have any? On our view spacetime has no physical properties either. How could spacetime have substantial properties not being a thing but rather a (relational) property of things? Still, it is often stated dogmatically that spacetime does have certain physical properties, in particular the following three. One is that spacetime is the "carrier" of physical fields such as the gravitational field – a function formerly assigned to the ether. This is suggested by the mathematical representation of a physical field as a tensor (or spinor or vector) field over a manifold. But a mode of representation should not be identified with the object represented. Physical fields need no carrier or substrate: they are things on their own – as first recognized by Einstein. Surely we shall continue to speak of, say, the gravitational field at a given place, without however attributing places an autonomous existence nor, a fortiori, physical properties.

Another property commonly attributed to spacetime is that any given event is connectible only with events lying inside its forward light cone (see Figure 6.9). This restricted connectibility is sometimes said to be a

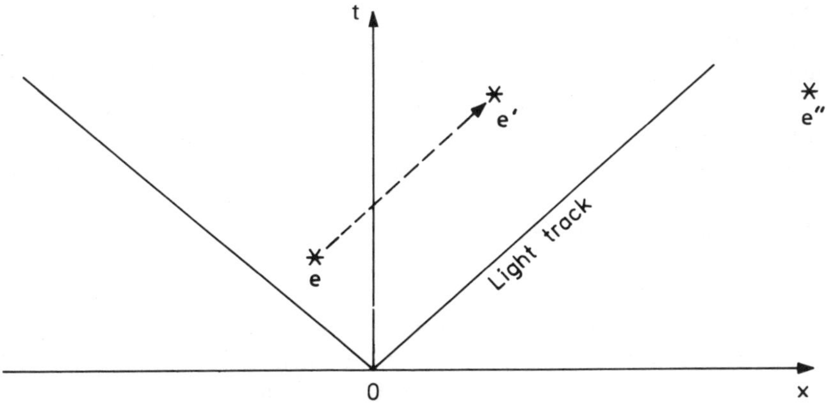

Fig. 6.9. Events e and e' are connectible, but e and e'' are not because there are no (known) superluminal signals.

fundamental property of spacetime. Yet note that (*a*) connectibility is definable as a ternary relation among physical entities, and (*b*) connectibility depends on there being things capable of moving. If superluminal entities (tachyons) were discovered we would cease to regard connectibility within just the light cone as a universal property of things and their changes

A third important instance of the attribution of physical properties to spacetime concerns the vacuum polarization postulated in quantum field theory. Actually this is not a property of spacetime but of the background field, e.g. of the electromagnetic field in its ground state, where it exists and fluctuates but has no quanta (photons).

We generalize the preceding considerations, in line with our relational view of spacetime, and state that the latter *has no properties*, and this because it is itself a property, namely the basic mesh of the sum total of changing things. It is things and their changes, as well as the patterns of either – i.e. the laws – that have spatiotemporal properties. For example a hydrogen atom in its ground state is spherically symmetric, and the development of an embryo is not time symmetric, or rather reversible – but these are properties of the things concerned not of spacetime.

An important spatiotemporal characteristic of a factual item is its absoluteness or the lack of it. Some facts, such as light emission, childbirth and industrial production, are absolute, i.e. independent of any reference frame. Others are not: for example a light detector, such as a photocell or a retina, tuned to respond to a certain wavelength band, will detect certain signals coming from a stationary source but will fail to detect the same radiation if the source moves either towards or away from the sensor. Light absorption depends then upon the reference frame. Needless to say a relative fact is as objective as an absolute one. The circumstance that the detector might belong to a living being, in particular a subjectivist philosopher, does not render the fact subjective.

Likewise while some laws are absolute others are relative to some frame. In particular all the basic laws are invariant under spatial (actually spatiotemporal) translations and rotations. That is, place and spatial orientation make no difference to the basic patterns. (Most laws are also reflection invariant, i.e. they do not distinguish between left and right, or East and West.) In other words, if a basic law obtains in a given region $m \subset M^4$ of spacetime, then it also holds in any other region $m' \subset M^4$. This principle of spatiotemporal invariance (or rather covariance) applies only to the basic laws: the derived laws need not have this property, since they may embody particular circumstances of a spatiotemporal character, such as initial conditions and boundary conditions. Another caution: the preceding principle does not entail that, if there are biological (or sociological) laws in a given region, then the same laws are also in force in any other region of spacetime, even if devoid of organisms and a fortiori of societies. The principle is satisfied vacuously by the regional laws that happen to obtain in a region but are not "exemplified" in another.

5.2. *Time Reversal and Process Reversibility*

The matter of invariance under time reversal deserves a special section because it is often misunderstood. Call π a net representing a linear chain of states of some thing, i.e. a nonbranching process (relative to some reference frame of course). Call $\tilde{\pi}$ the transpose of π, i.e. the same set of states in reverse order. For example, if $\pi = \langle s_1, s_2, s_3 \rangle$ then $\tilde{\pi} = \langle s_3, s_2, s_1 \rangle$. The given net and its transpose are related by the order reversing operator T such that $\tilde{\pi} = T\pi$. (Caution: given π and T, the transpose is uniquely determined. But in general T is not uniquely determined by π and π', i.e. the preceding equation does not define T uniquely. As it so often happens the inverse problem does not have a single solution.) The operator T is usually called the *time reversal* operator. This is a misnomer because T represents the inversion of the order of states not the reversal of time itself (see Figure 6.10). A better name would be that of *process reversal* operator.

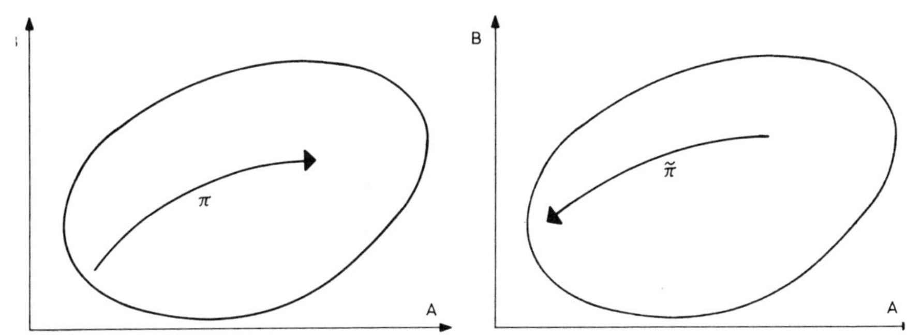

Fig. 6.10. A process π in the state space $S \subset A \times B$ of a thing and the transpose $\tilde{\pi} = T\pi$ of π. T maps the first onto the second.

The relation $\tilde{\pi} = T\pi$ between a process and its transpose can be interpreted in two different ways. A first interpretation is this: if we decide to count time intervals backwards, e.g. by using a clock whose pointers turn counter-clockwise, then we must invert the order of the states: the last state will come (temporally) before all the previous states, as shown in Figure 6.11. An alternative interpretation of the same mathematical operation of time (or rather state order) reversal is this. The transpose $\tilde{\pi}$ of π is not just a nonstandard representation of π but stands for a possible process going forward in time and traversing in reverse order the same stages as π – as when a film is projected backwards. However, this second interpretation of time reversal is legitimate only provided $\tilde{\pi}$ is lawful, which is not the case with every process. When a process happens to have a transpose going forward in time it is called a *reversible* process – otherwise it is said to be *irreversible*. (Recall Definition 5.21 in Ch. 5, Sec. 3.1.)

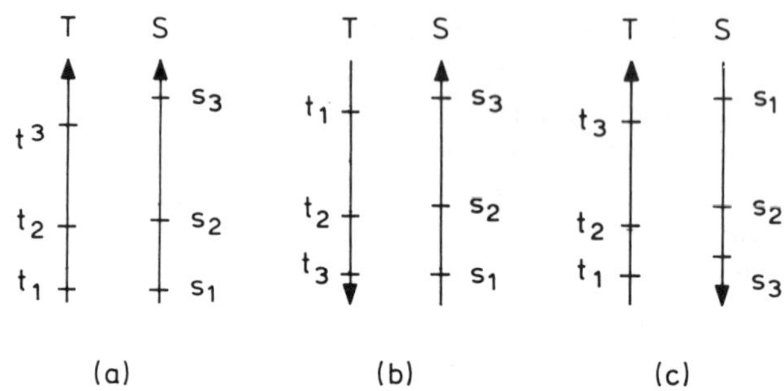

Fig. 6.11. (*a*) The standard instant-state correspondence. (*b*) The queer but possible representation of a process by its transpose. (*c*) The actual inversion of a process in case it is reversible.

If a process is reversible, the time coordinate values of its successive states afford a convenient if misleading representation. By Corollary 6.5 in Sec. 3.3, the metric duration function t_f is antisymmetric (odd) in the sense that, if s and s' belong to the same process, then $t_f(\langle s, s', g\rangle, u) = -t_f(\langle s', s, g^{-1}\rangle, u)$ for every frame f and every scale-cum-unit system u. Calling t the value $t_f(\langle s, s', g\rangle, u)$, as is usual in science, we discover the factual meaning of the inversion of the sign of t: it corresponds to an inversion of the order of appearance of the underlying states. That is, *time reversal represents process reversal*:

$$\textit{forward process } \langle s, s', g\rangle, \quad t_f(\langle s, s', g\rangle, u) = t$$

$$\textit{reverse process } \langle s', s, g^{-1}\rangle, \quad t_f(\langle s', s, g^{-1}\rangle, u) = -t.$$

Reversibility, a property of certain processes, must not be mistaken for the invariance of a basic law statement under time reversal, i.e. under the inversion of the sign of t (or, equivalently, under the T operation). Call $L(t)$ a law statement involving a time coordinate t. If $L(-t) = L(t)$, the law is said to be *T-invariant*, or invariant under time reversal. The basic electromagnetic laws (Maxwell's) are T-invariant, whereas the fundamental law of heat transfer (Fourier's) is not. The time inverse of a process described by a T-invariant law statement is just the reverse of the original (reversible) process – e.g. a motion with inverted velocities and spins. No "inversion of time" is involved even if the process can be described by exchanging t for $-t$ in the derived equations. The reverse or transpose of a reversible process goes forward in time just as much as the original process does. There is no *arrow of time* except as a misleading and overrated metaphor. The only "arrows" can be found in certain processes – namely the irreversible ones such as radioactive disintegration or ageing. Therefore the search for the origin of the arrow of time – in which many distinguished physicists have engaged – is as futile as the search for the philosopher's stone.

If a law statement fails to be T-invariant, i.e. if $L(-t) \neq L(t)$, then it concerns irreversible processes – provided the law is basic rather than derived. (E.g., whereas the equation of motion of a vibrating string is T-invariant, its elementary solution $u = a \sin(kx - \omega t)$, a derived law, is not.) The converse is false: some irreversible processes can be described with the help of T-invariant laws together with certain subsidiary conditions. For example the basic formulas of classical electromagnetism are T-invariant but, when conjoined with Sommerfeld's condition of outward radiation (exclusion of incoming waves), they describe the irreversible propagation of a retarded outgoing wave. In general a process or history is described jointly by a set of law statements and a set of constraints, initial conditions, boundary conditions, and possibly other subsidiary hypotheses representing particular circumstances of both the system and its environment. Thus radioactive processes are described by elementary quantum mechanics, which is T-invariant, together with the hypothesis that a nuclear potential barrier is semipermeable, in the sense that some nuclear particles can tunnel through it even if their energy is smaller than the height of the barrier. In sum the T-invariance of a bunch of law statements provides no indication as to the reversibility or otherwise of the process concerned: only the whole theory can supply such an indication (cf. Bunge, 1970b, 1972).

5.3. *Antecedence* ("*Causality*") *Principle*

Contemporary physics asserts explicitly, in various forms and under the misnomer *causality condition* (or *relation*), an overriding metaphysical principle, namely clause (vii) of Postulate 6.1. This principle states that causes precede their effects or, equivalently, that the present is determined (whether uniquely or stochastically) by the past, never by the future – this relation being moreover frame invariant unlike the mere succession of states. We proceed to state this principle in terms of the property concept and shall distinguish two cases, the stochastic and the non-stochastic ones (Bunge, 1968c).

Let P_1 and P_2 be two time dependent properties of things of a certain kind K. As usual we assume that the properties are represented by functions and, more particularly, that both $F_1 \triangleq P_1$ and $F_2 \triangleq P_2$ are defined on $K^n \times \{f\} \times T$, where f is a reference frame, T the range of the metric duration function attached to f, and $n \geq 1$ a natural number. (Actually there may be more than one kind of thing, so K may have to be interpreted as a thing genus or else K may have to be split into a cartesian product.) As far as pure mathematics is concerned there are three possibilities as to the relations between the two properties:

(*a*) P_1 and P_2 are mutually independent;

(*b*) one of them depends upon the other – e.g. P_1 determines P_2;

(*c*) both P_1 and P_2 depend upon a third property, so that they are indirectly related to one another.

Here we are interested in case (*b*), i.e. that of one-sided dependence. Again, as far as pure mathematics is concerned, there are no limitations on the form of dependence. But it so happens that actually the dependence forms are restricted by a general principle according to which the value of F_2 at any given time is the outcome of the P_1-history of the thing prior to that time. In other words the dependence of any one property on another is *retarded* or *nonanticipatory*: there is no action by the future – hence no precognition, i.e. no information flow coming from future events. This fundamental ontological hypothesis can be formulated as follows:

POSTULATE 6.12 Let P_1 and P_2 be properties of a system composed of things of kind K and assume that P_1 determines P_2.

(i) If P_1 and P_2 are representable by ordinary functions F_1 and F_2 respectively, both defined on the domain $K^n \times \{f\} \times T$, then the instantaneous value $F_2(x, y, ..., z, f, t)$ of F_2 for $x, y, ..., z \in K$, relative to the frame f, is a functional G of the P_1-history of the thing up to the given instant:

$$F_2(x, y, ..., z, f, t) = \underset{u=-\infty}{\overset{u=t}{G}} [F_1(x, y, ..., z, f, u)];$$

(ii) if P_2 is representable by a random variable F_2 (whether or not F_1 is one), then the probability that, at time t, its values lie in a preassigned interval $[a, b]$ is a functional of the entire P_1-history of the thing up to the given instant:

$$Pr[F_2(x, y, ..., z, f, t) \in [a, b]] = \underset{u=-\infty}{\overset{u=t}{G}} [F_1(x, y, ..., z, f, u)]$$

Simpler forms of the preceding axiom are obtained by forgetting that all substantial properties are defined for things and are often relative to a reference frame. I.e., by calling the property-representing functions x and y respectively, the antecedence principle simplifies to

$$\left.\begin{array}{r} y(t) \\ Pr[y(t) \in [a, b]] \end{array}\right\} = \underset{u=-\infty}{\overset{u=t}{G}} [x(u)].$$

Actually for every thing but the whole universe, the value of G will be nil between $-\infty$ and the instant t_0 of its coming into being.

About the simplest and most common functional dependence is this:

$$\left.\begin{array}{r} F_2(x, y, ..., z, f, t) \\ Pr[F_2(x, y, ..., z, f, t) \in [a, b]] \end{array}\right\} =$$

$$= \int_{-\infty}^{t} K(t, u) \cdot F_1(x, y, ..., z, f, u) \, du.$$

The particular case of instantaneous dependence of P_2 upon P_1, or instantaneous action, is obtained by setting $K(t, u) = \delta(t-u)$, where δ is the Dirac delta.

The antecedence principle has interesting consequences. One of them is

THEOREM 6.6 There is no return to the past.

Proof If there were (timelike) loops in spacetime it would be possible to modify the past states of things from the future – contrary to Postulate 6.12.

In other words, time machines are impossible. If they were possible then we could land in the past and mend it so as to obtain a better present. But then we would get two different presents. One seems enough.

Note that "time travel" to the past has nothing to do with reversibility. Even a strictly reversible universe would be subject to the restriction that the past cannot be undone. So much so that the antecedence or "causality" principle is part and parcel of quantum field theory and other stochastic theories dealing with reversible processes. Nor does Theorem 6.6 have any bearing on the frame dependence of duration as exemplified by the travelling twin "paradox". The latter consists in using the expansion of duration with motion in order to land in the future of a subject who has not taken part in the trip. This consequence of special relativity is counterintuitive to anyone brought up in the tradition of absolute time; but of course it is natural in the context of the theory and moreover it has been experimentally confirmed with great accuracy. But it is of dubious ontological value except in liberating us from the absolute theories of time.

Finally a few remarks. First, the antecedence principle is sometimes stated in this form: "Whereas the future can be influenced the past cannot". Actually neither past nor future can be influenced at any given time because they do not exist at that time (relative to a given frame). 'Influencing the future' is short and misleading for "Altering the odds of the present possibilities". There is no direct action of past on future: rather than jumping across a stretch of time we must let things unfold themselves relative to each other – or, as ordinary language puts it, in the course of time.

A second remark is this: the name 'causality condition' for Postulate 6.12 is mistaken on several counts (Bunge, 1959, Ch. 3, Sec. 3.2.) For one thing the postulate applies not just to input-output relations, which sometimes are genuinely causal, but also to certain pairs of functions mirroring different traits of one and the same thing – hence to situations in which there are no input and output terminals.

Third observation: although there is no experience of anticipation – but only hallucinations of such – some theories do allow for it at their own risk. The only great theory which is partially unfaithful to the antecedence principle is classical electrodynamics, according to which an electron is accelerated by an incoming wave before the latter actually strikes it. But this is usually regarded as one of the various unsatisfactory features of the theory, not just because it is unsupported by observation but also because it contradicts the general principle of antecedence. (See Bunge, 1967b, p. 167.)

Fourth: goal-directed behavior is only in seeming contradiction to antecedence. Indeed if an organism seems to be guided by its own future states it is only because the organism is equipped to behave in such a way that it will exploit its present circumstances so as to maximize its chances of survival or increase its welfare. Even human teleology is present-guided not future-guided: an anticipation (e.g. programming) of future events is nothing but the formation, at the present time, of a mental picture of certain possible future events, in particular those we wish to bring about.

5.4. *Action by Contact*

As far as logic is concerned a thing may act upon a separate (but of course connectible) thing either directly (*action at a distance*) or through the intermediary of a third thing (*nearby action*). And in either case such an influence could be exerted instantaneously or it could take some time. (See Figure 6.12.)

Action at a distance was assumed in all of the classical atomic theories, while nearby action has been a component of the plenistic cosmologies of Aristotle, Descartes and their followers. Until the

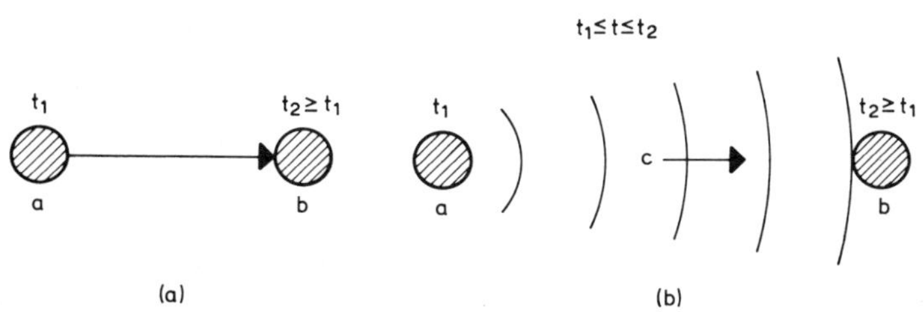

Fig. 6.12. (*a*) Thing *a* acts directly upon thing *b* (instantaneously or with some delay). (*b*) Thing *a* acts upon thing *b* through thing *c* (instantaneously or with some delay).

confirmation of the first successful field theory there was no compelling reason for preferring one of the alternatives to the other. Ever since then the hypothesis of nearby action has prevailed in every chapter of physics where interactions among different things are studied. True, corpuscularian metaphysics is still going strong among operationists who insist that we never measure anything but attributes of perceptible bodies such as the position of the hands of a clock. Here and there operationism raises its head and indicts the very concept of field, but only temporarily: it goes down every time because action-at-a-distance theories are definitely poorer than nearby action theories – in particular they cannot explain away processes which, like the propagation of an electromagnetic wave, are irreducibly plenistic. We bet then on nearby action. To formulate the principle we make use of the notion of place or bulk of a thing, introduced by Definition 6.6 in Sec. 2.4. The *principle of nearby action* reads then:

POSTULATE 6.13 If two things are at different places and one of them acts upon the other, then there is a third thing interposed between them. I.e.,

$$x, y \in \Theta \;\&\; \beta(x) \cap \beta(y) = \varnothing \;\&\; x \rhd y \Rightarrow$$
$$(\exists z)(z \in \Theta \;\&\; z \neq x \;\&\; z \neq y \;\&\; x|z|y).$$

A particularly appealing if specific version of this principle is the following. Let P_1 and P_2 be two substantial properties represented by functions F_1 and F_2 respectively defined on the space manifold M^3. Then for any point $x \in M^3$ there is a real valued function L on $M^3 \times M^3$ such that

$$F_2(x) = \int_{-\infty}^{\infty} dy\, L(x, y) F_1(y).$$

The principle of nearby action does not preclude the possibility that an action initiated in a thing remains confined to it. Such a possibility is denied by the strong version of the principle, which can be formulated thus: "Whatever happens at any given place is determined by what happens in its immediate vicinity and in turn influences every other thing". According to this principle no two things in the universe can be independent. We do not adopt it because there are cases of effective insulation and of spontaneous (uncaused) activity (Bunge, 1959).

5.5. *Spatiotemporal Contiguity*

The principles of antecedence and nearby action are logically independent: as suggested in the previous section one could assert the one while denying the other. But, since we have accepted both, we may as well merge them into a single principle, the more so since relativistic physics has taught us that spatial relations are enmeshed with temporal relations. A possible merger is the so called principle of *local causality*, which we prefer to call the *principle of spacetime contiguity*:

POSTULATE 6.14 Every action of one thing upon another satisfies both the antecedence and the nearby action conditions.

A possible but restricted formulation of this condition is as follows. If P_1 and P_2 are substantial properties of some thing (e.g. a field) representable by functions F_1 and F_2 respectively, defined on the spacetime manifold M^4, and if P_2 depends upon P_1, then relative to any frame f and for any quadruple $\langle x, t\rangle \in M^4$, there is a real-valued function N such that

$$F_2(x, t) = \int_{-\infty}^{t} dv \int_{-\infty}^{\infty} du\, N(x, u; t, v) F_1(u, v),$$

where

$$\langle u, v\rangle \in M_f^4$$

and $|x-u| \leqslant c|t-v|$, c being a positive real number. (To comply with special relativity c must be smaller than or equal to the velocity of light in vacuo. General relativity introduces certain modifications in the above formula, mainly the insertion of $g^{1/2}$ in the integrand, where g is the determinant of the metric tensor, and a slight liberation of the constraint on space and duration intervals, as c may now be a function defined on M^4, though one whose values will always be near the velocity of light in vacuo in the absence of gravitation.)

The previous principles, of nearby action and antecedence, are retrieved as special cases upon setting $N(x, u; t, v) = \delta(t-v) \cdot L(x, u)$ and $N(x, u; t, v) = \delta(x-u) \cdot K(t, v)$ respectively, where L and K are space dependent and time dependent functions respectively.

To realize that Postulate 6.14 underlies the whole of science suffice it to note that its denial would involve the following consequences. First, because the principle asserts the mutual independence of nonconnectible things, if it were false no thing could remain quasi-isolated from the rest of the universe even for a short while. Hence no thing could be known even approximately, and so the very concept of an individual thing would be idle. Far from being a system, the universe would be a block. Second, both the remote past and the remote future could pounce upon the present without notice and render it chaotic. This again would render knowledge utterly impossible and confusion would be even more dense and widespread than it is. Since we do succeed in knowing a thing or two we must infer that we are not at the mercy of things far removed in space and time. In particular, we may rest assured that events occurring in a distant space-like separated region of spacetime do not affect us.

True, it is occasionally claimed that Postulate 6.14 is at variance with quantum mechanics. But if this were true then the quantum theory would be incompatible with the special theory of relativity, which in turn embodies Postulate 6.14. And this conclusion is contradicted by the very existence of successful quantum relativistic theories such as Dirac's theory of the electron and quantum electrodynamics. Hence the results purporting to invalidate the so-called local causality – such as EPR inseparability and Bell's theorem – will have to be reinterpreted. In other words we do not have to give up ontological principles that are well entrenched in science just because they seem to be contradicted by an occasional and isolable fragment of scientific research. On the contrary, such strays have got to be reexamined in the light of scientific ontology. Thus when classical electrodynamics led to the conclusion that there are preaccelerations, hence events violating the antecedence principle (Postulate 6.12), most physicists doubted the physics rather than the metaphysics and made efforts to correct the former instead of the latter. (Recall Sec. 5.3.)

5.6. *The Causal Relation*

Finally let us elucidate the notion of causation and state the causal principle. We construe causation as a relation between events in different things – which of course may be parts of a bulkier thing. Furthermore we take it that causation subsumes antecedence and that, far from merely preceding the effect, the cause produces it (Bunge 1959). In sum, we make

DEFINITION 6.23 Let $e \in E(x)$ be an event in a thing x at time t and $e' \in E(x')$ another event in a thing $x' \neq x$ at time t', where the events as well as the times are taken relative to the same reference frame. Further, call $A(x, x')$ the total action or effect of x upon x' (Definition 5.31). Then we say that e is a *cause* of e' iff

(i) $t \leqslant t'$;

(ii) $e' \in A(x, x') \subseteq E(x')$.

A more precise condition than clause (ii) would be this: There is an energy transfer from x to x' during the interval $[t, t']$, and the quantity of energy transferred suffices for the occurrence of e'. But we do not need the extra precision.

The notion of causation allows us to formulate the strict *principle of causality*, namely as follows: "Every event is caused by some other event." More precisely: "Let x be a thing with event space $E(x)$. Then for every $e \in E(x)$ there is another thing $x' \neq x$, with event space $E(x')$ relative to the same reference frame, such that $e' \in E(x')$ causes e." But we shall not espouse this principle for, if we did, most of contemporary science would have to go (Bunge, 1959).

So much for the general principles involving the notions of space or time. Let us now approach matters of existence.

6. Matters of existence

6.1. *Existence in Space and Time*

It is usually assumed that, unlike ideas, physical things "exist in space and time". (For Aristotle only changing things exist in time: *Physics*, Bk. IV, Ch. 12. And for Kant ideas exist in time but not in space.) Moreover "existence in space and time" is sometimes taken to define the very concept of a physical object in contrast to that of a nonphysical object. In turn, what is meant by 'x exists in space and time' is that x occupies a region of space and endures throughout a time interval.

The statements we have just mentioned presuppose the autonomous existence of the spatiotemporal framework, which in turn would be a non-physical object. This is of course the absolute doctrine examined and rejected in Sec. 1.1. We do not accept the container view because our ontology does not assume any nonphysical entity – except of course as a fiction. In our view things do not float in a given spatiotemporal framework but hold spatiotemporal relations among each other – just as persons do not swim in a community but constitute it by holding connections with one another. Things "come" with their own (changing) spatiotemporal relations, and the latter are just relations (but not bonds) among things and their changes.

Consequently it makes as little sense, in our ontology, to say that a thing *exists in* spacetime, as to say that spacetime *exists in* a thing. The expressions 'x exists at a place y (or in a region y) relative to a frame z' and 'x exists at time t (or throughout time interval t) relative to a frame z' are elliptical. The location of a thing is always given by reference to another thing and the timing of a state or of an event is always referred to some other state or event – and in turn states are states of things, and events are changes of state of things. *No things, no nothing*, as the robust street corner philosopher put it. We can use those elliptical locutions because they save time, provided we remember what their purport is at the time of discussing foundations matters. In ontology and in the foundations of physics we have hardly any right to assume that spacetime preexists things, and that the spatiotemporal location of a thing or an event is in the nature of the part-whole relation, the part being the factual item and the whole spacetime. Only the bulk of a thing can be included in space, because that bulk is a set.

Finally, what about the *plenum* hypothesis held by Aristotle and his followers? (Cf. Lovejoy, 1936.) This hypothesis, that space is full of things, is favored by contemporary physics, in particular gravitation theory and quantum electrodynamics. In other words, there is no such thing as empty space or a perfect vacuum. (Particle void ≠ void.) We may then adopt this principle:

POSTULATE 6.15 There are things interposed between any two things located at different places. I.e.,

$$x, y \in \Theta \;\&\; \beta(x) \cap \beta(y) = \varnothing \Rightarrow$$
$$(\exists z)(z \in \Theta \;\&\; z \neq x \;\&\; z \neq y \;\&\; x|z|y).$$

Since this assumption entails Postulate 6.13, we could have gotten the latter as a theorem.

6.2. *Existence of Space and Time*

It is tempting to doubt the existence of space and time. After all, they are not things – not even imperceptible things – and they have no causal power. (It is sometimes said that, according to general relativity, spacetime guides the motion of things: that there are tracks in spacetime. But this is a mistake: what guides the motion of things is the gravitational field, which the theory represents as a departure of spacetime from flatness.)

As a matter of fact a number of philosophers and mystics have doubted the reality of space and many more have maintained that time is unreal. While the topoclasts have usually argued from the imperceptibility of space, the chronoclasts have often argued from the alleged immutability of a Reality behind appearances. In particular Kant held space and time to be phenomenal not real – which is the only reasonable view if the container doctrine is kept, not however on a relational doctrine of space and time (Lotze, 1887, Bk. II, Ch. I). And Russell held at one time that both space and time are fictitious because the *concepts* of space and time should be constructible out of "real entities", namely perceptual items. We cannot accept these phenomenalist views if only because they ignore the basic distinction made in contemporary physiological psychology between physical spacetime on the one hand and the various perceptual (e.g. visual and tactual) spaces and the perception of duration on the other.

In our view neither the topoclasts nor the chronoclasts are right. Although space and time are not things and have no being independent of things and their changes, they are no less real than any other substantial properties of real things. Indeed recall that a relation was called *real* if and only if it holds among real things – which is the case with spatiotemporal relations. (Cf. Definition 2.17 in Ch. 2, Sec. 5.1.) The *esse* of spatiotemporal relations consists not in their *percipi* but in their obtaining between real things and real events.

The role of space and time in science is not always well understood. Thus just because spatiotemporal concepts do not occur in the systematics of elementary particles, or because they play a subdued role in the black box theories of these entities – namely the scattering matrix and the dispersion relation theories – it has occasionally been claimed that they play no role at all (Chew, 1963). However, the fact that a dynamical theory of fundamental "particles" is not around the corner does not entail that we can dispense with space and time concepts. This is a case of sour grapes. It is no less glaring a mistake than that of assigning spacetime causal efficacy – which Aristotle did in his *Physics*. The locution 'x was knocked down because x was at place y' is just short for 'x was knocked down by some other thing z that happened to be at place y'. Likewise the phrase 'Time brought about the decline of x' is an ellipsis for 'Certain events brought about the decline of x'. Space and time are as harmless as they are unavoidable.

Another point that has not been well understood on occasion is the nature of coordinates. Because the general covariance requirement shows coordinate systems and coordinate values to be conventional or a priori, it is sometimes held that spacetime itself is not physically objective (e.g. Einstein, 1916). That coordinate values just label points in the four-manifold M^4 is true and even trivially so, as we saw in Sec. 4.2. But this does not entail that M^4 fails to represent objective relations among physical objects, as shown by the absolute character of certain invariants such as spatiotemporal distances (the values of the numerical partners of the qualitative spatiotemporal separation introduced by Definition 6.18). In fact they do not change under the substitution of coordinate systems; in particular they do not depend on the observer's point of view nor even upon his location and motion. Moreover the coefficients of the metric, though not covariant themselves (since they are the components of a tensor), are no less than the gravitational potentials and, as such, they occur centrally in the field equations of gravitation: they represent a feature of the thing called 'gravitational field'. If the metric were conventional (as Grünbaum (1973) has claimed), the laws of gravitation should be conventional as well. And if this were so we should be smart enough always to choose a metric that will prevent any nasty fall and even keep our enemies at bay.

In conclusion, space and time have no autonomous existence, hence no causal power either. But then no relation exists on its own, i.e. apart from its relata – as we know from Ch. 2. Although spatiotemporal relations are not entities, nor even couplings or connections on the same footing with mechanical or electrical or social bonds, they are real relations in the sense that they hold among real things. Spacetime is no more nor less than *the fundamental structure or basic framework of the world.*

In sum: mathematical spaces are multiple and unreal – neither objective nor subjective; physical space(time) is one, real, and objective (or subject-independent); and phenomenal (or perceptual) spaces and times are multiple, real, and subjective (or subject-dependent).

7. Concluding remarks

According to Cartesian mechanism all facts can and should be explained in terms of space and time only. This program, though wonderfully fruitful, has never been fully implemented in physics, let alone in other fields of research. Indeed some of the basic properties of matter, such as mass, charge, spin, and parity, are not explained in terms of spatiotemporal concepts. And even features that are dependent on spatiotemporal location, such as field strengths and cultural production, are not spatiotemporal themselves. In sum, things are not warts of spacetime. And spacetime, not being independent of things, is not illusory either. Reality is the aggregation of things holding spatiotemporal relations. Now, relations do not preexist their relata but hold among them. In particular spatiotemporal relations, such as "between" and "before", can be understood only in terms of the things, states, or events between which they hold. Hence we must try and understand spacetime in terms of changing things: this is what the present chapter is all about.

Once the spatiotemporal framework has been anchored to things we can invert the procedure and study things *in* spacetime – but this second, more detailed look, is taken by science not philosophy. (Likewise once we have formed the general concept of an event, belonging to ontology, we can start asking questions about events of special kinds.) In summary we have the situation depicted in Figure 6.13.

How do the various ontological theories of space and time fare in

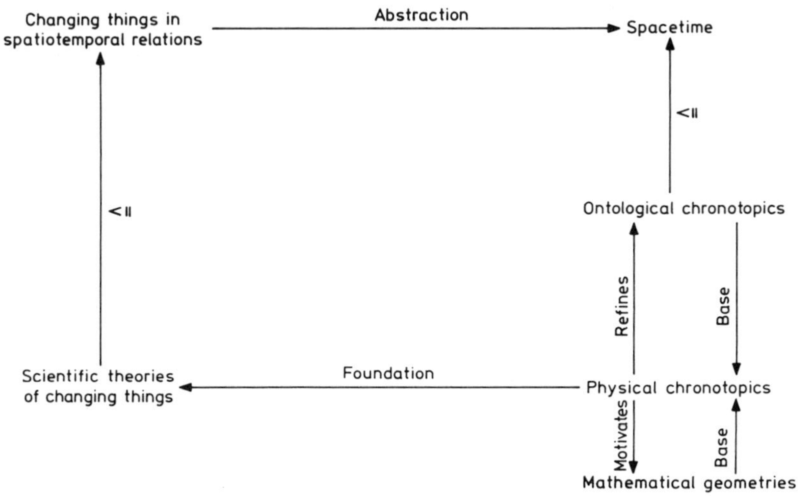

Fig. 6.13. Relations among items discussed in this chapter. The real thing is in the upper left corner: the rest are constructs.

contemporary science, i.e. what support do they get from the latter? This is not easy to say because any physical theory containing spatiotemporal concepts can presumably be married either to an absolute or to a relational theory of spacetime. Such an ambiguity, clear as it is on analysis, is obscured by history. Thus just because Newton assumed that space and time are self-existent, and in particular independent of matter in motion, most philosophers believe that Newtonian mechanics and gravitation theory are soaked in the doctrine of the absolute character of space and time. This is mistaken: it is possible to reformulate classical mechanics assuming a relational (though not relativistic) theory of space and time (Noll, 1967). In other words, the equations of motion and the constitutive equations of classical mechanics are independent of the way space and time are understood. The same holds for all other non-relativistic theories, in particular quantum mechanics.

Even relativistic theories are rather insensitive to the absolute-relational issue. In this case too either one may attempt to build spacetime out of changing things, or one may take it for granted and build thereupon. One may even get carried away by the second procedure and conclude that, since the gravitational field equations possess solutions even in the absence of matter and radiation, general relativity actually uses an absolute concept of spacetime. But this would be just as mistaken as claiming that Newtonian mechanics is indissolubly married to an absolute doctrine of space and time. The solutions for the hollow universe are physically meaningless because, by definition, physics is concerned with physical objects. In sum, the fact that physics is compatible with a given ontological theory of space and time does not endorse the latter but only renders it epistemically possible. In particular, the compatibility of our own chronotopics with the science of the day, though necessary and encouraging, is insufficient. We must look elsewhere for additional confirmation.

We cannot verify conclusively our relational theory of spacetime – nor indeed any other chronotopics. However we can dispose of its rival, namely the absolute theory, in any of its possible versions. We can do it not with the help of crucial experiments but by a semantic analysis. Indeed, if one admits the methodological principle that a physical theory, by definition of 'physical', must contain only physically meaningful concepts, and if one accepts the definition of a physically meaningful concept as one that refers to physical items (cf. Vol. 1, Ch. 2 of this treatise), then the concepts of absolute or autonomous space and time must be discarded for not being physical. By this criterion only the class of relational theories of space and time remain. And the only criteria for choosing among the latter are (*a*) mathematical cogency, (*b*) closeness of fit with contemporary physics, and (*c*) clarifying power. We leave the reader the task of deciding whether our own theory of spacetime meets these criteria.

So much for extension and duration. We have now all we need to handle systems in general and in particular. But this will be a task for the 4th volume of this work, namely *A World of Systems.*

BIBLIOGRAPHY

Agassi, J. (1964). The nature of scientific problems and their roots in metaphysics. In M. Bunge, Ed., *The Critical Approach*, pp. 189–211. New York: Free Press.

Agassi, J. (1975). *Science in Flux*. Dordrecht & Boston: Reidel.

Alexander, H. G., Ed. (1956). *The Leibniz–Clarke Correspondence*. Manchester: University Press.

Alexander, S. (1920). *Space, Time and Deity*, 2 vols. New York: Humanities Press.

Allen, J. A. (1870–71). On the mammals and winter birds of East Florida, etc. *Bull. Museum Compar. Zoölogy* (Harvard) **2**: 161–450.

Aquinas, T. (1259–64). *Summa contra gentiles*. Transl. A. C. Pegis. Garden City, N.Y.: Hanover House 1955.

Aristotle. *Metaphysics*. In *Works*, J. A. Smith and W. D. Ross, Eds., vol. VIII. Oxford: University Press 1928.

Aristotle. *Physics*. In *Works*, J. A. Smith and W. D. Ross, Eds., vol. II. Oxford: Clarendon Press 1930.

Ashby, W. R. (1956). *An Introduction to Cybernetics*. New York: McGraw-Hill.

Basri, S. A. (1966). *A Deductive Theory of Space and Time*. Amsterdam: North-Holland.

Bergmann, G. (1960). The philosophical significance of modal logic. *Mind* (N.S.) **69**: 466–485.

Bergmann, H. (1929). *Der Kampf um das Kausalgesetz in der jüngsten Physik*. Braunschweig: Vieweg.

Bergson, H. (1903). Introduction à la métaphysique. *Revue de métaphysique et de morale* **11**: 1–36.

Bernays, P. (1937). A system of axiomatic set theory. Part I. *Journal of Symbolic Logic* **2**: 65–77.

Blalock, H. M. (1961). Evaluating the relative importance of variables. *Am. Sociol. Rev.* **26**: 866–874.

Bohm, D. (1970). Further remarks on order. In C. H. Waddington, Ed., *Towards a Theoretical Biology*. Chicago: University of Chicago Press.

Bolzano, B. (1821). *Begriffe*. In E. Winter, *Die historische Bedeutung der Frühbegriffe B. Bolzanos*. Berlin: Akademie-Verlag 1964.

Bolzano, B. (1837). *Wissenschaftslehre*, 4 vols. Repr. Leipzig: Meiner 1929, and Aalen: Scientia Verlag, 1970.

Bolzano, B. (1849). *Was ist Philosophie*? Amsterdam: Rodopi 1969.

Bolzano, B. (1851). *Paradoxes of the Infinite*. Transl. by D. A. Steele. London: Routledge & Kegan Paul 1950.

Booij, H. L., and H. P. Wolvekamp (1944). *Catenary Processes, Master Reactions and Limiting Factors*. Bibliotheca Biotheoretica, Vol. I, Pars 4. Leiden: E. J. Brill.

Borel, E. (1948). *Le hasard*, 2nd ed. Paris: Presses Universitaires de France.

Bradley, F. H. (1893). *Appearance and Reality*. London: Oxford University Press 1969.

Braudel, F. (1969). *Écrits sur l'histoire*. Paris: Flammarion.

Brody, T. A. (1974). Probability: A new look at old ideas. Preprint, Instituto de Física, Universidad Nacional Autónoma de México.

Bruck, R. H. (1958). *A Survey of Binary Systems*. Berlin, Göttingen & Heidelberg: Springer-Verlag.

Buchdahl, G. (1969). *Metaphysics and the Philosophy of Science*. Cambridge, Mass.: MIT Press.

Bunge, M. (1959). *Causality*. Cambridge, Mass.: Harvard University Press.

Bunge, M. (1961). The weight of simplicity in the construction and assaying of scientific theories. *Phil. Sci.* **28**: 129–149.

Bunge, M. (1962). Cosmology and magic. *The Monist* **47**: 116–141.

Bunge, M. (1964). Phenomenological theories. In M. Bunge, Ed., *The Critical Approach*. New York: The Free Press.

Bunge, M. (1966). On null individuals. *J. Phil.* **63**: 776–778.

Bunge, M. (1967a). *Scientific Research*, 2 vols. New York: Springer-Verlag.

Bunge, M. (1967b). *Foundations of Physics*. New York: Springer-Verlag.

Bunge, M. (1968a). Physical time: the objective and relational theory. *Phil. Sci.* **35**: 355–388.

Bunge, M. (1968b). Physique et métaphysique du temps. *Proc. XIVth Intern. Cong. Philosophy* **II**: 623–629. Wien: Herder.

Bunge, M. (1968c). Conjunction, succession, determination, and causation. *Intern. J. Theoret. Phys.* **1**: 299–315.

Bunge, M. (1968d). Problems and games in the current philosophy of natural science. *Proc. XIVth Intern. Cong. Philosophy* **I**: 566–574. Wien: Herder.

Bunge, M. (1969). What are physical theories about? *Amer. Phil. Quart. Monographs* **3**: 61–99.

Bunge, M. (1970a). Virtual processes and virtual particles: real or fictitious? *Intern. J. Theoret. Phys.* **3**: 507–508.

Bunge, M. (1970b). The arrow of time. *Intern. J. Theoret. Phys.* **3**: 77–78.

Bunge, M. (1971a). Is scientific metaphysics possible? *J. Philos.* **68**: 507–520.

Bunge, M. (1971b). A mathematical theory of the dimensions and units of physical quantities. In M. Bunge, Ed., *Problems in the Foundations of Physics*, pp. 1–16. New York: Springer-Verlag.

Bunge, M. (1972). Time symmetry, time reversal, and irreversibility. In J. T. Fraser *et al.*, Eds., *The Study of Time*, pp. 122–130. New York: Springer-Verlag.

Bunge, M. (1973a). *Method, Model and Matter*. Dordrecht & Boston: Reidel.

Bunge, M. (1973b). *Philosophy of Physics*. Dordrecht: Reidel.

Bunge, M. (1973c). Review of P. Suppes' *Probabilistic Theory of Causality*. *Brit. J. Phil. Sci.* **24**: 409–410.

Bunge, M. (1974a). *Sense and Reference*. Vol. 1 of *Treatise on Basic Philosophy*. Dordrecht & Boston: Reidel.

Bunge, M. (1974b). *Interpretation and Truth*. Vol. 2 of *Treatise on Basic Philosophy*. Dordrecht & Boston: Reidel.

Bunge, M. (1974c). The relations of logic and semantics to ontology. *Journal of Philosophical Logic* **3**: 195–209.

Bunge, M. (1974d). Les présupposés et les produits métaphysiques de la science et de la technique contemporaines. *Dialogue* **13**: 443–453.

Bunge, M. (1974e). The concept of social structure. In W. Leinfellner and E. Köhler, Eds., *Developments in the Methodology of Social Science*, pp. 175–215. Dordrecht & Boston: Reidel.

Bunge, M. (1974f). Metaphysics and science. *General Systems* **19**: 15–18.

Bunge, M. (1974g). Things. *Intern. J. General Systems* **1**: 229–236.

Bunge, M. (1975). A critical examination of dialectics. In Ch. Perelman, Ed., *Dialectics/Dialectique*, pp. 63–77. The Hague: Nijhoff.

Bunge, M. (1976a). Possibility and probability. In W. Harper and C. Hooker, Eds., *Foundations of Probability Theory, Statistical Inference, and Statistical Theories of Science*, Vol. III, 17–33. Dordrecht and Boston: D. Reidel Publ. Co.

Bunge, M. (1977a). The GST challenge to the classical philosophies of science. *Intern. J. General Systems*, Vol. 4, No. 1.

Bunge, M. (1977b). States and events. In W. E. Hartnett, Ed., 1977.

Bunge, M. (1977c). The interpretation of Heisenberg's inequalities. In H. Pfeiffer, Ed., *Denken und Umdenken: Zu Werk und Wirkung von Werner Heisenberg*. München–Zürich: R. Piper & Co.

Bunge, M., and A. García Máynez (1977). A relational theory of physical space. *Intern. J. Theoret. Phys.*, Vol. 16, forthcoming.

Bunge, M., and A. A. L. Sangalli (1977). A theory of properties and kinds. *Intern. J. General Systems*: **3**: 183–190.

Burtt, E. A. (1932). *The Metaphysical Foundations of Physical Science*, rev. ed. London: Routledge & Kegan Paul.

Carnap, R. (1928). *Der logische Aufbau der Welt.* Transl. by R. A. George: *The Logical Structure of the World.* Berkeley & Los Angeles: University of California Press 1967.

Carnap, R. (1936–37). Testability and meaning. *Phil. Sci.* **3**: 419–471; **4**: 1–40.

Carnap, R. (1947). *Meaning and Necessity.* Chicago: University of Chicago Press.

Castañeda, H.-N. (1974). Thinking and the structure of the world. *Philosophia* **4**: 3–40.

Chaitin, G. J. (1974). Information-theoretic limitations of formal systems. *J. Assoc. for Computing Machinery* **21**: 403–424.

Chew, G. (1963). The dubious role of the space-time continuum in microscopic physics. *Sci. Progr.* **51**: 529–539.

Chew, G. (1966). *The Analytic S Matrix.* New York: Benjamin.

Chisholm, R. M. (1960). *Realism and the Background of Phenomenology.* Glencoe, Ill.: The Free Press.

Chwistek, L. (1949). *The Limits of Science.* London: Routledge & Kegan Paul.

Cocchiarella, N. B. (1972). Properties as individuals in formal ontology. *Noûs* **6**: 165–187.

Cocchiarella, N. B. (1974). Logical atomism and modal logic. *Philosophia* **4**: 41–66.

Collingwood, R. G. (1940). *An Essay on Metaphysics.* Oxford: Clarendon Press.

Costa de Beauregard, O. (1963). *Le second principe de la science du temps.* Paris: Ed. du Seuil.

Cournot, A.-A. (1851). *Essai sur les fondements de nos connaissances.* Eng. transl. with Introd. by M. H. Moore: *An Essay on the Foundations of our Knowledge.* New York: Liberal Arts Press 1956.

Cresswell, M. J. (1973). Physical theories and possible worlds. *Logique et analyse* (N.S.) **16**: 495–511.

Dantzig, D. van (1935). The fundamental equations of electromagnetism, independent of metric geometry. *Cambridge Phil. Soc. Proc.* **30**: 421–427.

Dehnen, H. (1970). The frame of reference in relativistic theories. *Intern. J. Theoret. Phys.* **3**: 509–511.

Dewitt, B. S. (1970). Quantum mechanics and reality. *Physics Today* **23**, No. 9, pp. 30–35.

DeWitt, N. W. (1954). *Epicurus and his Philosophy*. Minneapolis: University of Minnesota Press.

Ducasse, C. J. (1968). *Truth, Knowledge and Causation*. New York: Humanities Press.

Einstein, A. (1916). Grundlagen der allgemeinen Relativitätstheorie. *Ann. Phys.* **49**: 769–822.

Engels, F. (1878). *Anti-Dühring*. In K. Marx and F. Engels, *Werke*, vol. 20. Berlin: Dietz 1973.

Everett, H., III (1957). 'Relative state' formulation of quantum mechanics. *Rev. Mod. Phys.* **29**: 454–463.

Falk, G. (1966). *Theoretische Physik, I*. Berlin, Heidelberg & New York: Springer-Verlag.

Feibleman, J. and J. W. Friend (1945). The structure and function of organization. *Phil. Rev.* **54**: 19–44.

Feigl, H. (1967). *The 'Mental' and the 'Physical'*. Minneapolis: University of Minnesota Press.

Feller, W. (1968). *An Introduction to Probability Theory and its Applications*, I, 3rd. ed. New York: Wiley.

Fine, A. (1964). Physical geometry and physical laws. *Phil. Sci.* **31**: 156–162.

Finkelstein, D. (1973). A process conception of nature. In J. Mehra, Ed., *The Physicist's Conception of Nature*, pp. 709–713. Dordrecht: Reidel.

Fitch, F. B. (1971). Propositions as the only realities. *Amer. Phil. Quart.* **8**: 99–103.

Fokker, A. D. (1965). *Time and Space, Weight and Inertia*. Oxford: Pergamon Press.

Fraassen, B. C. van (1970). *An Introduction to the Philosophy of Space and Time*. New York: Random House.

Fréchet, M. (1939). The diverse definitions of probability. *Erkenntnis* **8**: 7–23.

Gaal, S. A. (1964). *Point Set Topology*. New York: Academic Press.

Gell-Man, M. (1956). The interpretation of the new particles as displaced charge multiplets. *Nuovo Cimento Suppl.* (S.10) **4**: 848–866.

Giere, R. N. (1974). Objective single-case probabilities and the foundations of statistics. In P. Suppes *et al.*, Eds., *Logic, Methodology and Philosophy of Science IV*, pp. 467–483. Amsterdam: North-Holland 1974.

Gold, T., Ed. (1967). *The Nature of Time*. Ithaca, N.Y.: Cornell University Press.

Gonseth, F. (1938). *La méthode axiomatique*. Paris: Gauthier-Villars.

Goodman, N. (1951). *The Structure of Appearance*. Cambridge, Mass.: Harvard University Press.

Goodman, N. (1956). A world of individuals. In *The Problem of Universals*, pp. 15–31. Notre Dame, Ind.: University of Notre Dame Press.

Grünbaum, A. (1967). *Modern Science and Zeno's Paradoxes*. Middleton, Conn.: Wesleyan University Press.

Grünbaum, A. (1973). *Philosophical Problems of Space and Time*. Dordrecht & Boston: Reidel.

Harary, F. (1971). Demiarcs: an atomistic approach to relational systems and group dynamics. *J. Math. Sociology* **1**: 195–205.

Harré, R. (1961). *Theories and Things*. London: Sheed and Ward.

Harré, R. (1972). *The Philosophies of Science*. London and New York: Oxford University Press.

Harris, E. E. (1965). *The Foundations of Metaphysics in Science.* New York: Humanities Press.

Hart, R. G. (1971). A ramiform view of unpredictable events. *Am. J. Phys.* **39**: 1336–1346.

Hartmann, N. (1938). *Möglichkeit und Wirklichkeit.* Berlin: W. de Gruyter.

Hartnett, W., Ed. (1977). *Systems: Approaches, Theories, Applications.* Dordrecht and Boston: Reidel.

Harvey, D. (1969). *Explanation in Geography.* New York: St. Martin's Press.

Hasenjäger, G. (1966). Logik und Ontologie. *Studium generale* **19** (3): 136–140.

Hebb, D. (1949). *The Organization of Behavior.* New York: Wiley.

Hegel, G. W. F. (1812–16). *The Science of Logic.* Transl. A. V. Miller. London: Allen & Unwin 1969.

Heidegger, M. (1927). *Sein und Zeit.* Eng. transl. J. Macquarrie and E. S. Robinson: *Being and Time.* New York: Harper 1962.

Heidegger, M. (1953). *Einführung in die Metaphysik.* Tübingen: Niemeyer.

Heisenberg, W. (1969). *Der Teil und das Ganze.* München: Piper.

Helmholtz, H. von (1873). *Popular Lectures on Scientific Subjects.* London: Longmans, Green & Co.

Hilbert, D. (1899). *Grundlagen der Geometrie.* Leipzig. Eng. transl. E. J. Townsend: *Foundations of Geometry.* La Salle, Ill.: Open Court, 1902.

Hintikka, J. (1969). *Models for Modalities.* Dordrecht: Reidel.

Hirsch, M. and S. Smale (1974). *Differential Equations, Dynamical Systems, and Linear Algebra.* New York: Academic Press.

Hiż, H. (1971). On the abstractness of individuals. In M. K. Munitz, Ed., *Identity and Individuation*, pp. 251–261. New York: New York University Press.

Hooker, J. D. (1853–55). *Flora Novae Zelandae*, being v. 2 of *The Botany of the Antarctic Voyage of H.M. Discovery Ships* Erebus *and* Terror *in the Years 1839–1843*, etc. London: Reeve Brothers 1844–1860.

Hughes, G. E., and M. J. Cresswell (1968). *An Introduction to Modal Logic.* London: Methuen.

Hume, D. (1739–1740). *A Treatise of Human Nature*, L. A. Selby-Bigge, Ed. Oxford: Clarendon Press, 1888.

Huntington, E. V. (1913). A set of postulates for abstract geometry, expressed in terms of the simple relation of inclusion. *Math. Ann.* **73**: 522–559.

Jammer, M. (1954). *Concepts of Space.* Cambridge, Mass.: Harvard University Press.

Jensen, U. J. (1972). Conceptual epiphenomenalism. *The Monist* **56**: 250–275.

Kant, I. (1781). *Kritik der reinen Vernunft.* Hamburg: F. Meiner 1952.

Keynes, J. M. (1921). *A Treatise on Probability.* London: Macmillan.

Klir, G., Ed. (1972). *Trends in General Systems Theory.* New York: Wiley-Interscience.

Kneale, W. and M. (1962). *The Development of Logic.* Oxford: Clarendon Press.

Körner, S. (1970). *Categorial Frameworks.* Oxford: Basil Blackwell.

Koyré, A. (1957). Galileo and Plato. In P. P. Wiener and A. Noland, Eds., *Roots of Scientific Thought*, pp. 147–175. New York: Basic Books.

Kripke, S. (1959). A completeness theorem in modal logic. *J. Symbolic Logic* **24**: 1–15.

Kripke, S. (1971). Naming and necessity. In D. Davidson and G. Harman, Eds., *Semantics of Natural Languages.* Dordrecht: D. Reidel Publ. Co.

Kulakov, Ju. I. (1970). The geometry of spaces of constant curvature as a special case of the theory of physical structures. *Sov. Math. Dokl.* **11**: 1055–1057.

Lakatos, I. (1969). Criticism and the methodology of scientific research programs. *Proc. Aristotelian Soc.* **69**: 149–186.

Lambert, J. H. (1770). Letter to I. Kant, 13.10.1770. In H. E. Fischer, Ed., *Briechwechsel von Imm. Kant,* 3 vols. München: Georg Müller 1912.

Leibniz, G. W. von (1704). *Nouveaux Essais.* Paris: Flammarion s.d.

Leibniz, G. W. von (1956). *Philosophical Papers and Letters,* L. H. Loemker, Ed., 2 vols. Chicago: University of Chicago Press.

Lejewski, C. (1967). A single axiom for the mereological notion of proper part. *Notre Dame Journal of Formal Logic* **8**: 279–285.

Leonard, H. S. and N. Goodman (1940). The calculus of individuals and its uses. *J. Symbolic Logic* **5**: 45–55.

Lévi-Strauss, C. (1958). *Anthropologie structurale.* Paris: Plon.

Lewes, G. H. (1879). *Problems of Life and Mind.* London: Trubner.

Lewis, C. I. (1923). Facts, systems, and the unity of the world. *Journal of Philosophy* **20**: 141–151.

Lewis, C. I. (1946). *An Analysis of Knowledge and Valuation.* La Salle, Ill.: Open Court.

Lewis, C. I. and C. H. Langford (1932). *Symbolic Logic.* New York: Century.

Lewis, D. (1973). *Counterfactuals.* Oxford: Basil Blackwell.

Ljapin, E. S. (1963). *Semigroups.* Providence, R. I.: American Mathematical Society.

Locke, J. (1689). *An Essay Concerning Human Understanding.* London: Routledge s.d.

Lotze, H. (1887). *Metaphysics,* 2 vols., 2nd ed. Transl. ed. by B. Bosanquet. Oxford: Clarendon Press.

Lovejoy, A. O. (1936). *The Great Chain of Being.* Cambridge, Mass.: Harvard University Press.

Lucas, J. R. (1973). *A Treatise on Time and Space.* London: Methuen.

Luce, R. (1960). The theory of selective information and some of its behavioral applications. In R. D. Luce, Ed., *Developments in Mathematical Psychology,* pp. 5–119. Glencoe, Ill.: Free Press.

Lucretius (55 B.C.?). *De rerum natura.* Eng. transl. by W. E. Leonard: *Of the Nature of Things.* London: J. M. Dent; New York: E. P. Dutton, 1921.

Ludwig, G., Ed. (1964). *Raum und Zeit als Probleme der Naturwissenschaften.* Freiburg: Karl Alber.

Łukasiewicz, J. (1970). *Collected Works,* L. Borkowski, Ed. Amsterdam: North-Holland.

Mach, E. (1872). *Die Geschichte und die Wurzel des Satzes von der Erhaltung der Arbeit,* 2nd ed. Leipzig: Barth 1909.

Mach, E. (1883). *Die Mechanik in ihre Entwicklung historisch-kritisch dargestellt,* 4th ed. Leipzig: Brockhaus 1901.

Mach, E. (1906). *Die Analyse der Empfindungen,* 5th ed. Jena: Gustav Fischer.

Marcus, R. B. (1968). Modal logic. In R. Klibansky, Ed., *Contemporary Philosophy,* vol. I, pp. 87–101. Firenze: La Nuova Italia Editrice.

Margenau, H. (1941). Metaphysical elements in physics. *Rev. Mod. Phys.* **13**: 176–189.

Margenau, H. (1950). *The Nature of Physical Reality.* New York: McGraw-Hill Book Co.

Margenau, H. (1966). Exclusion principle and measurement theory. In P. O. Löwdin, Ed., *Quantum Theory of Atoms, Molecules, Solid State,* pp. 81–91. New York: Academic Press.

Martin, R. M. (1965). Of time and the null individual. *J. Phil.* **62**: 723–736.

Maxwell, J. C. (1892). *A Treatise on Electricity and Magnetism,* 3rd ed. Oxford: University Press.

McGilvary, E. B. (1951). Space-time, simple location, and prehension. In P. A. Schilpp, Ed., *The Philosophy of Alfred North Whitehead*, 2nd, ed. New York: Tudor Publ. Co.
Meinong, A. (1904). *The Theory of Objects*. In R. M. Chisholm (1960) pp. 76–117.
Meinong, A. (1915). *Über Möglichkeit und Wahrscheinlichkeit*. Leipzig: J. A. Barth.
Mellor, D. H. (1971). *The Matter of Chance*. Cambridge: University Press.
Melvin, M. A. (1960). Elementary particles and symmetry principles. *Rev. Mod. Phys.* **32**: 477–518.
Mendelson, E. (1963). *Introduction to Mathematical Logic*. Princeton: Van Nostrand.
Mesarović, M. D., and Y. Takahara (1975). *General Systems Theory*. New York: Academic Press.
Minkowski, H. (1908). Space and time. In H. A. Lorentz *et al.*, *The Principle of Relativity*, pp. 73–91. New York: Dover s.d.
Misner, C. W., K. S. Thorne and J. A. Wheeler (1973). *Gravitation*. San Francisco: Freeman.
Montagu, W. P. (1925). *The Ways of Knowing*. London: Allen & Unwin.
Montague, R. (1960). Logical necessity, physical necessity, ethics and quantifiers. *Inquiry* **3**: 259–269.
Montague, R. (1974). *Formal Philosophy*. Selected Papers of R. M. ed. by R. Thomason. New Haven: Yale University Press.
Munitz, M. K., Ed. (1971). *Identity and Individuation*. New York: New York University Press.
Munitz, M. K., Ed. (1973). *Logic and Ontology*. New York: New York University Press.
Naess, A. (1972). *The Pluralist and Possibilist Aspect*. Oslo: Universitetsforlaget; London: Allen & Unwin.
Neumann, J. von (1932). *Mathematische Grundlagen der Quantenmechanik*. Berlin: Julius Springer.
Newton, I. (1704). *Opticks*. In *Opera quae exstant omnia*, vol. IV. London: Nichols 1782.
Nicod, J. (1923). *La géométrie dans le monde sensible*. Paris: Presses Universitaires de France 1962.
Noll, W. (1967). Space-time structures in classical mechanics. In M. Bunge, Ed., *Delaware Seminar in the Foundations of Physics*, pp. 28–34. New York: Springer-Verlag.
Ockham, W. (ca. 1320). *Summa totius logicae*. In P. Boehner, Ed., *Ockham: Philosophical Writings*. Edinburgh: Nelson 1957.
Ostwald, W. (1902). *Vorlesungen über Naturphilosophie*. Leipzig: Veit & Comp.
Padulo, L., and M. Arbib (1974). *System Theory*. Philadelphia: W. B. Saunders.
Peirce, C. S. (1892–93). *Scientific Metaphysics*. Vol. VI of *Collected Papers*, C. Hartshorne and P. Weiss, Eds. Cambridge, Mass.: Harvard University Press 1935.
Peña, L. de la (1969). New formulation of stochastic theory and quantum mechanics. *J. Math. Phys.* **10**: 1620–1630.
Peña, L. de la, and A. M. Cetto (1975). Stochastic theory for quantum mechanical systems. *Foundations of Physics* **5**: 355–373.
Penney, R. (1965). On the dimensionality of the real world. *J. Math. Phys.* **6**: 1607–1611.
Penrose, R. (1971). Angular momentum: an approach to combinatorial spacetime. In T. Bastin, Ed., *Quantum Theory and Beyond*, pp. 151–180. Cambridge: University Press.
Pielou, E. C. (1969). *An Introduction to Mathematical Ecology*. New York: Wiley-Interscience.

Plantinga, A. (1974). *The Nature of Necessity.* New York: New York University Press.

Popper, K. R. (1957). The propensity interpretation of the calculus of probability, and the quantum theory. In S. Körner, Ed., *Observation and Interpretation*, pp. 65–70. London: Butterworths Scientific Publications.

Popper, K. R. (1963). The demarcation between science and metaphysics. In P. A. Schilpp, Ed., *The Philosophy of Rudolf Carnap*, pp. 877–881. La Salle, Ill.: Open Court.

Popper, K. R. (1968). On the theory of the objective mind. Repr. in *Objective Knowledge*, pp. 153–190. Oxford: Clarendon Press 1972.

Popper, K. R. (1974). Autobiography. In Schilpp, Ed. (1974), Bk. I, pp. 3–181.

Putnam, H. (1969). On properties. In N. Rescher, Ed., *Essays in Honor of Carl G. Hempel*, pp. 235–254. Dordrecht: Reidel.

Quine, W. V. (1954). Reduction to a dyadic predicate. Repr. in *Selected Logic Papers*, pp. 224–226. New York: Random House 1966.

Quine, W. V. (1963). *Set Theory and its Logic.* Cambridge, Mass.: The Belknap Press of Harvard University Press.

Quinton, A. (1973). *The Nature of Things.* London & Boston: Routledge & Kegan Paul.

Raggio, A. R. (1969). Was heisst 'Bedingungen der Möglichkeit'? *Kant Studien* **60**: 153–165.

Rashevsky, N. (1960). *Mathematical Biophysics*, 3rd ed. New York: Dover.

Regge, T. (1967). Review of Chew's book (Chew 1966). *Physics Today* **20**: No. 9: 82–87.

Reichenbach, H. (1928). *The Philosophy of Space and Time.* New York: Dover 1957.

Reichenbach, H. (1956). *The Direction of Time.* Berkeley: University of California Press.

Rescher, N. (1975). *A Theory of Possibility.* Oxford: Basil Blackwell.

Rescher, N. and A. Urquhart (1971). *Temporal Logic.* Wien, New York: Springer-Verlag.

Robb, A. A. (1914). *A Theory of Time and Space.* Cambridge: University Press.

Robb, A. A. (1921). *The Absolute Relations of Time and Space.* Cambridge: University Press.

Robb, A. A. (1936). *Geometry of Time and Space.* Cambridge: University Press.

Rosenblueth, A. (1970). *Mind and Brain: a Philosophy of Science.* Cambridge, Mass.: MIT Press.

Russell, B. (1900). *A Critical Exposition of the Philosophy of Leibniz.* Cambridge: University Press.

Russell, B. (1914). *Our Knowledge of the External World.* London: Allen & Unwin 1952.

Russell, B. (1918). The philosophy of logical atomism. *The Monist* **28**: 495–527; **29**: 32–63, 190–22, 345–380. Repr. in *Logic and Knowledge*, R. C. Marsh, Ed., pp. 175–281. London: Allen & Unwin 1956.

Russell, B. (1940). *An Inquiry into Meaning and Truth.* London: Allen & Unwin.

Russell, B. (1948). *Human Knowledge: its Scope and Limits.* London: Allen & Unwin.

Rutherford, D. E. (1965). *Introduction to Lattice Theory.* Edinburgh & London: Oliver & Boyd.

Ryle, G. (1949). *The Concept of Mind.* London: Hutchinson.

Schilpp, P. A., Ed. (1974). *The Philosophy of Karl Popper*, 2 vols. La Salle, Ill.: Open Court.

Schlick, M. (1963). *Space and Time in Contemporary Physics.* New York: Dover.

Scholz, H. (1941). *Metaphysik als strenge Wissenschaft.* Darmstadt: Wissenschaftliche Buchgesellschaft 1965.

Schreider, Ju. A. (1975). *Equality, Resemblance, and Order*. Transl. M. Greendlinger. Moscow: MIR Publishers.

Schulz, C. M. (1967). *You're you, Charlie Brown*. New York: Holt, Rinehart & Winston.

Scott, D. (1970). Advice on modal logic. In K. Lambert, Ed., *Philosophical Problems in Logic: Some Recent Developments*. Dordrecht: Reidel.

Sellars, W. (1963). *Science, Perception and Reality*. London: Routledge & Kegan Paul.

Settle, T. (1974). Induction and probability unfused. In Schilpp, Ed. 1974, Bk. II, pp. 697–749.

Shea, W. (1972). *Galileo's Intellectual Revolution*. London: Macmillan.

Sinnot, E. W. (1963). *The Problem of Organic Form*. New Haven: Yale University Press.

Smart, J. J. C. (1963). *Philosophy and Scientific Realism*. New York: Humanities Press.

Smart, J. J. C., Ed. (1964). *Problems of Space and Time*. New York: Macmillan.

Sneath, P. H. and R. R. Sokal (1973). *Numerical Taxonomy*. San Francisco & London: Freeman.

Sobociński, B. (1954–55). Studies in Lesniewski's mereology. *Polskie Towarzystwo Naukowe na Obezyźnie Yearbook*, London.

Stapp, H. P. (1971). S-matrix representation of quantum theory. *Phys. Rev.* D **3**: 1303–1320.

Stapp, H. P. (1977). Theory of reality. *Foundations of Physics*, forthcoming.

Stiehler, G. (1967). *Der dialektischer Widerspruch*, 2nd ed. Berlin: Akademie-Verlag.

Strauss, M. (1970). Intertheory relations. In P. Weingartner and G. Zecha, Ed., *Induction, Physics, and Ethics*. Dordrecht: Reidel.

Strawson, P. F. (1959). *Individuals: an Essay in Descriptive Metaphysics*. London: Methuen. Repr. Cambridge University Press 1964.

Suppes, P. (1970). *A Probabilistic Theory of Causality*. Acta Philosophica Fennica XXIV. Amsterdam: North-Holland.

Suppes, P. (1974). *Probabilistic Metaphysics*, 2 vols. Uppsala: Filosofiska Studier.

Tarski, A. (1927). Foundations of the geometry of solids. In *Logic, Semantics, Metamathematics*, pp. 24–29. Transl. J. H. Woodger. Oxford: Clarendon Press.

Tarski, A. (1944). The semantic conception of truth. *Phil. and Phenomenological Research* **4**: 341–375.

Tarski, A. (1954). Contributions to the theory of models, I, II. *Indagationes Mathematicae* **16**: 572–588.

Teilhard de Chardin, P. (1964). *The Phenomenon of Man*. New York: Harper.

Teller, P. (1975). Essential properties: some problems and conjectures. *J. Phil.* **72**: 233–248.

Terletskii, Ya. P. (1971). *Statistical Physics*. Amsterdam: North-Holland; New York: Elsevier.

Trautman, A. (1965). Foundations and current problems of general relativity. In A. Trautman, F. A. E. Pirani and H. Bondi, *Lectures on General Relativity I*, pp. 1–248. Englewood Cliffs: Prentice-Hall.

Tripathi, R. K. (1969). The central problem of Indian metaphysics. *Phil. East and West* **19**: 39–43.

Truesdell, C. (1974). A simple example of an initial value problem with any desired number of solutions. *Rendiconti dell'Istituto Lombardo*, Classe di Scienze (A) **108**: 301–304.

Ujomov, A. I. (1965). *Dinge, Eigenschaften und Relationen.* Berlin: Akademie-Verlag.

Vuillemin, J. (1971). *La logique et le monde sensible.* Paris: Flammarion.

Wallace, A. D. (1941). Separation spaces. *Ann. Math.* (2) **42**: 687–697.

Walsh, W. H. (1963). *Metaphysics.* London: Hutchinson.

Weiss, P. (1963). The cell as unit. *J. Theoret. Biology* **5**: 389–397.

Weyl, H. (1940). The evasive ghost of modality. In M. Farber, Ed., *Philosophical Essays in Memory of Edmund Husserl,* pp. 278–303. Cambridge, Mass.: Harvard University Press.

Weyl, H. (1949). *Philosophy of Mathematics and Natural Science.* Princeton: University Press.

Wheeler, J. A. (1962). *Geometrodynamics.* New York: Academic Press.

Whitehead, A. N. (1919). *An Enquiry Concerning the Principles of Natural Knowledge.* Cambridge: University Press.

Whitehead, A. N. (1920). *The Concept of Nature.* Repr. University of Michigan Press 1957.

Whitehead, A. N. (1929). *Process and Reality.* New York: Macmillan. Repr. 1969.

Whitrow, G. J. (1961). *The Natural Philosophy of Time.* London: Thomas Nelson. Paperback: Harper Torchbooks 1963.

Williams, D. C. (1937–38). The realistic interpretation of scientific sentences. *Erkenntnis* **7**: 169–178, 375–382.

Williams, D. C. (1953). On the elements of being. *Rev. Metaphys.* **7**: 3–18, 171–192.

Wilson, N. L. (1955). Property and description. *Phil. Rev.* **64**: 389–404.

Wittgenstein, L. (1922). *Tractatus Logico-Philosophicus.* London: Routledge & Kegan Paul. Rev. ed. 1951.

Wittgenstein, L. (1969). *Über Gewissheit.* Ed. by G. E. M. Anscombe and G. H. von Wright. Oxford: Basil Blackwell.

Woodger, J. H. (1929). *Biological Principles.* London: Routledge & Kegan Paul.

Woodger, J. H. (1939). *The Technique of Theory Construction.* Chicago: University of Chicago Press.

Woodger, J. H. (1951). Science without properties. *Brit. J. Phil. Sci.* **2**: 193–216.

Yessenin-Volpin, A. S. (1970). The ultra-intuitionistic criticism and the antitraditional program for foundations of mathematics. In A. Kino, J. Myhill and R. E. Wesley, Eds., *Intuitionism and Proof Theory,* pp. 3–45. Amsterdam: North-Holland.

Yukawa, H. (1973). *Creativity and Intuition.* Transl. J. Bester. Tokyo: Kodansha International.

Zadeh, L. A. and C. A. Desoer (1963). *Linear System Theory; the State Space Approach.* New York: McGraw-Hill.

Zeeman, E. C. (1962). The topology of the brain and visual perception. In M. K. Fort Jr., Ed., *Topology of 3-Manifolds.* Englewood Cliffs, N.J.: Prentice-Hall.

INDEX OF NAMES

INDEX OF SUBJECTS

The companion of this book is Volume 4 of the

Treatise on Basic Philosophy
ONTOLOGY II

A World of Systems

Table of Contents